'Cambini and Rubino have produced an important book for researchers, institutions and students interested in exploring the creation and development of regional energy markets in the Mediterranean basin and South East Europe. This well-written edited volume provides coverage of a number of important issues and approaches not commonly treated in the recent debate and brings the Mediterranean Energy markets back in the forefront of the discussion'

Roberto Vigotti*, Secretary General RES4MED and Coordinator RE Industry Board of IEA*

'Imperative for the study of the regional integration of energy markets. Interesting and comprehensive input on the advantages and difficulties of the development of the Energy Community and MedReg. Different regulatory processes, renewable energy and infrastructure projects in the Mediterranean region are all analysed. Under the excellent leadership of editors who are themselves authors of the key chapters, the reader finds thought-provoking contributions from recognized European experts. Decisive for the work of the Mediterranean Energy Forum.'

M. Teresa Costa-Campi*, Professor of Economics, Chair in Energy and Environmental Sustainability, Barcelona University, Former President CNE 2005-2011*

'In this valuable volume, Carlo Cambini and Alessandro Rubino provide a careful review of the two most important energy initiatives in the Euro-Mediterranean area from a regulatory, technical and institutional point of view. They clearly set out the ways in which EU energy strategy has evolved within its wider neighborhood, allowing the reader to appreciate the complex interaction and the rapidly changing framework that characterize energy cooperation in the Mediterranean basin. The book provides also an important account of both MedReg and Energy Community experience. To my knowledge, this is the first systematic review of such initiatives and may represent a very useful reference for researchers and policymakers.'

Dr Matteo di Castelnuovo*, MaGER Director, Department of Economics, Bocconi University, Italy*

'Cambini and Rubino's book moves the debate from a Euro-centric approach to energy market integration and regulation, towards a Euro-Mediterranean framework that allows readers to grasp the new paradigm that is now penetrating the academic debate as well as policy making. This book offers a systematic and essential review of the dynamic at play in the Regional energy initiatives in the wider EU neighbourhood and represents a fundamental reference for those interested in understanding the origin, rationale and future of the forthcoming Euro-Mediterranean Energy free trade area.'

Michelangelo Celozzi*, Secretary General Med-TSO*

Regional Energy Initiatives

This book focuses on the two intra-regional initiatives created for the development and integration of energy markets: the Energy Community and MedReg.

The Energy Community and MedReg, apart from their common strategic role in providing a much-needed stable regulatory environment for energy markets in their respective reference countries, represent examples of a diverse development of regional energy initiatives. The former is initiated by external factors and is an example of a top-down approach, whereas the latter is a voluntary bottom-up initiative of the countries involved. The way the institutional framework is built is not without consequences on the functioning and organization of the two regional initiatives.

The book assesses these different approaches and their consequences in the framework of the development of the Energy Community and MedReg, with particular reference to their impact on regional integration, energy policy and institutional change. The analysis is enriched with several case studies on the role of independent regulatory agencies, the promotion of renewable energy sources, the infrastructure and interconnection development across the Mediterranean basin and the implications of exporting the EU institutional model. This book is aimed at policymakers, institutions, energy companies and academics to provide a better understanding of the economic and institutional ecosystems that characterize the Mediterranean area.

Carlo Cambini is Associate Professor of Industrial Organization at the Politecnico di Torino, Research Fellow at the Istituto di economia e politica dell'energia e dell'ambiente (IEFE – Bocconi, Milan), and Economic Member of the Scientific Committee of the Florence School of Regulation, Robert Schuman Centre for Advanced Studies – European University Institute (EUI). He received the Young Economist Essay Award at the 1999 EARIE conference in Lausanne, Switzerland. His research focuses on industrial economics, regulatory economics and competition policy with a strong background in telecommunications and energy markets. He has been published in leading scientific international journals and has advised national regulatory agencies in and outside Europe.

Alessandro Rubino is a Lecturer in Economics at Bari University. His research interests are in the fields of regulation, industrial organization, energy market integration and demand side participation. Alessandro obtained his PhD in Economics from Siena University. Before completing his PhD, he worked at Ofgem as Regulatory Economist. His main tasks were related to the analysis of the impact of the Large Combustion Plants Directive (LCPD, 2001/80/EC) on the generation portfolio in UK. In addition Alessandro has worked on the design and implementation of the incentive scheme for the national transmission system operator for gas and electricity. From June 2009 to July 2012 Alessandro worked as Research Assistant at the Florence School of Regulation (European University Institute) within the Robert Schuman Centre, a research unit specialized in applied economics. He has been junior expert in the EC founded project "Paving the Way for the Mediterranean Solar Plan" and coordinator of the International Energy Regulation Network (IERN). From July 2012 to January 2014 Alessandro was head of capacity building and knowledge dissemination at the Enel Foundation.

Regional Energy Initiatives

MedReg and the Energy Community

Edited by Carlo Cambini
and Alessandro Rubino

Routledge
Taylor & Francis Group
LONDON AND NEW YORK

First edition published 2014
by Routledge
2 Park Square, Milton Park, Abingdon, Oxon OX14 4RN

and by Routledge
711 Third Avenue, New York, NY 10017

Routledge is an imprint of the Taylor & Francis Group, an informa business

British Library Cataloguing in Publication Data

A catalogue record for this book is available from the British Library

Library of Congress Cataloging-in-Publication Data

Regional energy initiatives : MedReg and the energy community / [edited by] Carlo Cambini and Alessandro Rubino.
pages cm
Includes bibliographical references and index.
1. Energy policy—Mediterranean Region. 2. Energy development—Mediterranean Region. 3. Energy industries—Mediterranean Region. 4. Regionalism—Mediterranean Region. I. Cambini, Carlo. II. Rubino, Alessandro.
HD9502.M582R44 2014
333.7909182'2—dc23
2013048236

ISBN: 978-0-415-73723-4 (hbk)
ISBN: 978-1-315-77479-4 (ebk)

Typeset in Sabon
by Apex CoVantage, LLC

Contents

Abbreviations

GW	gigawatts
HVDC	high-voltage direct current
kWh	kilowatt hour(s)
m/s	meter(s) per second
Mt	million tons
Mtoe	million tonnes of oil equivalent
NTC	net transfer capacity
PPP	purchasing power parity
toe	tonne of oil equivalent
TWh	terawatt hour(s)
USD	United States dollar

Preface

There is almost no dispute in the economic literature and in the business practice that harmonized energy systems, and electricity markets in particular, produce positive effect in terms of aggregate welfare and cost reduction. Today modern energy-related dynamics extend well beyond national borders. Although national energy policies are still very important and strategic, these are nowadays widely affected by global trends and supranational dynamics. In this context, it is key to identify effective forms of governance that go beyond the state and to define ways to foster convergence in order to allow the creation of integrated energy systems able to guarantee cheap, sustainable and abundant energy for all.

Energy cooperation in the Mediterranean basin offers, in this perspective, a number of interesting live experiments. The region presents a variety of different situations when it comes to energy supply and demand, including, within its boundaries, energy exporting countries together with net importer and transit countries. In addition, energy issues interrelate heavily with climate change policies, in which EU countries are leaders on the global stage and provide an example of how participating nations can overcome different national priorities (and conditions) to pursue common targets.

This variety imposes the definition of a set of complex energy policies that is then reflected also into composite regulatory frameworks. National, regional and global policymakers need to address challenges such as provision of correct incentives to promote the deployment of the most efficient technology available, integration of renewable energy source (RES) generation in conventional electricity systems, security of supply in developing and emerging economies, matching energy demand in new urban areas, efficient use of the capacity currently installed, improvement/reinforcement of cross-border transmission capacity, demand side management and sustainability of the energy system, just to name few of them. Moreover, the additional challenge is to ensure that the solutions identified (from the global to the national level) are harmonized, consistent and robust to unpredictable circumstances.

To provide possible answers to these challenges, the main regulatory and governance questions has been for several years now how non-EU countries deal with the EU regulatory framework when it comes to energy trade and energy cooperation in the Mediterranean basin and, more recently, following the completion of the Mediterranean integrated electricity network, how non-EU countries interact.

The Energy Community and MedReg represent a living experiment on how two different approaches can provide answers to these complex interactions. The recipe provided by these two experiences is significantly different in their approach. The Energy Community proposes a top-down approach, calling for the adoption of the *aquis communitaires* outside EU borders; on the contrary, MedReg is based on a bottom-up approach, proposing a gradual integration of market-based principles with a voluntary development process of shared rules and regulations among associated countries. These two models have been carefully analyzed in this book, also offering a critical review of the lessons to be learned in order to foster regulatory convergence; such a process is vital for the development of integrated energy markets and represents a key element to guarantee the creation of efficient energy systems. We believe that the contributions collected in this volume will provide useful guidance for those interested in exploring the origin of energy cooperation in the Mediterranean region, along with an overview of the different dynamics that this complex process may trigger.

Simone Mori, Executive Vice President, Enel Group

Integrated and sustainable energy for the benefit of the people of the Mediterranean . . . and beyond

Peace, political stability, continued progress in the quality of life for the communities that belong to the Mediterranean basin are goals and aspirations of great relevance. A common history, a difficult present and the idea to build together a positive future constitute a set of demanding challenges particularly cumbersome for those who want to look with confidence to the future of energy cooperation in the Mediterranean Sea. The Mediterranean is a region full of different cultures, ideas, religions, lively exchanges in terms of social, economic and cultural dialogue. Moreover, it is the place where democracy was born and must be now further developed, where conflicts must be avoided by promoting constructive discussion, collaboration and cooperation with better and enduring integration in the political and socio-economic dimension of the countries of the region.

The countries considered are increasingly interdependent and interconnected. In this context, energy is playing an increasingly strategic role for all the countries in the area, and it also represents a key tool to promote the sustainable development of the entire region. The conditions for a promising future for the people of *Mare Nostrum* require strong commitment and dedicated effort to achieve a better integration of the national systems that should evolve toward a modern dimension:

1. Energy systems need to be safer, endowed with adequate capacity to cover proactively changes in demand and able to supply and exchange energy securely and efficiently thanks to an adequate set of reliable interconnections and infrastructure.
2. Energy systems have to offer their functions at a reasonable cost, be able to serve consumers at an affordable price and be able to guarantee, more and more, technically and commercially sophisticated services, designed to better protect the individual freedom of choice for each individual.
3. Energy systems need to be environmentally friendly and always consistent with the need to limit climate-altering emissions and increasingly efficient on the production and demand side; energy systems should pay special attention to the sustainable exploitation of primary resources, including renewable ones.

The integration of energy systems with this level of complexity and the creation of energy markets cannot be left unattended as wild savannahs, but require refined and shared rules, needed to dynamically follow the technological development and also to anticipate product and services requirements. Crucial to this approach is the necessity to open transparent consultation processes that should be implemented and enforced by regulators able to operate independently and autonomously, in order to ensure the promotion of healthy competition and consumer protection.

For the above mentioned set of reasons, all Mediterranean countries need to adopt/develop a solid legislative and regulatory framework at the national level, designed and implemented by responsible institutions (including Regulatory Authorities and Regulatory Commissions) able to work together to facilitate and accommodate the effort and the requirements of modern energy systems.

With this vision in mind, in 2005 the Italian Energy Authority (Autorità per l'Energia Elettrica e il Gas) took the initiative to promote a "dialogue platform" as a precondition for the creation of a real common *entity* for the existing independent Mediterranean regulators and representatives of the ministries in charge of energy matters in those countries where an independent regulator did not exist yet.

In order to face the immense challenges recalled so far, the Mediterranean regulators decided, voluntarily, to combine their national commitments with an open international dialogue through the creation of a platform targeted to the interests of the common areas of responsibility. According to this approach, after the first, positive experiences – e.g. Council of European Energy Regulators (CEER), Energy Regulators Regional Association (ERRA), African Forum for Utility Regulators (AFUR), Regional Electricity Regulators Association (RERA), Latin American Energy Regulators (ARIAE), and National Association of Regulatory Utility Commissioners (NARUC) – the Mediterranean energy regulators and ministries decided to gear up their contribution to the construction of a perspective, foreseeable energy free-trade area of the Mediterranean region, by setting up a new platform for their cooperation. Therefore, I invited my fellow regulators in 2005 to start a series of informal meetings aimed at the promotion of a body that could facilitate such collaboration. This initiative gave us the possibility to create a first "Working Group" which met several times in Rome in 2006. After these first positive experiences of dialogue and exchange of know-how, the regulatory authorities of the Mediterranean countries decided to expand their activities at this new regional dimension having well in mind the goal in the "Barcelona Declaration" in 1995 and pursuing a strong will to move towards a formal regional institution. Straight from the beginning, the participation in the various meetings was inclusive to the point that, it was decided to create the "Association of the Mediterranean Energy Regulators", MedReg,[1] with its own statute and regulation.

MedReg enjoyed, from the very beginning, an open collaboration and choral consensus around the strategic position and the activities defined by the Association. I had the honour to take over the first presidency of the new organization, whose permanent secretariat was entrusted to the Italian Authority. As for the organizational structure, it was defined by choosing a system of balances with a two-year rotation for the responsibilities of the executives involved in the steering committee, as well as for the organization of the General Assembly[2] meeting and for the participation of four working groups (Electricity, Gas, Institutional and Environment, Renewables and Energy Efficiency). All the participating countries, being from the north and south of the basin and from the eastern and western parts of the region were (and are) engaged with the same weight and according to criteria of periodic alternating mandates with a predefined time limit.

The governance structure made possible to determine, from the outset, a climate of constructive cooperation, but with a clear and transparent accountability, open to effective internal collaboration and with an outstanding institutional commitment. This new *entity* was also very welcome by the European Commission that offered support to the newborn association becoming a reliable partner, able to guarantee an important institutional dialogue, together with a valuable financial support built upon a constructive participation in the day-to-day work and on a long-term contractual agreement. At the same time, it is of the utmost importance to mention MedReg privileged relations with the Parliamentary Assembly of the Mediterranean (PAM) along with governments, financial institutions, consumer associations and with the operators of the region.

The key role played by MedReg in several international fora and its decisive contribution to the creation of the International Confederation of Energy Regulators (ICER), its continuous institutional consolidation as a focal institution in the region, as well as its current limits are explored in this remarkable volume. The present contribution is part of a publication that represents the result of careful research and studies carried out in the last five years and that will become a reference, I am sure, to the different stakeholders operating in the area of energy, environment, economic and social development of a region – the Mediterranean – that plays, and will continue to play, a strategic role in the global context.

My hope is that readers and scholars may also receive from this book useful information and valuable stimuli to support, at all levels of responsibility, the importance that multilateral cooperation and system integration can play as an essential tool for the future of peace, stability, development and to improve the quality of life for all communities belonging to the *Mare Nostrum* and everywhere.

Alessandro Ortis, Honorary President, MedReg

Notes

1. Officially registered in Rome, 15 November, 2007.
2. The General Assembly is the decision-making body responsible for defining MedReg's strategy and approving the action plan, operational guidelines, budget and any other official document.

Integrating energy

Lessons to be learned from the Energy Community

The Energy Community, which originally was supposed to have a reference to South-East Europe even in its name, was and sometimes still is associated with the Balkans. Historically, this is correct. When the Athens Treaty was signed in 2005, following a "testing phase" under two consecutive memoranda of understanding in 2002 and 2003, the main purpose was to help former socialist countries to attract investments by introducing market reforms and to prepare them for European Union (EU) accession. Geopolitics did still not play the main role, despite the fact that the territory covered by the Energy Community was about to be branded as the Southern Corridor. With Ukraine and Moldova joining in 2010 and 2011, respectively, not only the geographical scope of the Energy Community expanded. The Energy Community gained an additional political dimension, ensuring the security of supply of the wider Europe. If assistance and development were the decisive motives in the early, South-East European, years, the discovery of its strategic role and of concrete European Union interests characterizes the current phase. This is undoubtedly a positive development, as it establishes a certain balance of interests in the Energy Community between the Eastern and South-East European Contracting Parties on the one side and the EU on the other.

In any event, the Energy Community is more than the sum of the interests of its members. The treaty is a genuine attempt to translate Monnet's and Schuman's ideas of regional integration into the twenty-first century and to countries outside the European Union. When the Energy Community was born in the early 2000s, many compared it to the European Coal and Steel Community of 1952. Evidently, the approach chosen by the founding fathers and mothers of the Energy Community was one of linking this new community to the "mother ship" of European integration, the EU. EU legislation is the almost exclusive source of law for the Energy Community, its institutions are bound by homogeneity obligations to the EU judiciary (Article 94 of the treaty), and the EU institutions also assume a coordinating role in the Treaty's governance structure (the European Commission under Article 4) and a prominent role in the decision-making. Linking it with the European Union rather than only copying it was and is a central rationale of the Energy Community. That the EU actually would join *en bloc* as party to the treaty and take the far-reaching commitments under Title IV of the treaty was far from clear in the beginning. For the countries in South-East Europe, its membership was a strong sign of commitment.

The Energy Community is not the first case in which the EU integration model has been extended to third countries. The most consequential extension is the accession of third countries to the European Union, as has happened recently when Croatia became the Union's 28th Member State. In the case of the European Economic Area, third countries with a vital interest in access to the internal market accept submission under EU rules and enforcement institutions modeled on the Commission and the Court of Justice. In the bilateral relations with Switzerland, EU law is extended currently without an institutional package.

The Energy Community is the first multilateral agreement where both law and institutions are extended in one sector only. A multilateral treaty has obvious advantages vis-à-vis bilateral relations; once negotiated it is an open invitation for accession by new partners, thus shortening the process of finding tailored solutions for each interested country. In a world of multinational infrastructure projects and transactions, bilateral solutions also have their limits.

The common denominator of the examples just mentioned is the import of EU rules and, to some extent, its institutional model. Accepting those requires a strong and continued interest in EU law and institutions by the non-EU members. Countries opting for an integration agreement with the EU will usually have made some kind of cost-benefit analysis beforehand. In the energy sectors, submission under the EU liberalization program requires far-reaching changes and tough decisions, especially in countries where the state controls the sector, from public ownership to unlimited price regulation. Moreover, dreams of energy autarchy may have to be abandoned. Reasons for accepting a transition towards the EU's vision of open and integrated energy markets may be manifold. They can range from an interest in EU accession or at least association, to gaining market access, attracting investment or simply the expectation of increased sector efficiency by applying the EU programme. As might be expected, a bundle of all those and potentially more motive applies in the case of the Contracting Parties to the Energy Community. It is fair to say that in all existing and potential future Contracting Parties, a certain proximity (not necessarily through direct borders) with the EU is a relevant factor.

Binding, enforceable rules and an institutional architecture are the two pillars on which the EU, as well as the countries associated with it through integration agreements such as the Energy Community Treaty, is built. The difference they make for international energy relations, traditionally characterized by an explicit "hard politics" dimension, not subject to a tight legal framework and where disputes were often resolved by ad hoc diplomacy is easy to imagine. The Energy Community follows the path of the European integration process in the tendency to replace intergovernmental relations by the rule of law. The rules applied by both EU and Energy Community are first and foremost governance-related and grant direct rights (such as third-party access, eligibility) only as a reflex where their exercise contributes to achieving the overall *telos* of market liberalization and integration. Traditional international energy relations and their dispute resolution mechanisms in particular are based on the idea of balancing subjective interests in individual transactions, be it between corporations, between countries or between corporations and countries. This change in perspective from international to EU-style energy relations has interesting consequences. On the one hand, one may question whether the EU model protects private interest efficiently. This question is to be answered in connection with the enforcement procedures (see the following discussion). On the other hand, the implementation of governance-related rules, unlike transaction-related rules, can be monitored and measured. In the Energy Community, compliance by Contracting Parties is monitored by the Secretariat in an annual report. This allows for measuring the success of the organization, and intervening by the institutions if the progress achieved does not meet the expectations. This may include updating the applicable rules where needed, as has happened, for instance, in 2009 (European Union) and 2011 (Energy Community) with the so-called Third Package. An organization thus becomes and stays dynamic by constantly developing its yardstick.

The Energy Charter, albeit based on binding rules as the Energy Community, rather follows the traditional model of transaction-related international energy relations. Its rules are to be applied, not implemented. This makes it more apt for a global outreach but lighter in its governance. Other, less formal arrangements such as the Eastern Partnership or the Union for the Mediterranean, do not oblige their members to implement and apply rules in their domestic legal orders. In view of the advantage of measurability of progress and the possibility for dynamic evolution, it is no surprise that at some point, a paradigm change towards introducing legal rules is discussed in many of these contexts. As mentioned earlier, the Energy Community underwent precisely this evolution when changing from a community based on memoranda to a community under a binding treaty.

In integration agreements based on the EU model such as the Energy Community or the European Economic Area, upgrading the applicable legal rules serves another important purpose besides improvement of their quality, namely upholding homogeneity. As concluding integration and association agreements with interested third countries results in the co-existence of different integration circles in wider Europe, it must be assured that they do not drift away from the core, the EU itself. Homogeneity is to be maintained in terms of legislation, but also in terms of (administrative and/or judicial) enforcement. This aspect of European integration beyond the EU is particularly emphasized in the European Economic Area (EEA), less so in the Energy Community. The history of the EEA also shows that the EU has a certain dislike towards structural homogeneity: when the drafters of the EEA Agreement envisaged an EEA Court common for EU and European Free Trade Association (EFTA) member states, this idea was rejected by the Court of Justice in its Opinion 1/91.

In any event, pure homogeneity is difficult to maintain both in terms of legislation and of enforcement. With respect to the former, it may turn out that a set of rules developed for and agreed upon by some countries – the EU Member States – cannot be transferred to another set of countries without frictions. In the Energy Community, for instance, a debate has recently started about whether certain requirements of the *acquis* are not too ambitious for countries lagging behind with regard to the state of market development and their income structure. The treaty is characterized by a remarkable degree of flexibility; Article 24 allows for adaptations when incorporating EU law on account of "the specific situation of each of the Contracting Parties". Titles III and IV offer a wide range of legal bases for genuine Energy Community law. In practice, they have rarely been used. So far, the balance between homogeneity and flexibility in the Energy Community is in favour of the former. This "discipline" allowed for incorporation of rules even when their impact was far-reaching – the Third Package, the Renewables Directive or the Oil Stocks Directive may be mentioned in this context.

Regarding enforcement homogeneity, the EU enforcement institutions – the Court, the Commission and recently ACER – form the center towards which the institutions of an integration agreement orientate themselves when interpreting and applying the law. However, enforcement practice shows that no two cases are the same, and precedence in European energy law is surprisingly rare. The Energy Community Secretariat in all its cases so far had to break new ground, including for highly complex questions such as the existence, and the demarcation of, a principle of cost-reflectivity of network tariffs under the Electricity Directive.

Besides the law on substance, it is the institutions and procedures which make the quality of an integration agreement. In this respect, the Energy Community is by far no carbon copy of the European Union, despite being inspired by it. The structure of the treaty is more complex than the EEA's. It consists of three layers, of which the first one (Title II) enables the incorporation of EU law, the second one (Title III) envisages the creation of genuine governance rules for a regional market and the third one (Title IV) provides for common rules and principle for an internal market including both the Contracting Parties and the EU. In terms of institutions, it is fair to say that the bodies established by the Energy Community Treaty fall short of what the EU (and the EEA) has achieved. There is no judiciary and no parliamentary body, the only independent and permanent institution – the Secretariat – has no executive powers (which leaves an important gap in the enforcement of competition and state aid rules), and the Regulatory Board is modeled on a looser form of cooperation as compared to the European Union's energy regulatory agency ACER. Effective institutions, however, are crucial for the success of any energy integration agreement, especially where it is supposed to deliver results on par with the EU. Discussing and possibly reforming the institutional architecture under the Energy Community Treaty is be in the focus of the High Level Reflection Group established by the Ministerial Council in October 2013.

The effectiveness of procedures for decision-making, to turn to the final aspect of energy integration agreements discussed here, depends on whether they fulfill their purpose smoothly. With respect to incorporating new EU legislation, the Energy Community Treaty performs very well. The majority principle applicable in this area constitutes an important element of supranationalism, whilst the weighting of votes – one vote per Contracting Party – follows intergovernmental models. Safeguards such as the Commission's exclusive right to initiate the incorporation of EU law or veto rights where the Union's interest and territories are directly concerned make sure a balance between the interests of the third countries and the Union is maintained. But even when it comes to taking fundamental decisions on amending the treaty or the accession of new members, the procedural rules and practice in the Energy Community have functioned without the kind of frictions the European Union suffered between Nice and Lisbon, and which have prevented the EEA, for instance, from ever amending their agreement.

A specific procedure and a litmus test for the effectiveness of any international agreement based on the rule of law is the enforcement mechanism. Unlike the Energy Charter and despite its misleading denomination ("dispute settlement"), the founders of the Energy Community opted for a public enforcement procedure in Articles 90 *et seq.* of the treaty. In the way the principles set out in these provisions are spelled out further by procedural rules and applied by the Secretariat, they come close to the infringement actions known in the European Union. The Secretariat is usually the initiator of proceedings and carries out a bilateral preliminary procedure. The most striking difference to EU law is the lack of a court; the ministers assembled in the Ministerial Council assume the role of judges in cases in which a party to the Treaty is accused of violating Energy Community law. Evidently, they lack the independence and the legal expertise only a court of law or an arbitration tribunal can guarantee. To compensate for this deficit, the procedural rules introduce an Advisory Committee composed of three lawyers which delivers an opinion in each case submitted to the Ministerial Council. This committee may thus be compared to the Advocate General at the Court of Justice of the European Union. Despite having no decision-making power, its assessment constitutes an important element of professionalism and independence in the enforcement practice. The Committee gave its first opinion in a case against Bosnia and Herzegovina about the failure to implement the *acquis* on gas in October 2013.

This case was the first case to be decided by the Ministerial Council, but it was by far not the Secretariat's first one. Every year, some ten cases are scrutinized and considered for opening a procedure. Given the lack of a judicial decision maker, the Secretariat is committed to settle cases through a dialogue with the country concerned. Several factors support the Secretariat's success with this approach in many cases. Experience shows that governments may be more inclined to negotiate indirectly through an international organization than directly with private investors threatening arbitration. Countries aspiring to join the EU are afraid of falling behind in their negotiations for accession in case of non-compliance with the Energy Community Treaty. And the Secretariat can efficiently combine its role of guardian of the Treaty with providing assistance in rectifying breaches and solving problems. To some extent, enforcement practice thus turned a weakness into a strength, albeit at the cost of reintroducing a strong element of diplomatic dispute settlement in a community committed to the rule of law. This remains a deficiency in particular with respect to the protection of individuals and investors, who cannot challenge a decision by the Energy Community institutions and are limited to submitting complaints to the Secretariat where their rights are impinged upon by a party to the treaty.

Despite this, the contribution the Energy Community makes to organizing energy relations between countries committed to the same values and principles can hardly be overestimated. The Energy Community has been called a success model by the European Commission, a point of reference by the Council and the basis for establishing a fully integrated regional energy market by the European Parliament. Its duration was just extended until 2026. Seven years after its entry into force, the Energy Community has reached a degree of maturity and acceptance which allows it even to ponder possibilities of improvement and reform. The Energy Community can learn much from other organizations in that process. But its concept, structures and practical experience so far provide some interesting lessons for sectors and regions aspiring reform and market integration.

Dirk Buschle, Deputy Director and Legal Counsel,
Energy Community Secretariat

Regional energy initiatives in the Mediterranean basin

An introduction

Carlo Cambini and Alessandro Rubino[*]

The Mediterranean basin is the only region in the globe where three continents come together. A rich historical heritage and a variety of languages, cultures and religions meet over a deeply asymmetric economic distribution. Currently the region accounts for around 7% of the world population and 8% of the current world energy consumption together with 4.5% of proven oil reserves and 4.7% of the world's proven gas reserves.[1] Although oil has enjoyed the highest share in the region's energy mix (both in terms of production and of consumption), a turn to electricity is currently taking place. This transition increases the interdependence of energy markets. Much of this effort relies on the ability to promote renewable energy sources (RES) in the region together with new electricity interconnections.

The turn to electricity is grounded on the new political, economic and social transformations that are currently taking place in the region. The planning of the energy sector tends to follow closely the evolution of economic and population growth. In this perspective, there are two main trends which mainly characterize the Mediterranean region: first, there is a welfare gap between North Mediterranean countries (NMCs) and South and East Mediterranean countries (SEMCs). Second, most of the economic growth in the region is currently taking place in the SEMCs with NMCs experiencing persistent financial and economic crises. Both trends determine that the poorest regions in the Mediterranean are experiencing the greatest development and economic growth. This, by itself, poses serious challenges on how to finance the developments that will come to respond to the raising energy demand.

The implications of these developments are tremendous for the Mediterranean region as a whole. A great deal of the transformation in energy infrastructure will have to address the growing needs in the South and Eastern Mediterranean basins shaping an increase of energy consumption per capita in the region. Currently NMCs triple SEMCs' energy consumption per capita; however, it is expected that by 2030, this gap will be reduced to double. A great deal of this transformation in the energy sector is closely linked with broader political, economic and social events; hence, the appropriateness of the solutions adopted will have strong implications for the region. The strong relations between energy consumption and macro-economic growth are well known in the literature,[2] and it applies also to the Mediterranean region where energy has entered the political agenda with the highest of priorities.

The strong interaction existing between social condition and energy consumption[3] has multiple dimensions. Energy is a basic commodity that is required also to meet other basic needs, such as access to water, sanitation and health care. Together with this, energy has other indirect economic and social benefits. It enables the condition for job creation also in economic sector with traditionally low added values, such as agriculture and industry, that are widespread in rural areas. Energy access is necessary to allow access to primary education, giving the possibility to study also after the sunset, in rural and informal settlements. Finally, energy access reduces gender gap, by reducing in-house pollution and freeing women from traditional housekeeping and domestic tasks.

The development of the energy industry will affect countries exporting and importing fossil fuels in a different way. In the energy importing countries, economies are facing the difficulties created by inefficient and insufficient energy markets which contribute to creating energy poverty and poor living standards. Additionally, the volatility and high prices for fossil fuels have a negative impact on public finance and the overall economic performance. The picture is different for the exporting economies. These countries (mainly Algeria and Libya), where energy exports constitute a large share of their economic activities, have recently seen relevant political and social hurdles slowing down the development of the society. Notwithstanding this, all these countries are expected to have optimistic prospects for growth and the energy is a key sector for the success of this evolution.

SMCs have been at the centre of numerous initiatives targeted to improve the energy cooperation, in particular in the electricity sector, among the countries in the region. These initiatives are crucial to build regional and sub-regional energy integrated systems, able to generate the necessary condition to favor investment on critical infrastructure, such as electricity interconnection and gas pipelines, in addition to the much needed additional generation capacity. The existence of both intra-Mediterranean and Euro Mediterranean initiatives creates a favorable environment in order to promote a pan-Mediterranean area of energy exchange and to enhance security of supply. Moreover, SMCs also aim at increasing the diversification of primary sources in the region. Large-scale initiatives, such as Desertec and the Mediterranean Solar Plan, will now be accompanied also by small-scale pilot projects, which will be able to provide a useful illustration on the challenges related with the integration of renewable energy sources in the systems. These initiatives require the creation of an interconnected pan-European and cross-Mediterranean grid, relying also (but not exclusively) on renewable generation, that shall enable electricity production at the best sites for each technology, regardless of national or regional borders and distances to the central European load areas.

In the next twenty-five years, energy demand in these countries is expected to grow. Therefore, one of the major challenges Mediterranean countries will have to face will be to ensure energy supply in order to meet a fast-growing demand, especially in the Southern Mediterranean shore. In order to meet this challenge, increasing investment in Renewable Energy Sources is perceived as a pivotal strategy, not only to strengthen energy production, but also to create jobs. In fact, since the Arab Spring job creation is among the core objectives of many countries in the region. The current energy system, mainly based on fossil fuels, is perceived to become unsustainable in the near future, where the increased energy demand should be satisfied by increasing the penetration of RES, which in turn could be achieved by investing in RES technology. This should not only help mitigating CO_2 emissions, but will also lead to an efficient implementation of the energy systems. All these challenges could not be met by single countries alone. Fostering cooperation in the energy sector, as well as regulation among Mediterranean countries is considered an essential step towards a better-integrated energy market, although some other major obstacles remain, especially related to the political instability in many countries of the region. The contributions collected in this book try to shed a light on this process.

The situation in South-East Europe (SEE) presents a different picture and a diverse set of problems. The past decades of transition and conflict have left the Balkan region with a legacy of inadequate growth and declining living standards. In the region, most countries have recovered only 75% of their pre-transition income levels. This poor economic performance can be attributed to different factors. First, at the onset of the transition, the economic conditions in the region were particularly critical: the economies have had unbalanced industrial structures, weak institutions, and fragmented civil societies. Second, war, civil strife and ethnic conflict have had a devastating effect. In those countries directly involved in the conflict, the civil war has fragmented societies as well as destroyed institutions and infrastructure. In all other countries indirectly involved, regional conflicts have disrupted normal economic activities, creating an uncertain business climate. Third, macroeconomic stabilization policies have been inconsistent and this has further exacerbated uncertainty in the standard business activities. Finally, structural reforms have been weak; progress in trade liberalization, privatization, enterprise reform, competition policies and financial sector development have still to be realized. Similarly, disappointing progress has also been made in institutional, social and environmental policies.

To some extent, this opaque situation offers a more optimistic panorama when we look at the energy systems in the region. The normalization of the political situation led to an attempt to reintegrate the energy system, both in physical and in institutional terms.[4] The international community has played an active role in promoting assistance to this process, offering both technical support, allowing the reconnection of the network of the transmission system operator in the region, and at institutional level, promoting and supporting the creation of the Energy Community.

There is a long list of technical and economic characteristics of energy systems that differ widely in each country of the region. However, technical aspects are not the only significant factors: Institutions also play a major role. Starting from the first initiatives launched in 2002, Balkan countries have undertaken a major attempt to harmonize their energy systems, in both the technical and institutional dimensions, by adopting a framework based on the European Union *acquis* on energy. This poses the question whether a set of rules and legislative provision adopted in a specific context, can be efficiently implemented elsewhere.

The adoption of a set of legislative provision does not mean automatically that they will affect the behaviour of all different actors in the industry. Actually, those countries where formal law has not evolved spontaneously, but rather was introduced by transposition, usually show large scope for informal law. Thus, it is the contemporaneous existence of a feasible ground for a market reform and a credible commitment with regulatory strengthening that will ease the introduction and correct implementation of a market-based restructuring. How effective this process could be, in the medium and long run, to promote a stable and attractive investments environment? In this book we look into these complex dynamics, proposing a critical review of the existing processes.

******

The existence of a common past and geographic proximity alone is not enough to qualify the Mediterranean basin as region. A region is not a mere geographic area. It should have a well-developed and sustainable institutional structure in order to fulfil its primary function – that is to address political and economic problems before they escalate to global concerns. Energy markets are not exempted from these dynamics, and actually play a crucial role in the context of regional integration and cooperation between NMCs and SEMCs. This book focus on two intra-regional initiatives created for development and integration of energy markets: the Energy Community and Medreg.

The Energy Community and Medreg, apart from their common strategic role in providing a much-needed stable regulatory environment for energy markets in their respective reference countries, represent examples of a diverse development of regional energy initiatives. The former is initiated by external actors and is an example of a top-down approach, whereas the latter is a voluntary bottom-up initiative of the countries involved. The way the institutional framework is built is not without consequences on the functioning and organization of the two regional initiatives. The chapters in this book assess these different approaches and their consequences in the framework of the development of the Energy Community and Medreg, with particular reference to their impact on regional integration, energy policy and institutional change.

The first part of the book specifically looks at regional cooperation between Europe and Mediterranean and SEE countries and at how the economic, regulatory and institutional framework of the EU might be extended in these areas. In his chapter, Peter Polajnar stresses that support for multilateralism is a traditional feature of the EU external policy and is particularly strong in its relations with immediate neighbours. By stimulating neighbours to interconnect, the EU aims at improving hard and soft security at its borders. According to the author, EU policy is most successful when neighbours are candidates for membership and regional cooperation can be translated into conditionality for accession. However, in cases when the EU cannot use conditionality, it has to resort to other tools. Polajnar analyzes how a regional free-trade agreement (CEFTA) in the Western Balkans has been constructed and how it is governed. CEFTA has moved from an initiative justified by conditionality for EU accession into an economic development tool. Governance of CEFTA offers lessons to other EU-sponsored initiatives in the Western Balkans and in the wider neighbourhood, including the Energy Community.

As shown before, the Mediterranean region is composed of countries that show a significant variety in their levels and paths of economic growth rate and other economic fundamentals. This diversity is also mirrored by the energy landscape of the region. In the Mediterranean basin key hydrocarbon suppliers, transit countries from other regions to the EU and net energy importing countries co-exist. Consequently, energy policy in the Mediterranean area is characterized by a composite set of technical, economic and geopolitical factors that contribute to determine a multifaceted scenario. This poses the problem of how European countries should deal with non-EC neighbours. The chapter by Alessandro Rubino explores the regulatory challenges related to the development of a Euro-Mediterranean electricity area and identifies three pillars upon which it should be built and organized in the coming years. The first pillar is the foundation of a legislative and regulatory convergence based on the "energy corridor" specific approach; the second is the expansion of network infrastructures in the area; finally, the third pillar is represented by the establishment of an "energy free-trade area" in the Mediterranean region based on renewable energy sources.

The chapter by Francesca Pia Vantaggiato reviews the main policies enacted by the EU in order to tackle its external energy policy needs in its so-called neighbourhood, namely in Eastern Europe (Balkans and beyond) and in North Africa. A special emphasis has been placed on the most accomplished of these attempts (the establishment of the Energy Community) while not neglecting broader policy programmes containing energy clauses, such as the European Neighbourhood Policy. According to the author, the scant success in achieving a coherent external energy policy reported so far calls for an assessment of the reasons and an appreciation of the alternative routes energy market integration is taking. In this regard, and in line with the chapter by Cambini and Franzi, the crucial role energy regulatory authorities are coming to play in the construction of a common EU internal energy market and external energy policy is also highlighted. The creation of the Association of Mediterranean Energy Regulators (Medreg) is presented as a case in point to illustrate the significant changes occurring in the sphere of the EU energy policy and the growth of regulators as institutional actors.

The analysis to which extent projects of European interest are relevant for "outsiders" and whether the external EU energy policy can be used as incentive for defining and promoting regional interests is analyzed in the chapter by Ana-Maria Boromisa. Boromisa argues that EU policy is motivated primarily by energy security concerns and promoted by integration perspective. For countries in the Southern Mediterranean, Western Balkans, and Eastern Neighbours, the adoption of EU policy is seen as an opportunity to attract investment and to foster other economic interests on a national level, while capacity to formulate and implement regional interests is still missing.

The cooperation between EU and Mediterranean countries in renewable energy is recognized to be of fundamental importance. According to Miriam Bardolet the implementation of such an integrated EU–MENA power system requires a strong cooperation framework with a particular focus on renewable energy development. This framework should lead to progressive regulatory convergence as well as to the establishment of common institutions. An approximation of the regulatory frameworks is crucial in order to create a level playing field for investments and to ensure the operability of the system. Common institutions will be thus necessary for the implementation, monitoring and enforcement of this framework.

The need for a close cooperation is recognized also between EU and Western Balkans countries. The chapter by Milica Uvalic points out that such regional cooperation could substantially help economic growth and development of these countries in the future. The new EU policy for the Western Balkans launched after mid-1999 have promoted various initiatives of regional cooperation among the Western Balkan countries, including cooperation in the field of energy. The chapter considers the results achieved so far in regional cooperation among the Western Balkan countries in the energy sector, and illustrates why regional cooperation and integration of energy markets would have a series of beneficial effects.

Since energy represents a fundamental sector in ensuring the prosperity and the stability of the Mediterranean region, energy-related issues have always received considerable attention in the agenda of the Parliamentary Assembly of the Mediterranean (PAM). The chapter by Sergio Piazzi presents a summary of the activities carried out by the PAM on energy-related issues, underlining the importance and the key role of parliamentarians, in their capacity of policymakers, in support of cooperation in legislative matters at the regional level.

Finally, the chapter by Ernesto Bonafé points out that international energy cooperation must be based on a sound regulatory framework in order to increase stability and transparency in the investment environment. The chapter focuses on the role of the Energy Charter Treaty (ECT), of which the EU and its member states are already members. The treaty provides an international framework to enhance legal reform in transition countries. According to the author, the ECT is equipped to play a key role in the Euro-Mediterranean energy market.

The second part of the book focuses on how the EU institutional and policy model has been exported outside EU borders. The chapter by Josefine Kuhlmann analyzes the role of the European Neighbourhood Policy (ENP). The author shows that this policy is built on soft law instruments, and that using an additional instrument with legally binding commitments – such as a sectoral agreement – could facilitate the achievement of the ENP's objectives. Kuhlmann argues that the Energy Community provides such a legally binding framework and has already attracted attention amongst ENP countries. A more critical view on the role of the Energy Community is provided in the chapter by Natasha Hroneska. The author explores to what extent the Energy Community for South-East Europe is a viable model for the energy relations of the European Union with its southern and eastern neighbours, and discusses the perspectives and the limitations to the replication of the Energy Community in the wider EU neighbourhood. The author considers that by transposing the *aquis communautaire* in policy areas such as security of supply, energy efficiency, sustainable development and competition to the contracting parties of the Energy Community, the EU has been extending its energy norms outside its borders. The analysis focused on two important aspects: the deepening of the integration of the Energy Community and the diverse nature of the relations between the EU and the countries of its Eastern and Southern neighbourhood. The main findings of the analysis show that, even though the Energy Community should strengthen the implementation process of the EU *aquis*, it is still the most attractive model of energy cooperation with the EU and a potential model of norm diffusion in the EU neighbourhood. However, according to the author, two general limitations to the replication of the Energy Community in the EU wider neighbourhood exist. The first one is the level of integration between the EU and its neighbours. Indeed, the EU membership perspective plays a crucial role in the political commitment of the governments to import the EU energy norms and values. The second one is the geopolitics of energy supply, which is considered as an issue of highest national interest in some of the energy partners of the Caspian, North Africa and Middle East regions, rich with energy resources.

The third and last part of the book presents different case studies assessing the implications of the export of EU institutional model.

External objectives of the EU energy policy are being fulfilled through the Energy Community which embraces not only EU member states and candidate countries but also third countries without any perspective of membership in the EU. Roman Petrov's chapter analyzes the challenges of the process of Europeanization in Ukraine and Moldova through the application of the EU "energy acquis". Ukraine and Moldova were among the first eastern neighbouring countries to join the Energy Community and thereby to bind themselves to the EU energy acquis contained in the Energy Community Treaty. According to the author, the Treaty offers many legal challenges for the Ukrainian and Moldovan legal systems. Apart from the legal commitments of implementing the EU energy acquis the Treaty makes Ukraine and Moldova abide by fundamental principles of EU law, decisions of the Energy Community common institutions and the dispute settlement.

The Turkish Nabucco project is another interesting case for assessing the adoption of the EU set of formal laws in a non-EU country. The chapter by Bagdagul Kaya Caner revises this case study. The importance of Turkey's geo-strategic position in energy transit is acknowledged quite clearly in recent years, especially in the EU. It can be argued that possessing natural resources or being in a geo-strategic transit position is essential, though not enough, to trigger energy trade or attract investments. As the author argues, it is however necessary to maintain both political and legal stability in order to create a reliable investment climate. Turkey not only takes part in the ECT and is a member of the World Trade Organization. Turkey has also started accession negotiations to the EU already in October 2005, even before that the country initiated its extensive reforms in the energy sector that led to the liberalization of its energy markets. These fundamental steps made Turkey to be perceived as a reliable host for transit investments.

EU rules promotion in the southern Mediterranean countries has also a relevant effect on countries' regulatory framework, and in this perspective, a very peculiar role is played by the presence of a national independent regulatory authority. The chapter by Carlo Cambini and Donata Franzi shows how an independent regulatory authority constitutes a necessary condition to maintain the credibility of commitment to regulate natural monopolies. Independence, thus, improves the regulators' ability to make long-term commitments to regulatory decisions and contain risks of administrative expropriation. The authors thus examine both features of existing regulatory agencies of southern Mediterranean countries' regulators, stressing three dimensions of independence, as well as the perception of EU rules promotion by southern Mediterranean regulators. Findings show a low degree of agencies' independence from the existing national authorities, but potentials for a more incisive action of the EU as rules promoter and rules harmonization.

In order to fully understand the importance that progressive system harmonization will play between EU and Mediterranean countries, it is essential to assess the impact of the energy exchanges between the two regions. To this aim, the chapter by Angelo L'Abbate et al. presents a techno-economic evaluation of the impact of electricity generated in Northern Africa on the European system and the Italian grid in a 2030 horizon. The chapter also presents a closer assessment of the Italian power system as a potential electricity hub in the Mediterranean region, paying particular attention to the behaviour of the Italian transmission grid. The results confirmed that, in the long run, electricity will flow from south to north with practically fully loaded cross-Mediterranean interconnections. On the other hand, the analysis proves that further investments on the Italian national transmission network may be necessary.

Renewables play a fundamental role in the Mediterranean area reinforcing the external dimension of the EU energy policy. Rosa María Fernández analyzes the 2020 strategy and the increase role for negotiations with different partners. The author considers how bilateral agreements between Member States and Mediterranean countries can have a positive influence towards sustainable development, particularly through the promotion of renewable energy projects, while there may be doubts about the effect that these bilateral agreements can have over European integration. The 2020 strategy advocates for an increased use of these agreements, but the result will depend on the specific focus of each member state during the negotiations: its own energy needs or the general welfare of the parties involved. Finally, the chapter by Alessandro Sapio presents a detailed description of the renewable energy sources available around the Mediterranean area and outlines the main existing empirical findings about the diffusion of renewables and their effects on the Mediterranean electricity systems. This chapter offers new perspectives on the interdependent roles of regulatory and physical integration in driving the development of renewables in the area.

The Mediterranean is enjoying a revived centrality thanks to the burgeoning demand for green energy. European policymakers and businesspeople are increasingly alert to the investment prospects offered by that area. International energy cooperation, thus, must be based on a sound institutional and regulatory framework in order to increase stability and transparency to favour the investment environment. The comprehensive EU *acquis*, which may be considered as a benchmark for regional integration of energy markets, plays a fundamental role in promoting the convergence of energy policies with EU framework. This book provides a comprehensive assessment of this convergence process, pointing out the strengths and the limits of the current regional cooperation and integration and institutional change within the Mediterranean countries. We hope that this book will help policymakers, institutions, companies as well as academics to a better understanding of the economic and institutional ecosystem that characterizes the Mediterranean area, and will hopefully promote further research in this field.

Notes

* The findings, interpretations and conclusions expressed in this chapter are those of the authors and do not reflect the positions of the institution.

1. Observatoire Mèditerranèen de l'Energie (2011), "Mediterranean Energy Perspectives, 2011", Nanterre, France.
2. D. Stern (2011), "The Role of Energy in Economic Growth", *Annals of the New York Academy of Sciences* 1219, pp. 26–51; and P. Segal (2011), "Oil Price Shocks and the Macroeconomy", *Oxford Review of Economic Policy* 27 (1), pp. 169–185.
3. International Energy Agency, Organization of Petroleum Exporting Countries, Organisation for Economic Co-operation and Development, and World Bank (2010), "Analysis and Scope of Energy Subsidies and Suggestions for the G-20 Initiative", joint report, Toronto, Canada, June.
4. A. Rubino (2006), "Reverse-Balkanisation of Energy Markets in South East Europe", Working Papers SIDE, Rome.

Part I

Regional cooperation

Chapter 1

Multilateralism in the European neighbourhood

The case of the Central European Free Trade Agreement

*Peter Polajnar**

The chapter explores the growing role of regionalism in European Union (EU) external policies, which has become even part of EU accession conditionality for the candidates and potential candidates in the Western Balkans. As the countries are under pressure to engage regionally and as the EU is actively supporting the process, attention needs to be given to governance of newly designed initiatives of regional cooperation. Sections 1 and 2 establish the main principles of regionalism and governance. Section 3 presents examples of regional initiatives in the Western Balkans and in the eastern and southern neighbourhood. Section 4 assesses the functioning of the Central European Free Trade Agreement (CEFTA) through the prism of the main principles of governance. Section 5 examines the economic impact of CEFTA. The last section applies the experience of CEFTA to other similar initiatives in the European neighbourhood.

Regionalism in EU external policy

Over the last twenty years, regionalism has emerged as a growing factor in global and European politics. It is the process of political, economic and security integration in the European continent embodied in the EU.

Regionalism is defined as a process of regularized, significant political and economic interaction among groups of neighbouring states. The interaction can take place between national governments, local authorities, private businesses and civil society across a wide range of issues (Dwan, 2000). Tassinari (2005) defines three trajectories leading towards regional integration:

- Regions can be formed from "outside and above" at the initiative of countries, international institutions or, more generally, foreign policy elites.
- Regions can be formed from "within and below" as a result of networking by grass-roots movements in the civil society, firms or sub-national authorities.
- Regions can be formed for functional, interest-based or community-building purposes and can pursue goals as diverse as cultural cooperation, civil society development, trade or security.

Regionalism in today's Europe is highly influenced by the process of EU integration and has been extensively used in the EU external policy.[1] The concept emulates EU's own integration path, and "leads to national as well as regional stability and economic progress for all partners" (European Commission, 2002). The underlying rationale for the active role of the EU in multilateralizing relations in its neighbourhood is based in security concerns (Dangerfield, 2008). It facilitates top-level dialogue which prevents conflicts at EU borders, therefore providing hard security. It addresses socio-economic goals by stimulating trade, environment protection, energy supply and border management, which contributes to soft security. Due to the EU's political and financial resources, Pedersen (2002) describes the process as cooperative hegemony. Given the EU's strength in areas such as economy, technology, culture and ideology, the EU is a willing and a sought-after partner.

Cottey (2009) and Tassinari (2005) observe four functions which can be attributed to regionalism in EU external action:

1. A political, bridging function uses regionalism for overcoming historical divisions and preventing the emergence of new divisions.
2. An EU integration function promotes membership aspirations of countries involved in regionalism. It demonstrates their commitment to EU values such as trade liberalization, respect for supranational institutions and so on. It plays a key role in the enlargement process. The function is less pronounced in the neighbourhood policy, as its participants do not enjoy a clear EU perspective, which undermines their willingness to cooperate under the EU-drafted terms of reference.
3. A cross-border coordination function is built around areas of trade, environment, migration, energy, infrastructure, tourism and so on. Moreover, neighbouring states often have similar interests in these fields. The exchange of experiences can help can help in better designed policies and their implementation.
4. Finally, regionalism can function as a driver of national reforms. It supports policy transfer by providing the context. Regional initiatives are used as platforms for more efficient provision of financial and technical assistance from the outside.

For the EU, multilateral relations offer value added in comparison with the bilateral track. Bilateral relations are generally defined by conditionality, in which the EU demands reforms in return for membership, access to market financial assistance and so on. In regional cooperation, the EU is just one of many partners or often just a sponsor from the outside, so members have more say on the speed and the mode of cooperation. Regionalism has the advantage of bypassing issues which dominate the bilateral approach. It is a convenient tool to provide assistance without compromising the EU's stand in some specific bilateral situations, for example Belarus (Tassinari, 2005).[2]

The EU has constructed a "sophisticated toolbox" to support regionalism, with tools ranging from socialization and institutional cooperation to opening of the internal market or accession to full membership (Behr, 2010). Part of this toolbox relies on efficient governance, which is usually designed by the EU but needs to be implemented by the participating countries. Governance is particularly important in cases in which other incentives for cooperation are low, such as gaining the access to the internal market or having a very distant membership perspective.

Governance of regional initiatives

As regional cooperation enters mainstream policy and attracts more funds from donors, its effectiveness in improving political stability and economic situation falls under increased scrutiny. Policies and projects look for robust governance models to manage complex situations. However, there are no one-size-fits-all road maps. Governance is path dependent, conditional on the ambition, the capacity and the existing arrangements in the region and beyond.

Andreev (2009) analyzes the effectiveness of regional cooperation initiatives in South-East Europe and proposes four determinants of success, one being good institutions, expressed by the degree of institutionalization. He qualitatively measures this degree through the following variables: effectiveness of decision making bodies, the use of professional administration (secretariats), access to regular and adequate sources of finance and strong links with auxiliary bodies.

Alternatively, Cottey (2009) defines the degree of institutionalization by (a) the existence of a framework for meetings on regular basis and procedural rules and by (b) the allocation of powers and resources to supranational bodies. He measures the first variable by the number of meetings and their impact on lower levels of administration. When analyzing the functioning of supranational bodies, he balances the powers of secretariats (i.e. supranational bodies) against the power of intergovernmental arrangements (i.e. high-level decision-making meetings). The outcome of this balance is often dependent on relations with outside sponsors, who tend to prefer supranational bodies.

Willenberg (2009) asserts that the existence of supranational bodies is the key indicator of efficiency. As the second indicator, she proposes the commitment to regional cooperation defined as the percentage of funding for cooperation secured by the region as opposed to outside donors. She is the only author to use inclusiveness (participation of all regional countries) as a determinant of success.

In addition to the factors of governance proposed by Andreev (2009), Cottey (2009) and Willenberg (2009), we add the involvement of civil society. It provides legitimacy and public scrutiny and is a source of ideas. Its presence is often demanded when the EU provides its finance. Moreover, we distinguish between the financial and the political role of the EU.[3]

Table 1.1 summarizes the preceding theoretical proposals and constructs a heuristic which can be applied to analysis at the policy or the project level.

Table 1.1 Determinants of governance in EU-sponsored regional initiatives – the case of CEFTA

Initiative	*Status*	*Members/ inclusiveness*	*Decision making process*	*Number of meetings per year*	*Secretariat*	*Role of the EU*	*Regional funding vs. EU funding*	*Support of International financial institutions, bilateral donors*	*Involvement of civil society*
CEFTA	International agreement	Western Balkans with EU perspective plus Moldova	Only consensus	1 decision making. 5 to 10 technical.	Permanent based in Brussels. Employs 4 people.	EU not a member, but the biggest donor and initiator of the initiative.	Region 30% for the secretariat. 70% EU and donors. Trade agreements covered by the EU and donors. Events covered by the region.	Intelligence of international financial institutions through trade agreements.	Chambers of commerce

Examples of regionalism in the European neighbourhood

The origins of modern regionalism in EU's external relations lie in the 1990s in Central and Eastern Europe Countries (CEECs) (Cottey, 2009). The process emerged from "within and above", accompanying liberalization and democratization and replacing the obsolete structures built around the Soviet bloc. In transitioning from a planned to a market economy, regional groups were set as a framework to channel expertise and to share resources and experiences. By creating functional economic- and law-based regional initiatives, CEECs wanted to portray themselves as good candidates for EU integration. The CEFTA signed in 1992 is a primary example of endogenous regionalism of that time (Hyde-Price, 1996).

The willingness and the ability of CEECs to cooperate with one another regionally was viewed by the EU as an indicator of responsible, good neighbourly behaviour. However, the EU never embraced the principle as a core of its enlargement policy. The Copenhagen accession criteria set in 1993 defined that progress be assessed bilaterally. Own merit became the guiding principle in EU–CEECs relations (Bechev, 2006). Only parts of EU Copenhagen conditionality indirectly advanced the interstate cooperation. These parts concerned direct neighbours rather than larger groups of countries; for example, each country needed to ensure protection of its minorities; neighbours needed to define borders.

The nature of the enlargement process, in fact, weakened multilateralism in CEECs (Dangerfield, 2008). Although proposing that countries cooperate, the EU could not award any real incentives for regional cooperation. There even appeared to be certain tensions between regionalism and the pursuit of EU membership: more advanced candidates feared that regional groups may be proposed as alternatives to full integration into the EU; the principle of own merit injected an element of competition in the race to join the EU, thus hindering cooperation.

However, by the end of the 1990s, the regional approach was fully inserted in the new enlargement policy targeting the Western Balkans. It also entered the neighbourhood policies towards southern and eastern neighbours.

Stabilization and association process in the Western Balkans

Regional and multilateral approach to the Western Balkans has been introduced since the very start of the EU engagement with the region in 1996 (General Affairs and External Relations Council, 1996). The approach was then taken up as a condition for accession in the council conclusions of April 1997 and of June 1999 when the Stability Pact for South Eastern Europe was launched.[4] Regional cooperation was projected as both the recipe for growth and a generator of political stability. The Stability Pact organized frequent ministerial meetings and engaged the lower levels of public administrations. The pact developed special relations with the International Financial Institutions (IFIs). In June 2003, the European Council adopted the "Thessaloniki Agenda for the Western Balkans" (General Affairs and External Relations Council, 2003), which endorsed a number of regional initiatives to be pursued, including a multilateral free trade agreement.

Besides exogenous EU ideas, cooperation had well-developed roots in the region (Bechev, 2006). Politically, the region knew that it had to rebrand its image of being the black hole of Europe. Its own initiative, the South East Europe Cooperation Process (SEECP), worked along these lines during the 1990s. In parallel, cooperation was pursued by the businesses which sought to consolidate and expand. The exogenous and endogenous ideas were brought together in 2008 when the Regional Cooperation Council was established, a regional supranational organization replacing the Stability Pact and serving as the secretariat for the SEECP. Today, the EU continues to observe regional cooperation as a condition in the accession process and devotes approximately 10 per cent of its pre-accession assistance to regional activities.

Neighbourhood policy

The European Neighbourhood Policy (ENP) is built on the "ring of friends approach", which promises the neighbouring countries in the south and in the east "everything but institutions" (Prodi, 2002), that is, integration without membership.

South

Regional approach to the southern neighbourhood is defined in the Euro-Med Partnership launched in 1995 to address security threats and the worsening socio-economic situation. The partnership is officially described as regional forum with an added bilateral component. It perceives the fragmentation in the South as a major hurdle to development and proposes to build a Pan Euro-Med zone of free trade (Behr, 2010).[5] Since 1997, the countries are working on a south–south trade zone. However, only four advanced to the point that they could sign a modern free-trade agreement (FTA) in 2004 (Agreement Setting Up a Free Trade Area between the Arab Mediterranean Countries – Agadir Agreement, 2004). The EU supports the Agadir Agreement which it sees as a predecessor of the Pan-Euro-Med FTA.

In 2008, the process in the south was institutionalized in the Union of the Mediterranean, which shapes its governance. The process was given an annual political summit, a shared presidency (a country from the region and a country from the EU), a permanent secretariat and a permanent joint committee. The EU supports regionalism in the south with 10 per cent of all funds invested in the region.

East

The ENP obtained a new regional identity in 2009 through the establishment of Eastern Partnership (EaP), which proposed a more ambitious agenda for the east when compared with the south. Its underlying rationale is to blur the dividing lines at the eastern border following the enlargement to CEECs, to build bases for potential EU perspective and to function as a counterbalance to Russia's influence (Cottey, 2009). The partnership deepens bilateral cooperation and introduces new mechanisms for regional cooperation, which are often inspired or extended from the enlargement process.

The economic cooperation in the east largely mimics the approach of the Stability Pact. A large number of platforms and panels are introduced and other actors than governments are institutionalized (e.g. civil society, parliamentarians, local authorities and businesses). The EU is opening its community programmes for participants from the Balkans and the east. In the energy sector, the Energy Community has expanded from the Balkans to the east. Trade-wise, the EaP falls short of proposing a regional FTA but starts modestly with a regional trade panel.[6] One of the panel's ambitions is to help countries build a network of modern FTAs, however, only at later stage.[7]

The EaP process is steered by an EU-EaP summit every two years, an annual ministerial meeting and a number of *ad hoc* ministerial conferences in sectors of most intense cooperation. The EaP envisages frequent high-level dialogues in the areas of the platforms and the panels. Donors and IFIs are invited to join the process. The European Union supports these processes in the East with 25% of all funds invested in the region.

The case of CEFTA

CEFTA 1992 versus CEFTA 2006

CEFTA was initiated by the three Visegrád[8] countries in 1992 demonstrating their commitment to economic liberalization, cross-border cooperation, the rule of law and so on. Although the economic benefits of the agreement were perhaps not fully exploited and the liberalization of agriculture failed (Dangerfield, 2006), the agreement soon became a norm for regional cooperation and was extended to the current candidate countries and the potential candidates for EU accession.

The idea of a free-trade zone in the Western Balkans was introduced to the region through the Stability Pact and was endorsed by the EU Thessaloniki agenda. When launched in 1999, the Stability Pact had strong economic component, emphasizing reconstruction, cross-border infrastructure, foreign investment and engagement of IFIs, among other items. Linking countries with free-trade agreements was at the base of these plans. The Sarajevo declaration (1999) committed the countries to "work together to remove policy and administrative barriers to free flow of goods and capital in order to increase economic cooperation, trade and investment in the region." By 2005, a hub-and-spoke network of bilateral FTAs was accomplished. As a consequence, trade flows among countries increased absolutely and relatively in their total trade balance (Petreski, 2011). This was, on one side, a result of the agreements but, on the other side, it was also a spill-over of reconciliation, which led to re-establishment of old commercial ties and of improved statistics (Pjerotić, 2008).

The bilateral network of FTAs had a number of deficiencies. It liberalized industrial goods but exempted some agriculture and, in general, did not deal with services and public procurement. Implementation of the agreements was in practice often obstructed. Some countries suspended parts of the agreements, but many did not have sufficiently organized customs services which could cope with large number of varying procedures and forms that these agreements imposed (Pjerotić, 2008). The bilateral agreements were not uniform. They differed significantly from the template agreed on in 2001. Their annexes on rules of origin were different; therefore, they could not develop a system of diagonal cumulation which was needed to attract multinational investors, which operate via cross-border supply chains (Zirojević and Jelisavac, 2009).

Subsequent talks within the round table for trade of the Stability Pact aimed at concluding with a multilateral agreement. The European Commission insisted on this approach (Wittenberg, 2009). Despite some hesitations linked to fears that regional groups could be an alternative for EU accession, the decision to sign a multilateral FTA by amending and modernizing the CEFTA agreement was adopted by trade ministers in Sofia in 2005. The agreement was signed in Bucharest in December 2006 and went into force for all countries by 22 November next year. The new CEFTA, nicknamed CEFTA 2006, replaced 22 FTAs in the region.

In the area of trade policy, the new CEFTA introduced much more ambitious approach than was the original. Full liberalization of industrial goods was prescribed within three years, as opposed to eight years in the original 1992 agreement. A schedule for liberalization in agriculture was enshrined in the treaty. It introduced competition rules based on EU law. It envisaged opening public procurement markets by 2010. The parties committed to work on technical barriers to trade and to eventually liberalize services.

CEFTA 2006 brought a number of novelties regarding governance. The liberalization schedule was enshrined in the treaty without any room for future negotiations. Regarding institutions, the role of the joint committee remained the same – to supervise and administer the agreement and meet regularly. CEFTA 2006 established consultation, mediation and arbitration mechanisms which are implemented by the joint committee. A permanent and professional secretariat of the agreement was established in Brussels, which makes CEFTA 2006 a basis of an international organization (Pakowski, 2008).

Status, members and inclusiveness

CEFTA is an intergovernmental agreement among eight parties from South-East Europe.[9] The group is homogeneous in the sense that it includes seven countries which had to sign a regional free-trade treaty as part of their obligations under the EU Stabilisation and Association Agreement. CEFTA is thus a condition of EU accession, which secures participation and engagement of these seven parties. Another factor adding to the homogeneity is economic history of the parties. Six parties had been part of the single Yugoslav economic space for 70 years. They remain economically integrated. On average, CEFTA represents 20 per cent of their trade balance, which makes is the second-most important trade partner after the EU.

Two parties stand out, namely Moldova and United Nations Interim Administration Mission in Kosovo (UNMIK), which signed the agreement on behalf of Kosovo under United Nations Security Council Resolution (UNSCR) 1244. Moldova is the only member without an EU perspective. Its relations with the EU are governed in the context of an EU neighbourhood policy. CEFTA is economically irrelevant for Moldova. It represents 1 per cent of its trade. However, because CEFTA is perceived as an EU vehicle, the country has a strong interest in participating in the agreement and hopes that participating in the agreement will increase its chances of gaining EU perspective. The other outlier is UNMIK on behalf of Kosovo. Economically, Kosovo is highly integrated with its neighbours. However, the representation and the denomination of Kosovo have been disputed by the government in Pristina, on one side, and by three other parties which do not recognize an independent Kosovo, on the other side.[10]

Another source of heterogeneity in the agreement is different speed of the accession process. It leads to complying differently with EU technical standards and, as a consequence, can slow regional trade expansion. As countries individually converge towards EU norms, this creates new non-tariff barriers to trade in the CEFTA zone. For example, Bosnia-Herzegovina, which is highly integrated in the region economically but lags in the EU accession process, has been most affected by this situation. When its neighbours upgrade product rules to EU norms, Bosnia will not be able to export into their territories anymore.

Meetings

According to the agreement, the intergovernmental joint committee performs the executive function. The joint committee meets once a year. Its decisions are prepared by three subcommittees (customs and rules of origin, non-tariff barriers to trade and agriculture and phyto-sanitary standards). These bodies also meet once a year. Additionally, there are two subcommittee-like working groups on services and logistics, which also meet once a year. Every year, there are several meetings of technical assistance (TA) projects, which feed the work of the subcommittees.

All meetings are prepared for by the national contact points and the secretariat, who are in daily contact. The high number of meetings ensures the intensity of relations and demands significant resources in national administrations.

All meetings are occasionally affected by the Kosovo participation and denomination dispute. To this aim, CEFTA developed a number of solutions; for example, there are no country nameplates, representatives of UNMIK appear but leave the room as soon as possible and high-risk meetings are organized in Brussels.

The secretariat

In 2007, during its first meeting, the joint committee mandated that the secretariat support the committee and the three subcommittees by preparing background materials, keeping the minutes and following up on the decisions and preserving archives (Decision of the Joint Committee, 2007). More important, it mandated the secretariat to coordinate with donors and providers of technical assistance. Over the years, the role of the secretariat grew, particularly because of weak national administrations, which have little experience in multilateral commercial policy. Its supranational character became more pronounced by the appointment of a Slovenian director and a Turkish expert, two nationalities which are not part of CEFTA. The role of the secretariat is enhanced by its position in Brussels. Unlike the parties, it enjoys better access to the European Commission and IFIs.

The growing importance of the secretariat seems to be acceptable to the parties. Only Kosovo has complained that the secretariat was too political when it actively engaged in finding solutions on how to accommodate participation of UNMIK/Kosovo.[11]

Decision making, disputes and the role of the EU

All CEFTA decisions are adopted by consensus. This fact combined with the dispute over Kosovo's denomination/participation blocked progress of the treaty for several years. The two most affected areas were the liberalization of agriculture and the Kosovo custom stamp recognition. A solution for agriculture was found in bilateral agreements within the CEFTA context. However, the customs stamp dispute led to a mini trade war. Even if the dispute settlement mechanism, as defined by the CEFTA treaty, was triggered, the parties did not use this forum to discuss substance (i.e. import bans) but the underlying politics (i.e. denomination/participation). After more than a year, the parties and the international community[12] perceived that the

EU-sponsored bilateral dialogue was a better forum for finding a solution. Indeed, the solution was found there.

Other minor disputes were resolved quickly and no other party sought to use the dispute mechanism. To facilitate quick resolution of irritants, a dedicated matrix was developed within a technical assistance project.

The work of CEFTA is closely followed by the EU. The EU is not a party to the CEFTA,[13] so it does not intervene in its formal structures.[14] Nevertheless, the European Commission influences the process through its political reporting. In one of these reports, the commission systematically praised the good economic results of CEFTA, but demanded compliance from Serbia concerning the treatment of Kosovo and pragmatism from Kosovo regarding the participation (European Commission, 2010). Serbia's acceptance of Kosovo participation without the additional presence of UNMIK in the regional initiatives, including CEFTA, and the end of the trade dispute over the customs stamp were proposed as conditions for Serbia to be granted the status of a candidate country and to open accession negotiations (European Commission, 2011).

Donors and IFIs

Donor, EU and IFI contributions secure the funding of the secretariat and pay for studies and seminars which guide the implementation of the agreement.[15] The EU funding for CEFTA and other trade-related regional activities amounts approximately to €5 million in the 2011–2013 period. Individual donors contribute around 10 per cent of the budget of the secretariat. EFTA countries host the secretariat in its premises. The IFIs are associated through technical assistance projects which they implement on behalf of the European Commission or bilateral donors. Among IFIs, the Organisation for Economic Co-operation and Development (OECD) plays the most important role. In recent years, the World Bank has intensified its engagement in trade integration of the region and is now implementing a major project on customs and logistics.

Civil society

The CEFTA agreement envisages a role for the civil society only in the area of protecting intellectual property. In 2007, its role was upgraded; on the initiatives of the chambers of commerce, the joint committee established a "CEFTA Forum" to promote better dialogue between businesses and the governments within CEFTA structures. Under the CEFTA Forum, the chambers of commerce established working groups that mirror the CEFTA subcommittees. The forum is a major actor in "CEFTA Week", a set of public seminars organized annually by the secretariat and the chairing party, which reviews the implementation and examines future agenda in dialogue with civil society, academia and other stakeholders.

The impact of CEFTA 2006

Economic impact

During the wars of the 1990s, the relative share of regional trade in the total trade balance of the region fell sharply. In the 2000s, after the conclusion of bilateral FTAs, the importance of regional trade in total trade balance increased, but it stabilized around 2005.[16] The entry into force of CEFTA in 2007 coincides with the last year of economic expansion before the global financial crisis of 2008 affected the region. The global crisis negatively impacted trade flows within the region, with the EU and with the rest of the world.

As Table 1.2 shows, CEFTA was successful in preventing a disproportionate contraction in trade during the crisis.[17] Contraction of intra-CEFTA trade is on average similar to contraction with the rest of the world. It is lesser than contraction with the EU. In 2009, CEFTA trade contracted slightly less than did trade with the rest of the world, and now it is returning less swiftly to its previous levels than is trade with the rest of the world. It is interesting to observe that since the financial crisis of 2008 and the CEFTA entry into force, regional trade became relatively more important for Serbia and Croatia, the two most developed and competitive economies of the region.[18]

The preceding developments can be explained by the composition of trade flows in CEFTA.[19] A quarter of CEFTA trade is in agriculture or in processed food, which is more resistant to recession but grows slowly. Trade in this sector is supported by cross-border supply chains, and it is the main beneficiary of intra-regional foreign direct investment (e.g. Croatia investing in Serbia and Serbia investing in Bosnia). The parties continue to support it with agreements on phyto-sanitary recognitions and by converging with EU standards. Further convergence with EU standards is key if the parties wish to expand their regional competitive advantage in agriculture towards more trade with the EU or with the rest of the world. Further consolidation of the sector and cross-border mergers will play an important role.

The other three quarters of trade are dominated by commodities and low-value-added products. OECD (2009) analysis reveals that CEFTA parties enjoy overlapping comparative advantages in product categories with low level of export sophistication. To support trade in other goods, the countries would need to move away from commodities into mid-tech products. The subcommittee on technical barriers to trade is working on proposals to harmonize standards in product areas which could be the basis of this upgrade. However, the OECD (2009: 4) remains very critical about the results so far: "There is lack of administrative capacity and fragmented institutional responsibility. The business community lacks technical knowledge to identify and document different non-tariff barriers to trade in their sectors."

A recent issue dominating CEFTA's economic agenda is liberalization in services, which is envisaged in the treaty, but its schedule and modalities remain to be defined. To this aim, a number of studies are being prepared;

Table 1.2 The impact of the economic crisis on trade flows within CEFTA and with the rest of the world

Country	*Imports from CEFTA (in €)*			*Exports to CEFTA (in €)*		
	2008	*2009*	*2010*	*2008*	*2009*	*2010*
Albania	125,000	106,975	119,066	298,000	203,690	316,855
Bosnia-Herzegovina	1,529,000	1,169,566	1,302,049	2,916,000	1,601,972	1,883,276
Croatia	2,253,000	1,461,957	1,665,440	1,051,000	918,867	811,698
Macedonia, FRY	922,000	701,739	771,021	511,000	444,369	475,081
Moldova	0	4,916	7,143	0	8,766	9,017
Montenegro	159,000	146,195	130,521	1,227,000	630,453	669,999
Serbia	2,458,000	2,163,697	2,590,000	1,216,000	970,921	1,096,740
UNMIK/Kosovo	93,000	87,717	70,934	705,000	652,489	799,013
Total	7,539,000	5,842,743	6,656,174	7,924,000	5,431,527	6,061,679

Country	*Total trade CEFTA (in €)*			*Total trade Rest of the World (in €)*		
	2008	*2009*	*2010*	*2008*	*2009*	*2010*
Albania	423,000	310,647	435,921	38,717,000	374,575	4,202,779
Bosnia-Herzegovina	4,445,000	2,771,538	3,185,324	9,040,000	6,373,622	7,404,404
Croatia	3,304,000	2,380,824	2,477,138	30,255,000	20,368,662	21,552,442
Macedonia, FRY	1,433,000	1,146,108	1,246,101	5,712,000	4,427,727	5,370,544
Moldova	0	13,682	16,160	5,695,000	3,610,672	4,061,899
Montenegro	1,386,000	776,648	800,521	1,466,000	1,154,533	1,187,175
Serbia	3,674,000	3,134,618	3,564,889	20,937,000	14,283,235	16,825,025
UNMIK/Kosovo	798,000	740,206	869,948	2,124,000	1,321,077	1,569,018
Total	15,463,000	11,274,271	12,596,002	79,100,000	55,285,103	62,173,286

Source: CEFTA Secretariat.

however, no concrete steps have been taken yet. If implemented, liberalization in services could bring about significant benefits and could pave the way towards establishing of a genuine single market in the Western Balkans. The obvious next step would be the establishment of free movement of labour.

Policy impact: CEFTA contribution to regional cooperation

Of all regional initiatives in the Western Balkans, CEFTA has attracted most public attention (Delević, 2011). There seems to be a consensus across businesses and consumers that CEFTA is a beneficial agreement. Rather than appearing weary of deepening political integration, the parties obviously cherish deeper economic integration.

However, the figures presented in Table 2.1 point out that the economic effects of CEFTA are only marginal and comparable with other ongoing trade integrations. In the eyes of the parties, the consumers and the business, CEFTA must have performed well in areas other than trade. The analysis of governance offers anecdotal evidence of how CEFTA successfully managed political disputes and of how it furthered its economic agenda. CEFTA offers a number of lessons to other regional initiatives:

- Regional cooperation will be systematically applied as EU membership conditionality. The EU will apply it by examination of the functioning of regional initiatives. It will support these initiatives financially and politically, but in return, it will demand their functioning and inclusiveness.
- Pragmatism is needed. Regional initiatives can fall hostage to lingering bilateral disputes. However, the economic benefits can outweigh the political agenda. Parties not directly involved can move the initiatives forward by proposing pragmatic solutions.
- Supranational bodies have added value. National administrative structures have limited capacities, which help bring acceptance and strength to supranational institutions. Their role is further enhanced when supranational bodies are recognized by donors and are in position to secure funding.
- Regional integration and EU integration can be conflicting. The EU portrays regional cooperation as mutually reinforcing with EU accession. Although this is correct politically, various paces of EU integration can hamper regional engagement. Solutions going beyond the joint alignment with the *acquis communautaire* need to be developed to accommodate interests of all countries.

Conclusion

This chapter has analyzed a multilateral initiative in the Western Balkans through the prism of governance. It first established the key determinants of good governance, such as inclusive relations between members, regularity of meetings, a functioning supranational secretariat and the capacity to deal with disputes and relations with external actors, particularly support from international community and engagement with civil society. These determinants were then examined on the functioning of CEFTA.

The practical experience from CEFTA's economic agenda and its track record in dealing with disputes suggest that good governance plays an important role. However, the evidence also points out that there are limits to good internal governance. The role of the external actor – the EU – remains key. Arguably, only a few other regional initiatives have the economic importance of CEFTA. Therefore, they cannot expect that international community be equally responsive to their needs. However, other initiatives can learn lessons on the internal organization, the evidence-based agenda or the importance of obtaining legitimacy from the civil society. By doing the practical things right, good internal governance can help bring about the external political support when needed.

Notes

* The findings, interpretations and conclusions expressed in this chapter are those of the author and do not reflect the positions of the institution.

1. The EU is simultaneously engaged in regionalism in the EU by creating its own so-called macro-regions. In 2008 and in 2010, it adopted the Baltic Strategy and the Danube Strategy, respectively. It is currently preparing grounds for the Ionian-Adriatic Strategy.
2. Belarus is temporarily excluded from the Eastern Partnership, but it can be engaged in and benefits from regional actions.
3. The role of the EU in decision making varies from legal exclusion (no vote) to the right of veto.
4. An explicit prospect of EU membership for the countries of the Western Balkans was first made at the European Council Summit in Feira in June 2000.
5. A striking example of economic fragmentation of the region is that only 5 per cent of all trade flows have regional character, which is one of the lowest rates in the world in a region united by language and cultural identity.
6. The panel discusses issues such as custom cooperation and trade facilitation, IPRs, SPS issues and public procurement. Further topics on the agenda of this trade panel include technical regulations and standards, services and establishment, investment, capital movement and payments, trade in goods and competition.
7. On their own initiative, countries nevertheless signed a regional CISFTA in November 2011, which is perceived as old style (Institute for Economic Research and Policy Consulting, 2011).
8. The Visgrád countries include Czechoslovakia, Poland and Hungary.
9. The eight parties involved in CEFTA include Albania, Bosnia-Herzegovina, Croatia, Macedonia, Montenegro, Moldova, Serbia and UNMIK on behalf of Kosovo under UNSCR 1244/99.
10. The countries not recognizing an independent Kosovo are Serbia, Bosnia-Herzegovina and Moldova.
11. See Syla and Arifaj (2010) for a negative assessment of the agreement and the role of the secretariat.
12. Even if the commission assisted in drafting of the agreement, it never called on the parties to use its dispute settlement mechanism as the means to solving the customs stamp dispute.
13. The commission is officially represented in the steering board of the secretariat, which assembles the donors of the secretariat. The commission is the biggest donor of the secretariat, although this does not give it any special role in the steering committee. The issues in the management of the secretariat are contracted in a grant agreement.

14. On the working level, the Commission secures presence in high-level CEFTA events and in technical assistance projects when its expertise is needed.
15. Ongoing studies to support the work of CEFTA in 2011 were as follows: SEE Regional Trade Logistic (World Bank), CEFTA Trade portal (German Development Agency), Barriers to liberalization in services (World Bank), Monitoring and elimination of non-tariff barriers to trade (OECD), Evaluation of the impact of regional integration on investment location (OECD), Quality infrastructure (European Standards Committee) and Government procurement (OECD/Sigma).
16. For more details, see Petreski (2011).
17. The low number of irritants also demonstrates that the parties did not resort to any protections measures, which are otherwise more frequent in the crisis.
18. The importance of CEFTA for Serbia and Croatia has increased as these two countries export a high share of agriculture and food produce to the region and much less to the EU. In the crisis, the relative importance of agriculture increased.
19. For more details on the structure of trade and trade flows, see www.cefta.int/statistics.

References

Agreement on Amendment of and Accession to the Central European Free Trade Agreement. Bucharest, 19 December 2006.

Agreement Setting Up a Free Trade Area between the Arab Mediterranean Countries. Rabat, 25 February 2004.

Andreev, S. (2009) "Sub-regional Cooperation and the Expanding EU: The Balkans and the Black Sea Area in a Comparative Perspective." *Journal of Balkan and Near Eastern Studies*, 11(1): 83–106.

Bechev, D. (2006) "Carrots, Sticks and Norms: The EU and Regional Cooperation in Southeast Europe." *Journal of Southern Europe and the Balkans*, 8(1): 27–43.

Behr, T. (2010) *Regional Integration in the Mediterranean: Moving out of the Deadlock?* Notre Europe Research Series 77. Paris: Notre Europe. Retrieved 15.2.2012: www.notre-europe.eu/uploads/tx_publication/Etud77-MediterraneanIntegration-TBehr-en.pdf

Cottey, A. (2009) *Sub-regional Cooperation in Europe: An Assessment.* Bruges Regional Integration Global Governance Papers 3/2009. Bruges: College of Europe.

Dangerfield, M. (2006) "Subregional Integration and EU Enlargement: Where Next for CEFTA?" *Journal of Common Market Studies*, 44(2): 305–324.

Dangerfield, M. (2008) "The Visegrád Group in the Expanded European Union: From Preaccession to Postaccession Cooperation." *East European Politics & Societies*, 22(3): 630–667.

Decision of the Joint Committee of the Central European Free Trade Agreement No. 7/2007: Mandate of the Secretariat. Adopted 28 September 2007. Retrieved 15.2.2012: www.stabilitypact.org/trade/Final%20CEFTA%20Secretariat%20mandate.pdf

Delević, M. (2011) "CEFTA 2006 – an Economic Pillar of the Process of Political Stabilization and Regional Cooperation in the Western Balkans." In Međak (ed.), *CEFTA 2006 – Challenges and Opportunities: Collection of Essays*, 7–12. Belgrade: International and Security Affairs Centre.

Dwan, R. (2000) "Subregional, Regional and Global Levels: Making the Connections." In G. Herolf (ed.), *Subregional Cooperation and Integration in Europe*, 81–100. Stockholm: Utrikespolitiska Institutet.

European Commission. (2002) *Mediterranean: Regional Strategy Paper 2002–2006, and Regional Indicative Programme 2002–2004*. Brussels.

European Commission. (2010) *Enlargement Strategy and Main Challenges 2010–2011*. COM (2010) 660. Brussels.

European Commission. (2011). *Enlargement Strategy and Main Challenges 2011–2012*. COM (2011) 666. Brussels.

General Affairs and External Relations Council. (1996) *Conclusions*. Brussels, 26 February.

General Affairs and External Relations Council. (2003) *The Thessaloniki Agenda for the Western Balkans: Moving towards European Integration, Conclusions*. Thessaloniki, 16 June.

Hyde-Price, A. (1996) *The International Politics of East Central Europe*. Manchester: Manchester University Press.

Institute for Economic Research and Policy Consulting. (2011) *Spotlight Ukraine: Monthly Economic Monitor November 2011*. Retrieved 15.2.2012: www.dokumente.ios-regensburg.de/publikationen/SpotlightUkraine/SpotlightUkraineNovember2011.pdf

Organisation for Economic Co-operation and Development. (2009) *Aid for Trade: Implementing CEFTA: Analysis of Technical and Administrative Barriers to Trade*. Roundtable on Non-Tariff Barriers in the CEFTA area. Budapest. Retrieved 15.2.2012: www.oecd.org/investment/psd/44180621.pdf

Pakowski, G. (2008) *From Krakow to Budapest: Central European Free Trade Agreement CEFTA 1992 and 2006 – Outlines and Comparison*. Central European Case Studies, Vol 2. Gyor, Hungary: University of Gyor.

Pedersen, T. (2002) "Cooperative Hegemony: Power, Ideas and Institutions in Regional integration." *Review of International Studies*, 28(4): 677–696.

Petreski, M. (2011) *No Nation Was Ever Ruined by Trade: South Eastern European Trade Analysis*. FREIT Trade Policy Working Paper. Washington: Forum for Research in Empirical International Trade. Retrieved 15.2.2012: www.eiit.org/WorkingPapers/Papers/TradePolicyRegional/FREIT371.pdf

Pjerotić, Lj. (2008) "Trade Liberalization in the South East Europe Effects and Controversial Issues." *Panoeconomicus*, 55(49): 497–522.

Prodi, R. (2002) *A Wider Europe's Proximity Policy as the Key to Stability*. Speech at Peace, Security and Stability International Dialogue and the Role of the EU Conference. Brussels. Retrieved 15.2.2012: http://europa.eu/rapid/press-release_SPEECH-02-619_en.htm

Sarajevo Summit Declaration of the Heads of State and Government of the participating and facilitating countries of the Stability Pact and the Principals of participating and facilitating International Organizations and Agencies and regional initiatives. Sarajevo, 30 July 1999.

Syla, D. and Arifaj, A. (2010) *CEFTA – Free Trade Disagreement*. Periskopi no. 2. Pristina: Foreign Policy Club. Retrieved 15.2.2012: www.fes-prishtina.org/wb/media/Publications/2010/periskopi%202010%20Free%20Trade%20Disagreement%20(English).pdf

Tassinari, F. (2005) *Security and Integration in the EU Neighbourhood: The case for regionalism.* CEPS Working Documents No. 226. Brussels: The Centre for European Policy Studies.

Willenberg, A. (2009) *The Promotion of Regional Economic Integration in the EU's Neighbourhood: CEFTA 2006 and the Agadir Agreement.* Bruges Regional Integration Global Governance Papers 5/2009. Bruges: College of Europe.

Zirojević, M. and Jelisavac, S. (2009) "Cooperation in South East Europe within the Framework of the CEFTA Agreement." *Megatrend*, 5(1): 99–121.

Chapter 2

A Mediterranean electricity cooperation strategy

Vision and rationale

*Alessandro Rubino**

The Mediterranean region is characterized by the presence of important hydrocarbon producers and energy abundant countries (such as Algeria, Libya, Syria and Egypt).[1] According to the classical International trade theory approach (Helpman & Krugman, 1985), this would suggest that we should expect significant volumes of energy exchanged between energy-rich countries in the southern and eastern rim of the Mediterranean basin and the industrialized demand hubs of the north.

However, we observe persistent inconsistency between this theoretical prediction and the empirical evidence; the volume of energy traded is significantly lower than the level that theoretical model predict. This is all the more so when it comes to electricity trade. Along with the differences in factors endowment that, over the years, would have certainly justified the development of the infrastructure and would have been consistent with the forecast of a significant energy exchange volume between the two shores of the Mediterranean Sea, a number of other features are crucial in explaining the determinants of energy trade.

After two decades of regulatory practice, it is now clear that institutional diversity and the existing legal and regulatory frameworks play a fundamental role in explaining the volumes of energy exchange between countries. Notwithstanding the recent progress towards global economic and financial integration, national borders in the Mediterranean area still show a significant and depressing effect on electricity trade. This is even more evident in the south–south route (direction), where physical connection is already in place but where the rate of utilization of the existing capacity is extremely low (refer to Table 2.1). North–south interconnection is even more limited and barely consists of the Spain–Morocco interconnection.

Currently the electricity exchange within North African countries and Eastern Mediterranean countries is scarce. This is not only due to the inadequate capacity of the interconnection (that is, anyway, underutilized) or to constraints inside the countries. The limited use of interconnections, where they exist, mostly depends on the lack of commercial agreements that hinders larger energy flows (MED-EMIP, 2010, p. 64). Interconnectors are mainly used for mutual aid, and often remuneration is in kind. Therefore,

Table 2.1 Average load factor

	2008	*2007*	*2006*	*2005*	*2004*
Syria–Jordan	0.70%	0.50%	2.40%	27.50%	4.30%
Jordan–Syria	8.00%	5.20%	0.00%	0.00%	0.00%
Syria–Lebanon	32.90%	53.40%	41.60%	20.10%	8.20%
Syria–Lebanon	29.90%	52.70%	53.30%	26.10%	12.80%
Jordan–West Bank	90.20%	80.50%			
Egypt–Gaza	90.00%	82.60%	19.50%		
Turkey–Syria	4.40%	31.40%	4.30%		
Libya–Egypt	4.70%				
Egypt–Libya	7.90%				
Jordan–Egypt*	0.90%	1.00%	0.50%	0.10%	0.10%
Egypt–Jordan*	18.80%	12.00%	17.00%	24.20%	29.50%
Algeria–Tunisia	7.80%	7.00%	8.10%	7.90%	5.40%
Tunisia–Algeria	9.10%	8.20%	8.60%	9.00%	5.60%
Morocco–Algeria	10.20%	9.50%	10.40%	8.10%	9.30%
Algeria–Morocco	11.10%	9.10%	9.70%	8.20%	6.20%
Morocco–Spain	0.30%	0.40%	0.50%	1.50%	0.20%
Spain–Morocco	74.20%	61.50%	35.70%	23.90%	44.50%

Source: MED-EMIP – MEDRING update, Final Draft April 2010 (p. 70).

* The ratio between the yearly energy exchanged (MWh) and the Maximum Capacity (NTC value multiplied by the number of hours in a year) for the existing interconnections.

most countries in the region have to rely on their domestic generation (which is not necessarily an efficient arrangement).

When we look at the economic fundamentals that characterizes electricity generation, we see that energy trade occurs in a thick legal and regulatory framework that largely depends on the micro-technical characteristics of the commodity traded. Indeed energy markets are characterized by (1) large variations in demand over the course of the year, (2) the need to balance physically the supply and demand at every point of the network, (3) limited storability (in particular for electric power), (4) the inability to control energy flows to most individual consumers, and (5) the limited use of real-time pricing by retail consumers.

The considerable level of coordination between subsequent stages of the energy supply chain, in the last century, has produced almost everywhere two main variants:

- a vertically integrated public monopoly or
- different local monopolies for the distribution-retailing phase for small customers and a national firm in the production-transportation phase, integrated by a long-term contractual relationship.

Both options have been widely adopted in most countries belonging to the Mediterranean region and are still in place in the countries belonging to the south rim of the Mediterranean basin. In Europe this paradigm has been questioned in the 1990s with the promotion of the single European Union (EU) market. The starting point was the opening up of the transportation network to TPA – "third party access". This first step aimed to promote the wholesale market, breaking up the exclusive right of supply for the owner of the net.[2] Twenty years after the EU have produced a fairly integrated internal electricity market that currently trade cross border around 7% of the total gross consumption,[3] especially in continental Europe. This radical shift towards a competitive approach hinges on a model that foresees the separation of the energy supply chain in different stages with the establishment of competition where possible (production/generation, wholesale and retail market and related services) and regulation where necessary (typically transmission and distribution networks). This market paradigm is crucially based on the consideration that the creation of a functioning competitive market is possible, and most importantly desirable, from a welfare perspective.

However, this market model (based on liberalization of potentially competitive markets) has not been universally adopted. Actually, the contrary is true. Apart from several experiences in Europe, where a variety of different regulatory approaches emerged over the last decades (forming the so-called Regulatory Patchwork) and North America and Australia, the so-called neo-realist approach (Escribano, 2010) is widespread in other countries and regions. According to this approach, the relationships between consumers and producers are defined through bilateral long-term contracts, and national systems are interconnected with point-to-point connections, most of the times only for mere network security reasons. Typically, in this system, "national champions", traditionally supported by governments, dominate the national markets, in which there is only a narrow, or no role, for competition. In the Mediterranean region these two paradigms coexist. Energy cooperation in the area today needs to be formulated according to a model that conjugates those different approaches in a stable framework.

To respond to these compound needs Mediterranean countries are developing a common regulatory and legal framework. To date, 22 states, being members of the Association of the Mediterranean Regulators for Electricity and Gas (Medreg), are building a shared view on the regulatory framework for the area, coupled with a progressive move towards integration of the energy markets.[4] A peculiar feature of this process is the voluntary nature of this regional cooperation.

Medreg is a forum for exchange of know-how and expertise in the field of energy regulation for the benefit of its members and promotes the overall welfare of the Euro-Mediterranean area. Acknowledging the diverse historical background of the Mediterranean region, energy regulators have a central role to play in promoting a "bottom-up" process for harmonization of energy markets and legislations. The same accounts for progressive market integration in the Euro-Mediterranean region.[5] In front of these strategic energy challenges, Medreg's duty is to constitute a strong institutional basis to address economic, social and cultural development, as well as the need to conciliate a rising energy demand with sustainable development conditions.

The regulatory framework within Medreg countries is not imposed. It is discussed and evaluated in the light of potential mutual gains and possible (prospective) complementarities between diverse countries (exporting vs importing countries) and regulatory frameworks (liberalized market vs vertically integrated monopolistically organized ones). Within this forum the existing regulatory institutions and rules within members' countries are analyzed in an attempt to highlight possible harmonization strategies and road maps for convergence.

Private actors (such as Medreg) have begun to play an increasingly important role in regulatory policymaking. Although most of these policy arrangements follow a "top-down" logic, there is also an increasing number of "bottom-up" modes of inclusion of private actors. These modes are characterized by the emergence of voluntary self-regulation of individual firms or industry sectors that are recognized and integrated into policymaking and/or formal law at a later stage. In addition, more inclusive approaches, such as multi-stakeholder initiatives, became common in public policymaking, adding a participatory spin to the classic top-down logic.

In particular in the North African regions (Maghreb and Mashreq), where the power systems are generally smaller and scarcely interconnected, participation in the evolving Euro-Mediterranean market is a driving rationale for the restructuring of the sector. The relatively small size of the domestic market is a major constraining factor for the entry of international investors into the sector, together with the incomplete reforms and the low level of government commitment to the issues that investors perceived as priority when planning investment in developing countries.[6] In general, reform in the region has been slow. But the exporting potential of the area raised the European Commission (EC) and potential investors' interest.

Historically, concerns over security of supply, coupled with the significant transaction costs related to the technical difficulties connected with the considerable level of coordination required between subsequent stages of the electricity supply chain, dictated the shape of the power system that currently sees most Mediterranean countries being self-sufficient in electricity. When we look at electricity dependency in the Euro-Mediterranean area, we can see that almost no electricity is imported to or exported from the EU. Only Finland (with 13% net import of domestic demand from Russia), Italy (17% from France), The Netherlands (15% from various countries) and Morocco (10% from Spain) rely heavily on continuous electricity imports (Eurostat, 2010; International Energy Agency, 2010).

But this situation is varying rapidly. Regional integration in the Mediterranean area occurs in markets that are changing quickly under the confluence of globalization and growing links between national economies and major regional players (typically EU members states). The adoption of the EU climate and energy package and the stricter emission obligation targets (also known as the triple twenty targets[7]), and the growing competition to access secure and affordable energy sources and fuels from developing countries is likely to increase the integration between electricity markets in the area. The current power systems in Europe, North Africa and Middle East region are very different in most respects due to the large differences in economic development and the abundance of oil and gas in North Africa and the Middle East (that dictates the generation portfolio), but nevertheless, they are expected to show significant convergence in terms of electricity demand in the future. Actually, their variance constitutes an important added value when it comes to system risk minimization. In this framework, two forces appear to represent the driving rationale in shaping the sector evolution: climate change and energy security.

Despite these emerging circumstances, Mediterranean integration, so far, is unimpressive in the formal international institutions that students of integration normally have in mind. Compared to the development and the market potential observed when analyzing the integration of Euro-Mediterranean energy markets, the relative weakness of formal political institutions is very notable. The Mediterranean area, or any of its sub-regions, lacks equivalents to the panoply of European-wide institutions, foremost the European Union. In the establishment of formal institutions, Mediterranean regionalism during the last decades has experienced a series of false starts. Recently a joint communication to the European parliament, the council, the European economic and social committee and the committee of the regions[8] has reinforced the need to enhancing cooperation with neighbouring countries with the aim to establishing a complementary "EU-Southern Mediterranean Energy Community".

Renewable energy sources (RESs) are at the core of this harmonization process, for various reasons. They are abundant in the area (the Mediterranean Solar Plan alone planned to deliver 20 GW of new renewable electricity capacity by 2020 across Mediterranean countries[9]) and provide a sustainable solution in the long run. RESs also respond to the more stringent environmental policies European countries are adopting in the framework of the stricter emission obligation targets included in the "Green package".

In order to deploy secure renewable electricity and to provide a reliable market integration in the area three main actions will guarantee the backbone of the future Euro-Mediterranean energy area: (a) gradual regulatory and legislative convergence based on a "Corridor approach", (b) the expansion of the electricity network in the area and (c) the establishment of a energy free trade area in the Mediterranean basin.

The establishment of a Mediterranean specific electricity paradigm will be the result of an evolutionary process from the conventional Euro-Mediterranean Partnership (EMP) and European Neighbourhood Policy (ENP) approach that failed to achieve "convergence of interests between

European [and South-Mediterranean] consumers and producers and transit countries" (Darbouche, 2011, emphasis added). Those past initiatives were plagued by an "uncooperative approach by the EU" (Darbouche, 2011),[10] based on the mere reproduction of the *aquis communitaire* on a larger scale with the accompanying thick normative and regulatory dimension. Most South Mediterranean countries (SMCs) were sceptical on the market-oriented approach proposed by the EU about energy relations, especially when based on a narrow concept of strategic energy relations, mostly based on EU security of supply needs.

The evolutionary process involves the convergence of legal and regulatory framework between the two shores of the Mediterranean basin, together with a better physical and institutional integration. All these do not entail a giant leap in terms of policy and technological arrangements, but require that some fundamental pillars are posed in order to develop a sustainable electricity exchange in the Mediterranean area. These three main pillars will be the necessary conditions to sustain the establishment of an electricity market in the Euro-Mediterranean area. Let's illustrate the main characteristics of the three fundamental actions that constitute the required pre-condition for the Euro-Mediterranean electricity area to exist.

Corridor approach toward market convergence

Energy markets, and electricity market in particular, are still largely dominated by national champions that determine the characteristics of the electricity system, with their own set of preferences. Large market players are normally able to influence the set of policy options locally available and to make them country (or sub-regional) specific. In addition electricity market tended to be closed system (in order to ensure self sufficiency),[11] but the increasing interdependence of the national electricity systems (due to the RES penetration and thanks to the reduced reliance on indigenous fuel for generation requirements) requires stricter integration between adjacent and interconnected systems.

Adjacent countries, physically interconnected and endowed with complementary electricity system (in terms of portfolio mix and load profile) tend to develop a regulatory framework that maximizes some sort of joint welfare (or profit) function (Chao & Peck, 1996) and that drives many countries to coalesce in clusters, or "corridors".

Energy corridors arrangements are different from simple point-to-point connection, because the transit of energy from one system to another requires that the physical interconnection and the institutional and regulatory framework in the area interested are able to sustain and favour the necessary exchange across its borders. It also implies that the energy system interested manage to integrate, harmonizing rules and managing (without disruption) the physical and legislative constraints stricter integration requires.

Whereas the peculiarity of the Mediterranean region impede that a region-wide arrangement is reached in the short to medium run, and in fact also in EU a variety of different regulatory approaches has emerged over the last decades (with the emergence of a regional approach), clusters of countries with complementary electricity market and convergent interest normally manage to emerge, even in complex institutional environment as in the Euro-Mediterranean area, and normally group together supply hubs with demand centre.

The corridor approach allows for taking into consideration convergence within specific corridor while introducing flexibility to take into consideration country and region specific preferences.

Currently three main electricity corridor are emerging in the Mediterranean region in the north–south direction, but they require significant reinforcement of the corresponding south–south interconnection in order to sustain the significant flow that will be determined along the main directions.

The three corridors are the West corridor (Morocco–EU through Spain), Central Corridor (Maghreb–EU through Italy) and the East corridor (Middle East–EU through Turkey). The existence of energy corridors alone, while allowing for a sub-regional harmonization of the regulatory framework, does not guarantee in itself the creation, in the long run, of a regional energy market. The process of generating consistent and mutually compatible systems is largely devote to the implementation of the third pillar, the creation of the Mediterranean energy free-trade area, that requires that neither institutional arrangement is mutually exclusive.

Network expansion

The Mediterranean power systems are split into various sub-systems: the continental European grid,[12] and the three western North African grids, which are synchronous with the continental European grid, whereas the Libyan and Egyptian grids are synchronous with Jordan, Lebanon and Syria. The Turkish grid is now connected and synchronized with continental Europe through South-East Europe, all those blocks compose a number of different national or regional market and control areas. Interconnection capacity is rather small, even inside EU, and congestion is a problem at most national borders, a problem exacerbated by the regional expansion of intermittent renewable power. Power grids, along with generation portfolios, have been constructed from a national perspective. Therefore, there was no urgent need for interconnections in the past, and only few were built. Over time the benefit of stronger interconnections, in terms of efficient sharing of control capacities and imports during capacity shortages, became apparent and reinforced that interconnections are becoming a priority. However, interconnections are not strong enough to unify the continental European markets, not to mention the even more fragmented situation in the rest of the Mediterranean basin.

Limited interconnections cause market distortion and system-balancing problems. Increasing cross-border trade coupled with very strong growth in wind and concentrated solar power (CSP) in geographically limited areas, cause local overload and system instability. Negative prices in some EU national markets have already raised the necessity of increased grid interconnections.

A pan-European cross-Mediterranean super grid is key in enabling the efficient deployment of the growing wind generation in the north and CSP and photovoltaics (PV) in the south.[13] It will also require a reinforcement of the existing HVAC grid in the area and the wide introduction of smart technologies and smart grids. For example, in particular, all artificial bottleneck-limited cross-border trade caused by national or legislative borders needs to be removed.

It is well known that RES power cannot be installed close to demand hubs but must be located where it is more efficient to generate it. This means that a geographical specialization of RES generation will prevail in the long run, with large wind-generation areas located in the north; hydropower generation plants in the mountain areas in continental Europe; and CSP and photovoltaic generation located in the south part of the Euro-Mediterranean region and North Africa in particular. This implies that better interconnection between those remote areas, and with the meshed network around them, will be needed to secure adequate RES generation to be dispatched.

A reinforced super grid will serve two primary functions: it will transport renewable energy from production sites in North Africa and peripheral regions in Europe to load centres and to evacuate electricity away from areas with momentary excess supply to areas with momentary excess demand.

Euro-Mediterranean countries have strong incentives to improve the network interconnection, and this is true for various reasons. First, because it will lower the dependency of power sector from imported fuels,[14] allowing also for a greater diversification of the countries from which imports stem; second, the very nature of RES generation that, different from oil and gas (to a lesser extent), cannot be stocked from exporters countries to profit from increased prices and more favourable market conditions at a later stage. Cutting electricity supply (to exercise market power over importers) will lead to substantial foregone profits that the exporter will suffer and that cannot be recouped in subsequent timeframes. Therefore, neighbouring countries that trade substantial renewable power are tied very strongly to each other and tend to have aligned concerns over security of supply (SoS).

On the other hand, better interconnections, especially with the presence of significant share of RES generation transferred across countries, require the increase of the geographic scale of network operation (at least at subregional level), the effective promotion of system integration at corridor level, and a significant improvement in the institutional capacity. In particular this latter aspect will prove to be key when large amount of electricity is exchanged, because the coordination between systems will be fundamental to secure the delivery of electricity power at each point of the network. Therefore, the promotion of sub-regional (corridor) system integration, in presence of significant RES generation, requires stronger and better harmonized interconnections. These open issues have been acknowledged by the EC. In order to stimulate the coordinated construction and more efficient use of new interconnectors, as well as increase the cooperation and coordination between transmission system operators (TSOs), two new institutions were established in 2009:[15] Agency for the Cooperation of Energy Regulators (ACER), which will oversee the development of common standards and approaches for cross-border trades, and European Network of Transmission System Operators for Electricity (ENTSO-E), which aims to promote the reliable operation and optimal management of the European transmission system. A similar positive step in this direction is the creation of an analogous dedicated body that will oversee the development of cross-country interconnections in the Mediterranean region (Med-TSO).[16]

Establish an energy free-trade area in the Mediterranean region

The vision of an interconnected pan-European and cross-Mediterranean grid, based on a "corridor" approach and relying mostly (but not exclusively) on renewable generation, would enable electricity production at the best sites for each technology, regardless of national or regional borders and distances to the central European load centres. This provision both increases the efficiency of the system and reduces the intermittency problems.

Normative convergence together with physical infrastructure within different (geographical and source) energy corridor surely entails some sub-regional dynamics along a particular corridor. However, at a regional level, it may imply the fragmentation of the Euro-Mediterranean energy space, at least in the short run.

In order to avoid the risk of another uncooperative set-up and the proposal of a top-down imposition and extension of the *aquis communautaire* to a larger region, that has proved to be ineffective and scarcely attractive for the SMCs, the Euro-Mediterranean energy space will be based on the establishment of an "Energy free-trade area", to be set up through a process based on a mandatory sets of minimum requirements rather than on best practices. Regional institutions involved in the regulatory process, such as Medreg and Med-TSO, have to identify the minimum requirement in their respective areas of action such as market structure, regional TSO engagement, institutional capacity, governance structure, legal and regulatory framework.

These are obviously major challenges to be addressed and need the simultaneous action of a plurality of players, such as national and regional regulatory authorities, governmental bodies, TSOs, investment and financial institutions.

Without denying the complexity of this task, it is important to understand what the approach for creating a regional free trade area needs to be rather than describing the final outcome of this effort.

The binding constraint in this process is the voluntary convergence towards the minimum set of measures. They must ensure compatibility with the existing set-up and the institutional framework of neighbouring countries. The process will necessarily require gradual and stepwise evolution towards a regional market and will see different levels of engagement in different sub-regions. The main driver will be to align the interests of the consumers and producers in the northern and in the southern basins of the Mediterranean Sea.

The "energy free-trade area" will be based on an intervention that must guarantee longevity of energy policy, transparency, co-development and regional view.

The life span is a key attribute of government policy. In order to require and propose a long-lasting transformation of the industry, the new policy should have a long-term life span. These policies should be proposed by government and policymakers in an open and transparent way, with plans that are communicated to investors and to the industry in a clear and timely manner. In developing the legislation supporting the various schemes, it is vital that there is a dialogue among governments and interested parties, because the success and effectiveness of the policies proposed will be directly related to how much input there has been from the market that ultimately is expected to use them. This clearly includes consumers. Finally, a regional view requires that a region-wide goal is shared and is incorporated into individual countries' action plans.

The final aim for the creation of a regional free-trade area is to provide a legal and regulatory framework that would allow each state of the region to trade energy with its immediate neighbour. This minimal step is not yet in place everywhere. As explained earlier, self-sufficiency and network isolation are the rule, rather than the exception, in many non-EU countries in the region.

An obvious example (and model) is the European Coal and Steel Community (ECSC) established in the aftermath of the Treaty of Paris in 1951 and creating a common market for coal and steel between the signing countries. The ECSC was the first supranational institution where the decision-making process takes place in a multinational environment for a sector-specific policy area.

Conclusion

The EC has promoted over time an approach that proposed to extend the EU's energy *aquis communitaire* to its neighbouring countries (European Commission, 2006), and has often highlighted the importance of creating strategic energy partnership with producing and transit countries, which should be based on the EU's energy regulations and policies (European Commission, 2007). The EU external energy policy has been traditionally based on a "Europeanization" strategy of exporting EU regulations to non-EU states and in particular to partner countries. Although this approach, from an EU perspective, can be desirable, and represents a means for modernizing partner countries' energy sectors and for securing energy supplies, it is not always easy or possible, because of at least two separate factors.

The first obvious reason is that the progress of the internal European energy market is making slow progress. In fact, currently there is no single model yet and no single energy market. European countries' preferences over energy policy and existing regulatory and legislative tools are very disperse. The level and the structure of tariffs and taxation of energy products, as well as of energy prices, differ widely across Europe.

Second, the introduction of the third energy package will add considerable variability to the existing situation through the different levels of unbundling models applicable in each member state.

Currently, in this diverse scenario, there is a strong unifying energy policy across member states: the imposition of climate change policies. The triple twenty target has been able to drive most member states' energy policies towards the development of renewable energy sources.

Taking into account the substantial changes that this provision is imposing to EU member states and to its neighbouring countries, we analyzed the pillars on which the Mediterranean Electricity paradigm should be based (see Figure 2.1). They are the development of a corridor approach towards

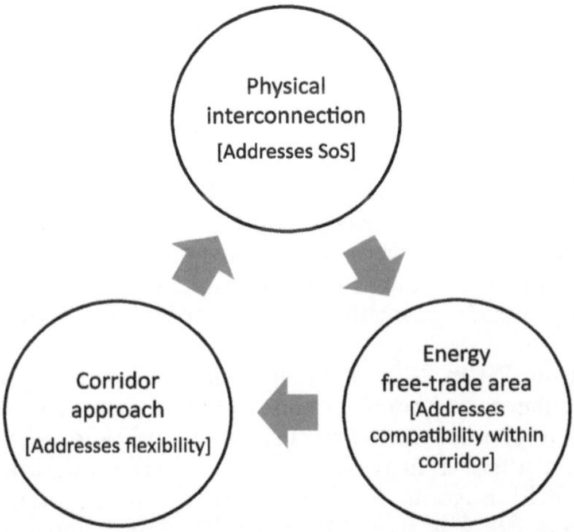

Figure 2.1 Energy trilemma revisited

Source: Author's elaboration.

convergence, a capillary network expansion in the region and the creation of an energy free-trade area in the Mediterranean basin.

These three main action points will enable the region to adopt a flexible approach towards regulatory convergence through the application of a corridor approach to market convergence. They will also address SoS issues by developing an integrated network, based on (but not limited to) the exchange of RES generation. Finally, the establishment of a Mediterranean energy free-trade area will guarantee that compatibility between the different corridors is ensured, allowing different countries to converge although at a differentiated speed.

The tools to deliver these main goals are several and depend on various factors: on the governance and the legal and regulatory frameworks in place in each market, on the integration and on the level of diffusion of renewable generation at present and, in the coming years, besides the financial instruments and risk mitigation mechanisms that each system will be able to implement. Many factors influence the development of the Mediterranean electricity sector, but we believe they are crucially based on the three pillars we have indicated. These three pillars are integrated policy goals; they mutually reinforce and sustain each other. The adoption of only a subset of these provisions might lead to an unstable system or to a highly fragmented electricity market. Neither is desirable.

The integration of the Euro-Mediterranean energy market will face significant challenges in the coming years. The adoption of unifying approach developed according to a bottom-up process will enable the energy potential of the area to be effectively deployed, avoiding unilateral policies that have repeatedly failed to deliver the necessary progress towards the development of an integrated Mediterranean energy community.

Notes

* The findings, interpretations, and conclusions expressed in this chapter are those of the author and do not reflect the positions of the institution.

1. See Eurostat (2010) for Euro-Mediterranean statistics.
2. The introduction of the directive 90/547/CEE introduced the possibility of free transit between producers and wholesale traders of electric energy.
3. For total cross-border monthly physical flows/gross consumption, see EC (2010).
4. Medreg was founded in 2006 with the primary aim of elaborating proposals for the development, harmonization and integration of Mediterranean energy markets and for promoting the exchange of know-how and expertise in the field of energy regulation. In November 2007, in the occasion of its fourth General Assembly held in Rome, Medreg was turned into a permanent association. The Constitutive Act was signed by the following countries: Albania, Algeria, Bosnia-Herzegovina, Croatia, Cyprus, Egypt, France, Greece, Israel, Italy, Jordan, Lebanon Malta, Montenegro, Morocco, Palestinian Authority, Portugal, Slovenia, Spain, Tunisia and Turkey. Libya joined the Association in 2013.

5. The essential role of energy regulators has been acknowledged by Euro Med Energy Ministers on 17 December 2007 in Limassol, in order to implement the Priority Action Plan agreed for 2008–2013:

    1. to guarantee greater harmonization of the energy markets and legislations and to seek progressive market integration in the Euro-Mediterranean region;
    2. to foster sustainable development in the energy sector and;
    3. to develop initiatives of common interest in key areas such as infrastructure extension, investment financing, research and development.

6. See Lamech and Saeed (2003).
7. The triple twenties refer to, respectively, a reduction of greenhouse gas by at least 20% by 2020, a reduction of energy consumption by 20% by 2020, and, finally, an increase in the production of renewables up to a portion of 20% of total production by 2020. See the EU climate and energy package available at http://ec.europa.eu/clima/policies/package/index_en.htm.
8. European Commission (2011).
9. MED-EMIP, 2010, p. 9.
10. ENP was launched in March 2003 and was established to build a wider energy community (European Commission, 2003).
11. This is not the case, for instance for other energy market, such as the gas and oil market, that are regional or wider.
12. The world largest synchronous working grid covers the entire continental Europe with a total consumption of 2,600 TWh/a (Union for the Coordination of Transmission of Electricity - UCTE, 2008).
13. Mediterranean countries are currently expanding the diffusion of RES in electricity system. Motivations for this policy are probably different between EU and non-EU countries. The former aims at reducing their carbon footprint together with their (fuel) import dependency and the latter, and in particular the gas and oil exporting countries, intend to reduce the domestic fuel consumption in order to free valuable resources for external markets.
14. Whereas most countries are self sufficient in relation to electricity supply, almost all European and most of the Mediterranean countries are dependent on imported fuels for their power generation. In 2008, EU 27 countries imported more than 60% of their energy (International Energy Agency, 2010).
15. Third package was firstly proposed at the Commission's Energy Policy for Europe (EC, 2007).
16. The launch of the Association of the Mediterranean Transport System Operators (Med-TSO) was officially announced in Naples, Italy, on the occasion of the international high-level conference on Mediterranean energy highways: an urgent need, organized by the Parliamentary Assembly of the Mediterranean (PAM) on 17 November 2011. Please refer to the PAM newsletter available at www.pam.int/welcome.asp?m=documents&a=cat&cid=141. Med-TSO was then established on April 19, 2012, at the Terna headquarters, as an international non-profit association on a voluntary basis with the mission to promote facilitate the dialogue and cooperation between the Transport System Operators and the Institutions in order to promote a coordinated and pro-active development of the electricity grids and of the Mediterranean connections.

References

Chao, H.-P., & Peck, S. (1996). A Market Mechanism for Electric Power Transmission. *Journal of Regulatory Economics*, 10 (1), 25–59.

Darbouche, H. (2011). Third Time Lucky? Euro Mediterranean Energy Co-operation under the Union for the Mediterranean. *Mediterranean Politics* 16 (1), 193–211.

Escribano, G. (2010). Convergence toward Differentiation, the Case of Mediterranean Energy Corridor. *Mediterranean Politics* 15 (2), 211–229.

European Commission. (2003). *Communication from the Commission to the European Parliament and the Council on the Development of Energy Policy for the Enlarged European Union, its Neighbours and Partner Countries*. COM (262) 262 final/2.

European Commission. (2006). *Energy Green Paper. European Strategy for Sustainable Competitive and Secure Energy*. COM 105 final.

European Commission. (2007). *An Energy Policy for Europe*. COM 1 final.

European Commission. (2010). Directorate-General for Energy, Market Observatory for Energy, Quarterly Report On European Electricity Markets. Market Observatory for Energy, 3 (4).

European Commission. (2011). *Joint Communication to the European Parliament, the Council, the European Economic and Social Committee and the Committee of the Regions. A new response to a changing Neighbourhood*. COM 303 final.

Eurostat. (2010). *Eurostat Pocketbook, Energy Transport and Environment Indicators*. Luxembourg: Publications Office of the European Union.

Helpman, E., & Krugman, P. (1985). *Market Structure and Foreign Trade: Increasing Returns, Imperfect Competition, and the International Economy*. Cambridge, MA: MIT Press.

International Energy Agency. (2005). *Projected Costs of Generating Electricity*. Paris: Nuclear Energy Agency – IEA.

International Energy Agency. (2010). *World Energy Outlook*. Paris: International Energy Agency.

Lamech, R., & Saeed, K. (2003). *What Investors Look For When Investing in Developing Countries*. Energy and Mining Sector Board Discussion Paper Series Paper No. 6, World Bank, Washington, DC.

MED-EMIP. (2010). *MEDRING update. Overview of the power systems of the Mediterranean Basin*. Brussels: MED-EMIP.

Union for the Coordination of Transmission of Electricity - UCTE. (2008). *Annual Report*. Brussels: UCTE.

Chapter 3

Mechanisms and outcomes of the EU external energy policy

An alternative approach

Francesca Pia Vantaggiato[1]

Introductory remarks

Until the Treaty of Lisbon, no EU treaty recognized any competence for the commission on energy issues. Energy was considered as any other economic commodity and therefore provisions for the internal market and competition rules were deemed to apply.

Kanellakis et al. (2013) provide an overview of the European Union (EU) energy policy since its very beginning, when the Economic Coal and Steel Community (ECSC) was formed in 1951. As they recall, at that time the six member states involved looked at nuclear power as the driver of future economic growth. For this reason, the European Atomic Energy Community was created in 1957. It can be argued, therefore, that the very origins of the EU lie in energy issues. Molle (1997) confirms their importance and traces their history starting from the 1962 Memorandum on Energy Policy, followed by a 1968 "First Orientation to a Common Energy Policy". Another important milestone in the construction of an EU energy policy is represented by the November 2000 Green Paper concerning security of supply (Prange-Gstöhl, 2009). At its informal meeting held at Hampton Court in October 2005, the European Council asked the European Commission (EC) to formulate a concept of EU energy policy. This was followed by the March 2006 Green Paper titled "A European Strategy for Sustainable, Competitive and Secure Energy" (European Commission, 2006). In it, the EC argued for the necessity of achieving energy security through a "pan-European energy community" and the creation of a "common regulatory space" around Europe. Following the Green Paper, the EC's Communication of 2007 titled "An Energy Policy for Europe" (European Commission, 2007) made explicit the three objectives of the EU energy policy: security of supply, sustainability and competitiveness. In November 2008, an energy security and solidarity action plan was tabled by the commission and made the external energy policy one of the main topics of the Second Strategic Energy Review (Prange-Gstöhl, 2009).

The Treaty of Lisbon, entered into force on 1 December 2009, includes a chapter on energy issues, thus embodying EU competence in energy issues in the main legal basis of the EU. The chapter consists of only one article, number 194. The article states the EU energy policy shall aim to ensure a functioning energy market, security of supply, energy efficiency and renewable energy development and to promote the interconnection of energy networks. The council and the parliament should establish the necessary measures. However, these should not affect a member state's right to determine the conditions for exploiting its energy resources, its choice between different energy sources and the general structure of its energy supply.[2]

Article 194 introduces shared competence in energy for member states and EU institutions. However, these are constrained by the national sovereignty over energy issues. The tension between the supranational and the national level of governance is a hot topic of the EU integration literature. Nowhere is this tension sharper than in energy issues.

In fact, the different resource endowments of member states and their very geographic location influence greatly their stances and their practices in external energy policy. Moreover, the size of their "national champions" and domestic resource endowment do matter in their stances for or against integration (Pointvogl, 2009).

The EU external energy policy is, therefore, a sensitive topic. The EC presents it on both the European External Action Service (EEAS) website and on Directorate General (DG) Energy website. The former states that "[t]he EU's key energy partners are Russia, Norway, the US, India, China and OPEC. Energy dialogues are underway with each. Memoranda of Understanding have also been signed with key energy partners, while regional cooperation initiatives such as the Black Sea Synergy, the Eastern Partnership, the Union for the Mediterranean and the Central Asia Strategy also contain energy components".[3]

The latter divides the topic of "energy from abroad"[4] into various subheadings. "Security of Supply" concerns the gas diversification strategy the EU is attempting to implement with regard to the famous "gas corridors". "Bilateral cooperation" redirects to the website of the EEAS, where the European Neighbourhood Policy (ENP) is presented. Also, a list of countries with which the EU has established bilateral agreements regarding energy is presented. These vary in depth and continuity, ranging from Memoranda of Understanding and broad ministerial dialogue (i.e. with Japan, South Africa and India) to more structured partnership (i.e. with Ukraine and the US) to very recent developments (i.e. a Memorandum of Understanding with Algeria). Among these energy partners, we find Algeria and Norway, which, with Russia, are the principal gas suppliers of the EU.

Russia, however, has been awarded a dedicated policy for the management of its energy relations with the EU, the only one with an exclusive energy focus in this list.

Following the "EU – Russia energy relations", the lists of policy spaces continues with "The Energy Community", "the Eastern Partnership" (EaP) and "the Mediterranean", to end with "Central Asia".

The focus of this chapter is on the EU External Energy Policy in its immediate neighbourhood. Therefore, attention is focused on energy relations with countries to the east and the south of the EU. These relations are elaborated through the mentioned ENP, of which the Black Sea Synergy, the Eastern Partnership and the Union of the Mediterranean represent area-specific sub-programmes (Prange-Gstöhl, 2009).

The previously mentioned energy-related policy programmes represent but a snapshot of the wide range of energy-related policy programmes that the EU has started. The full list of EU–third countries energy or energy-related dialogues, partnerships, organizations, and memoranda is astonishingly long. These are detailed in the Commission Staff Working Paper titled "Key Facts and Figures on the External Dimension of the EU Energy Policy" (European Commission, 2011a) accompanying the 2011 EC Communication titled "The EU Energy Policy: Engaging with Partners beyond Our Borders" (European Commission, 2011b).[5]

The EU external energy policy in the "neighbourhood"

Therefore, the main objectives of the EU external energy policy regard ensuring security of supply through diversification. Prange-Gstöhl (2009) explains how the EU external energy policy has been studied accordingly to a notion of "rings of energy cooperation" around the EU consisting of three circles of countries. Circle 1 consists of the Western Balkan countries (i.e. those having membership perspective). With them, the EU established a highly institutionalized framework of market integration following the hierarchical governance (Lavenex and Schimmelfennig, 2009) pattern typical of enlargement procedures; Circle 2 includes countries with no membership perspective but with potential to become members of the Energy Community soon (i.e. Ukraine, Moldova, Armenia and Georgia, which in fact are members or observers to the Energy Community at present); Circle 3 includes "the neighbours of the neighbours", that is countries with no membership perspective but with potential to become members of the Energy Community in the long run (Central Asian/Caspian Sea region, Mediterranean, Gulf region). Among the reasons that would push countries with no membership perspective to adhere to the EU relevant legislation and market practice in energy, Prange-Gstöhl quotes an "identification motive" (willingness to demonstrate capability to become part of the club), an "independence motive" (from another regional hegemon) and an "economic motive" (commonality of rules would enable market transactions with the EU).

The Energy Community Treaty (ECT) had been planned as a legal tool to initiate the integration of prospective EU members in the Western Balkans into the EU internal market, starting from the energy sector. Through the establishment of the Energy Community, the EU took the initiative to start the process of integration of South-East European (SEE) countries' energy infrastructure both across SEE territories and with the EU energy market. The EU provided for the legal and regulatory framework in the form of a Treaty binding its signatories to implement relevant pieces of the EU energy legislation. The significant costs of adaptation would be, in this case, rewarded by membership (Karova, 2009). Two Athens memoranda that would have brought to the creation of the Energy Community were signed in 2002 and 2003. The Community was officially set up in 2005.

In time, however, the concept of the ECT seems to have evolved. As stated on its website, "In Dec 2009 the Energy Community Ministerial Council decided on the *accession of Moldova and Ukraine.* With the decision, the geographical concept of Western Balkans, with which the process was linked initially, lost its validity. Today, the leitmotiv behind the Energy Community Treaty is rather the import of the *EU energy policy into non- EU countries*" (emphasis in the original).[6] Prange-Gstöhl (2009) confirms that the ECT has become, in the eyes of the EC, the legal tool to export its internal energy markets norms, rules and institutions.

Karova (2009) outlines the mutual advantages the EU and the countries of South-East Europe enjoyed by entering the legally binding framework of the community itself. In the aftermath of the 1990s wars, the EU was keen in providing financial assistance to the SEE in order to ensure its stability and create scope for trading with the region. In an area devastated by war and deep divisiveness, several regional forms of cooperation were launched, such as the Stability Pact for the SEE. The region's potential of hydropower plants and potential for energy efficiency (totally unexplored due to the subsidized cost of energy) coupled with its scarce gasification met the EU needs in several ways. An important aspect is the possibility of building gas storage facilities in the region, seen that the planned new pipelines meant to bring gas from the Black Sea and the Caspian region (such as the much-disputed Nabucco pipeline and the Trans-Adriatic pipeline, connecting the Turkey–Greece Interconnector to Italy via Albania) to the EU would pass beneath the SEE countries territories. Last but not least, the SEE energy market could help some of the member states to meet their renewable energy targets.[7]

Several authors (Schimmelfennig, 2009; Smith, 2005), however, have pointed to the likely exhaustion of the so-called enlargement method. EU enlargements in 2004 and later have proved to be a rather costly and onerous endeavour. The 2004 enlargement brought 10 more members into the EU, being at very uneven stages of economic development, requiring much assistance in implementing the reforms required by the *acquis*, complicating the decision-making process of the EU while internalizing many of their unsolved regional conflicts (i.e. with Russia). Full membership seems, therefore, unlikely to be the integration pattern to be followed in the future (Schimmelfennig, 2009).

At the same time, the EU's wish to ensure stability and security at its borders pushes it to pursue deeper market integration with the neighbourhood, especially in certain, vital economic sectors, such as, critically, the energy and, more generally, infrastructural one. The way out of this conundrum has been found in the design of a broad cooperation/integration programmes, such as the already mentioned ENP.

The ENP was launched in 2004. This framework was proposed to 16 countries, called "the EU's closest neighbours".[8] The ENP had been proposed as a way to address the constant dilemma of how large should the EU become that has been daunting since the end of the Cold War (Smith, 2005).

The SEE countries were not included in the ENP. In fact, the Energy Community countries' acknowledged eligibility for membership made them unfit to be inserted into the periphery of the EU. Therefore, including a country in the ENP has been equivalent to declare that it had very slight hopes of being offered EU membership (Cardwell, 2011). As Browning and Christou (2010) point out, clearly stating the fact that ENP partners have no membership perspective has marked a radical departure from previous narratives of the EU, which depicted itself as a peace project aiming at stabilizing the area through its openness to enlargement. Coupled with this stance was the spread of "European values" and the unwillingness to draw definitive borders of exclusion.

The ENP, instead, uses the leverage of the market size of the EU and of its financial support in return for transformations in accord with the EU *acquis:* "the previous conditionality logic behind enlargement therefore remains intact, except the carrot is now smaller".

The peace project the EU represents is still an integral part of its external action. For instance, the ECT is presented as not only "about *investments, economic development, security of energy supply* and *social stability*; but – more than this – the Energy Community is also about *solidarity, mutual trust* and *peace*" (emphasis in the original).[9]

Russia refused to be one of the partners of the ENP and demanded (and obtained) a dedicated strategic partnership, to underline its weight in international relations.

The ENP consists of action plans and association agreements being established bilaterally between the EU and each partner country. These plans are adapted to the specific areas of interest of both parties. They cover subjects ranging from civil society development to electricity interconnection plans.

The ENP is complemented by three regional or multilateral policies: the Black Sea Synergy, the Eastern Partnership (EaP) and the Euro-Mediterranean Partnership (EUROMED).

The Black Sea Synergy, launched in 2008 in Kiev, was preceded by the so-called Baku Initiative.[10] The focus of the initiative was on regulatory approximation, on extension of the current infrastructure and the attraction of investments to build new infrastructure, particularly interconnections, thanks to the technical and financial assistance of the EU through the TACIS and INOGATE programmes. This policy seems to be at a standstill. Probably, the participation of Russia to the strategy has not made progress easy to achieve for the EU.

The Eastern Partnership was launched in 2009 in Prague. The countries involved in the partnership are Armenia, Azerbaidjan, Belarus, Georgia, Moldova and Ukraine. Its founding document, a joint declaration issued after the Prague summit, clearly mentions "political association" as one of its ultimate objectives.[11] However, the uneven progress made with regard its overarching goals pushed for a re-orientation of the EaP towards forging "[d]eep and Comprehensive Free Trade Areas (DCFTAs) where regulatory approximation can serve to strengthen the positive effects of trade and investment liberalisation, leading to convergence with EU laws and standards" (European Commission, 2012).[12] The main step forward of the ENP in the politics of the energy sector concerned the accession of Ukraine and Moldova to the ECT as members. Georgia and Armenia joined it as observers.

The Euro-Mediterranean Partnership, instead, has a longer history: the first attempts to forge an external EU policy towards these countries date back to the 1970s, when the Global Mediterranean Policy was put forward by the then European Community, in the wake of the oil crisis (Cardwell, 2011).[13] However, only with the Maastricht Treaty and the stated objective of increasing the EU presence abroad was a structured policy planned for the Mediterranean. It took the form of the so-called Barcelona Process, initiated with the Barcelona conference of 1995 and bearing the official name of Euro-Mediterranean Partnership.[14] In 2008, the then French president Sarkozy announced the intention to establish a Union for the Mediterranean (UfM), foreseeing closer cooperation in economic issues between the two borders of the Mediterranean. The initiative should have concerned only EU members bordering the Mediterranean. The EC opposed the initiative by pointing out that an economic cooperation programme that excluded most EU member states was inacceptable. The idea was therefore repackaged as another re-launch of the EUROMED Partnership/Barcelona Process; its full name became "Barcelona Process: Union for the Mediterranean". The UfM has several objectives, ranging from promotion of human rights and democratic development to support to small businesses, to the establishment of a Mediterranean solar plan.

The events of the "Arab Spring" triggered a reaction in March 2011 consisting in a communication by the EC titled "A Partnership for Democracy and Shared Prosperity with the Southern Mediterranean" and a support programme called SPRING (Support for Partnership, Reforms and Inclusive Growth).[15] The Communication put forward the idea of creating "a form of EU-Southern Mediterranean Energy Community. Extending the Energy

Community Treaty with the Union's Eastern and South-East neighbours, or building on its experience, this community should cover relevant parts of the EU's energy legislation with a view to promoting a real and reliable convergence of South Mediterranean partners' energy policies with EU policy" (European Commission 2011c, p. 11). Importantly, the May 2011 ENP review introduced more explicit "carrot-and-stickism" in the policy by stating the adoption of the so-called more-for-more principle. Basically, the more progress each partner country makes in converging with the EU regulatory stances, the more financial support it will receive from the EU, the European Investment Bank (EIB) and the European Bank for Reconstruction and Development (EBRD). The "more-for-more" principle has been included into the EaP and into the SPRING programme.[16]

The May 2011 review of the ENP, moreover, restated the desirability of establishing an "EU-Southern Mediterranean Energy Community" as a complement to the SEE Energy Community.[17] This important statement seems, however, to have got lost into subsequent reviews of the ENP.

Despite the attractiveness of the idea of establishing an EU-Med energy community, the main and obvious obstacle to such a plan is the modality of its establishment. As mentioned, the Energy Community was a fully fledged top-down initiative, whereby energy market integration was achieved in an asymmetric, hierarchical governance (Lavenex and Schimmelfennig, 2009) form; such a deep change in the energy markets structure of SEE countries was made possible by the membership perspective.

Mediterranean countries, as mentioned, have no prospect of membership. Some of them (Algeria) are regional powers of their own, besides being endowed with fossil fuels. The view of the "economic motive" as being sufficient to convince a set of widely different countries to passively accept the EU *acquis* seems to be rather one-sided and incomplete.

Normative power Europe?

The concept of the EU as a normative power has been first proposed by Manners (2002). He maintained that the best understanding of the international identity of the EU is to be found in its ability to act as a normative power. Laffan (2001) echoes his opinion when she proposes to look at the EU as a regulative, normative and cognitive power. In her view, the EU, however, is a social construction that finds its legitimacy in legal, moral and identity values, differently from traditional nation-states, which found their legitimacy in other forms of power, specifically, military power. The EU is described as regulative, in that it is created by law, and it is a source of law at the same time; normative, in that it represents a community of values (the fundamental freedoms Manners also refers to in his article); cognitive, in that it attempts to prevail over national symbols and identification in the effort to create a sense of "Europeanness" in its citizens. Therefore, the legal body of the EU represents its main source of internal legitimacy. For this reason, compliance with EU law (which has precedence over national law) by member states is fundamental in supporting the credibility of the EU itself.

By looking at the EU as a normative power, both Manners and Laffan refer, specifically, to the commitment of the EU to the promotion of certain key values in its external action: promotion of democracy and respect of the rule of law, fundamental rights and human freedoms. These objectives represent a declaration of commitment to values enshrined in the Universal Declaration of Human Rights of the UN. This conceptual framework pinpoints the narrative of the EU as a peace project with goals that are ultimately social and democratic. By way of this discourse, market integration programmes are not to be disconnected from democratization objectives.

Some observers have criticized the legitimacy of EU normativeness. These authors tend to see the EU external action in more traditional power dynamics, whereby the EU leverages its economic weight to establish itself as a regional hegemon (Haukkala, 2011)

A stronghold of the EU's external action is, indeed, promoting alignment with its regulatory and normative practices in both market contexts and social contexts, as the two are presented as intimately connected.

In most EU umbrella programmes, in fact, sociopolitical objectives are conflated with market integration ones. Promotion of democracy and respect of human rights sit next to market-based principles and market integration and development.

Haukkala (2011) reckons, in her turn, that the EU is using its economic and normative clout to build up a set of highly asymmetrical bilateral relations that help to facilitate an active transference of its norms and values. However, her analysis argues, the EU's ability to project its power is crucially dependent on the perceived legitimacy of its actions in the eyes of its partners. This legitimacy was found in the open-ended nature of the Union's membership, as already mentioned. The ENP can be seen as a policy at the intersection between domestic and foreign (Lavenex and Stulberg, 2007). Its scarce progress, not last in the energy sector, can be connected to several reasons: member states privilege relations with key fossil fuels suppliers, such as Russia and Algeria, thus constraining the support that can be given to transit countries or other, less crucial energy partners in the neighbourhood; the ENP itself, moreover, includes eastern neighbours with membership perspective and southern neighbours, with no such perspective at all. The same difficulties are likely to be faced if an attempt to build up a pan-European Energy Community should be made.

MedReg

Analyses on the role of regulation and of Regulatory Authorities (RAs) within the EU member states abound (Gilardi, 2002, 2005; Thatcher and Coen, 2008; Coen and Thatcher, 2008; Thatcher and Stone Sweet, 2002). The recent creation of EU agencies for the cooperation of utility regulators, including one dedicated to energy regulators (ACER, which stands for Agency for the Cooperation of Energy Regulators), spurred a remarkable strand of literature on the role of these agencies within the EU institutional architecture (Keleman, 2002; Gehring and Krapol, 2007; Thatcher, 2011) and what can be expected from them in terms of delivering the single market objectives. The role of market regulation as the instrument to introduce competition in highly sensitive sectors (such as the utilities) while preserving a commitment to consumer protection and ensuring economic growth has emerged as a key aspect of EU governance. Correspondingly, utilities regulators have come to occupy a crucial role in the EU market integration path. In the energy sector, as well as in others, national regulators are called on to perform several important functions, which could be summed up in three main roles: mediators between multiple political and economic principles (i.e. national ones and EU ones), gatekeepers of their national markets' functioning and agents of integration at inter-European level.

The role of regulatory networks and their "informal governance" practices have also been explored by scholars (Thatcher, 2002; Weinmann, 2007; Maggetti and Gilardi, 2011) who recognized in these networks regulators' efforts to bypass the procedural complexity of decision making at EU level and to achieve closer and faster harmonization of regulatory practices, using them also as precious sources of updated information on the market developments of other member states.

The extent of the expansion and importance of the role of regulators in the EU is such that it has brought important observers to argue about the emergence of a "regulatory state" in the EU (Majone, 1994, 1999).

Nicolaides (2004, 2005) also touches upon a key point when describing the reasons why RAs were created in EU member states: their importance stems from the need to eliminate the ambiguity of the rules that are agreed on at EU political level.

The heterogeneity of preferences and objectives of the member states, and the lengthy procedures of EU decision making cause that, in order to achieve agreement, common rules have to be left purposefully broad. This is also why the instrument of the directive is most commonly used to promulgate EU law. The directive sets out the overall objectives of the policy; policy implementation measures, thereafter, can be decided at national level according to the peculiar legal, institutional and economic features of member states.

This approach, however, has important shortcomings. Although the agreement on common rules is "beneficial to all partner countries because their companies avoid the costs of complying with multiple rules" (Nicolaides, 2005), the fact of having to deviate from their national optimum is costly for member states. Therefore, vague common rules create a compliance problem afterwards and can incentivize member states to deviate from them and to go back to their national optimum. To obviate to this risk, national Regulatory Authorities were established. In this sense, national RAs are effectively agents of integration on behalf of the European Commission.

Majone (1994) also underlined the fact that the EC has a very small budgetary endowment, which must serve to accomplish a wide variety of policy objectives. This resource constraint acted as an incentive for the EC to expand the extent of its action through the expansion of its rule-making activity and the creation of subsidiary bodies to which it can delegate part of its administrative burden. In this logic, the creation of EU agencies in many sectors, ranging from pharmaceuticals to food to energy, finds a straightforward explanation.

The national level of regulatory practice, however, is at least as important in that national RAs only can help the European Commission filling the information gaps it has with regard to the market developments of each single member state. In fact, the EU Agency for the Cooperation of Energy Regulators is primarily an institutionalized platform for national RAs interaction.

The regulatory approach described earlier is, by now, well established and familiar to EU member states. It foresees a specific place for RAs in the institutional framework of the EU, which implies they should be endowed with the necessary autonomy, independence, discretion and decision-making power to be detached from the political short-term circumstances of their national contexts and regulate markets in a broader perspective of sustainability, security of supply, and market integration. Part of the requirements of the EU *acquis* concern, indeed, the establishment of state-level independent regulatory authorities to promote (or enforce) market liberalization. In most Eastern European countries and in several North African ones, such RAs have indeed been created.

MedReg (the association of Mediterranean electricity and gas regulators) was established as the Mediterranean Working Group on Electricity and Natural Gas Regulation in May 2006 within the frame of the Euro-Med partnership.[18] It is funded by the European Neighbourhood and Partnership Instrument (ENPI),the main financial mechanism through which assistance is given to ENP countries, plus Russia.

The MedReg "Concept and Strategy Paper"[19] highlights the parallel between the Energy Community and MedReg itself. It also stresses the possible involvement of Mediterranean countries into an energy cooperation relationship on the basis of market integration, but with a different approach: "Considering the strategic energy challenges of the Mediterranean region, Mediterranean Energy Regulators consider as their duty to constitute a strong institutional basis promoting a "bottom up" approach [. . .] to guarantee greater harmonization of the energy markets and legislations and to seek progressive market integration in the Euro-Mediterranean region". In contrast to the top-down, strongly institutionalized approach of the ECT, MedReg proposes itself as a bottom-up initiative to constitute a network of RAs from the countries facing the Mediterranean.

MedReg gathers subsets of regulators from the EU, from Maghreb and Mashreq countries, from the Middle East and from most of the Energy Community member countries. Its approach is self-regulatory: it has its own statute and procedures, its own internal organs; its membership is varied and does not reproduce the membership patterns of other organizations.

Its mission, as can be read from its website, reflects, however, most of the EU external energy policy goals. It seeks to promote a greater harmonization of the energy markets and legislations and progressive market integration in the Euro-Mediterranean region, besides serving as a platform for cooperation, information exchange and assistance among its members.[20]

MedReg is clearly inspired by an EU agenda, besides being financed by the ENP financial arm and supported organizationally and strategically by CEER (Council of European Energy Regulators). CEER is the voluntary association of European energy regulators. It is not financed by the EU nor forms part of its institutional structure. However, its mission and values coincide with the EU ones. Before the creation of the ACER, CEER was the sole cooperative network of EU regulators and it provided for the "informal governance" arrangements previously mentioned. Furthermore, as stated on its website, in 2012 the CEER has started annual workshops at which EU meet EaP regulators.

One may acknowledge in these developments the beginning of a possibly new approach to energy issues on the side of the EU. Such an approach seems intended to carve out a specific space for energy market issues to be handled, by relying on regulators as agents of integration not only in the EU internal space, but also in its external one.

Energy governmental departments' representatives are allowed to take part in the assemblies and the discussions. MedReg's network involves institutions; industrial associations such as MedGrid, Med-TSO, Desertec, Medelec;[21] and financial organizations such as the EBRD and the World Bank.

During its 15th General Assembly held in Egypt, a representative from the Ministry of Electricity and Renewable Energy of Libya joined the MedReg Assembly and officially expressed the will of his country to become a member of MedReg.

This piece of information raises important issues regarding the famous "capture" problem: in the EU, the issue of regulators' independence from both politics and industry is a very sensitive one. Regulators are not supposed to cooperate with government or to allow government representatives to take part in their dealings. MedReg is an informal tool of networking and cooperation, yet it has been conceived as a regulators' network. However, several countries in the Maghreb do not have regulatory institutions to take part in such dialogues, thus risking being excluded from the "club" and missing important economic development opportunities. Moreover, the EU would also miss economic opportunities by insisting on the export of its institutional features and market practices as they are.

These instances seem to point to a shifting conception in the EU governance of energy matters: the attempt to create a sort of collaborative governance (Bohne, 2011) between all the stakeholders involved seems to be running parallel, or complementing, political summit meetings and policy papers.

These developments might have to do with different aspects that characterize the energy field: in particular, the intertwining political and technical aspects of it. Aspects of mutual convenience in the integration of energy markets between the EU and its closest neighbours, in the face of the scarce results obtained by applying heavy conditionality with too small a carrot, shape these new practices. Increasingly, normative stances regarding broad social objectives are sidelined by more down-to-earth considerations regarding market feasibility, economic development and mutual energy security.

Initiatives such as MedReg show how networked forms of governance (Lavenex and Schimmelfennig, 2009) are actually a favoured outcome for energy stakeholders. In particular, they show the institutional entrepreneurship of EU regulators.

Within the Energy Community, the parallel action of politicians and regulators is actually enshrined into its institutional structure. In it, a ministerial council steers policy action and decides on the admission of new members, whereas a regulatory board works as a coordination body of the national regulators for exchanging knowledge and developing common best-practice solutions for implementing the treaty in a harmonized way. However, the regulatory board has mainly advisory powers.

MedReg, instead, prides itself with having received political endorsement and having been identified as one of the key actors for energy cooperation in the region.[22]

One could possibly recognize in the initiative (strongly supported by the regulators of France, Italy and Spain), both the fact that EU national regulators are indeed agents of integration and a sign that the agent (regulator) has reached a new maturity in its institutional development. This may push RAs to transform into policy-making entities (rather than mere rule-setting ones) and to pursue their own institutional goals. This widely interesting topic is ripe for future research.

Concluding remarks

The distinction between positive and negative integration has been first made explicit by Tinbergen (1954) and then applied by Pinder (1968) to the analysis of economic integration in the EU. In short, negative integration consists in removing obstacles to market creation, whereas positive integration implies the setting up of common policies. Clearly, spotting discrimination and removing it from the treaty is much easier than finding agreement on common policies, which can take many different forms (Laffan, 1992). Majone (2009a and 2009b) sees a return to the principle of negative integration as enshrined in the Treaty of Rome as inevitable for managing a future EU comprising thirty or even more members. He envisages, therefore, that even countries that currently have no membership perspective will eventually become members. Managing such an enlarged EU comprising such a wide variety of national features, he argues, will be impossible unless the current top-down community method is abandoned. Majone envisages an EU political arena that is always less political and more regulatory:

> Under a negative-integration regime, most regulatory responsibilities would be left with the people who are most directly affected by a given problem, and who have to bear the cost of regulation. The tasks of the European institutions would primarily consist in monitoring the behaviour of national regulators to make sure that they do not abuse their autonomy for protectionist purposes, or to violate rights guaranteed by European law [. . .] Centralized, top-down harmonization would become a measure of last resort. (2009a, p. 191)

In his view, therefore, regulators will bridge across national differences and interests. Whether this forecast will turn into reality or not, it seems useful for now to borrow these ideas and apply them to the external energy policy of the EU.

Initiatives such as MedReg, which have the support of the EC, or the establishment of the information exchange mechanism on agreements with third countries, seem to point in this direction already. As Pointvogl (2009) convincingly shows, an energy market integration policy that does not take into explicit account the security of supply concerns of each member state is doomed to fail. Moreover, it is difficult to draw the line between external and internal energy policy: resource scarcity means a constant concern for security of supply; this implies the internal and external dimensions of each member state, and therefore of the Union as a whole, cannot be neatly split.

Furthermore, the action of the EU at political level could be positively complemented by the action of RAs in matter that concern narrower, technical issues that will necessarily exclude narratives of peace and communities of values. This, however, needs not imply that such objectives would disappear from the general external policy of the EU.

We would argue, however, that initiatives in support of democratization and civil society building should not be mingled with market integration objectives.

The centrality of energy to economic development makes it unfit to be tackled under umbrella programmes that set themselves many, varied and ambitious social objectives and put these next to more strictly economic ones. The final outcome of this strategy is a human rights record that remains "*strikingly disconnected from the EU energy policy*" (Youngs, 2009).

Achieving a transparent management of energy policy at internal level, one that takes into account member states' real issues and interests and allows for a disaggregated, negative integration approach to energy relations would be beneficial for the overall security of the EU energy supply and for the credibility of the EU in both energy and other, broader social objectives. A community approach to market creation that relies on the inputs of both EU and non-EU stakeholders would likely achieve more concrete outcomes. In the multi-level governance structure of the EU, regulators are best placed to act as effective brokers.

MedReg could represent a starting point towards an inclusive, collaborative and productive external action in the energy field. However, this implies a much broader and "heavier" role for regulators than the one traditionally assigned to them, that is, technical competence in narrow sectors. Whether initiatives such as MedReg represent the beginning of an important shift in the features regulatory authorities should have or will have is a crucial topic that calls for further exploration.

Notes

1. The findings, interpretations and conclusions expressed in this chapter are those of the author and do not reflect the positions of the institution.
2. http://eur-lex.europa.eu/LexUriServ/LexUriServ.do?uri=CELEX:12008E194:EN:HTML (last accessed on 22/02/14).
3. www.eeas.europa.eu/energy/ (last accessed on 22/02/14).
4. http://ec.europa.eu/energy/international/index_en.htm (last accessed on 22/02/14).
5. http://eur-lex.europa.eu/LexUriServ/LexUriServ.do?uri=SEC:2011:1022:FIN:EN:PDF (last accessed on 22/02/14).
6. www.energy-community.org/portal/page/portal/ENC_HOME/ENERGY_COMMUNITY (last accessed on 22/02/14).
7. See www.balkaninsight.com/en/article/the-underwater-cable (last accessed on 22/02/2014) and www.balkaninsight.com/en/page/specials-montenegro-energy-focus-home (last accessed on 22/02/2014).
8. They are Algeria, Armenia, Azerbaijan, Belarus, Egypt, Georgia, Israel, Jordan, Lebanon, Libya, Moldova, Morocco, Occupied Palestinian Territory, Syria, Tunisia and Ukraine.
9. www.energy-community.org/portal/page/portal/ENC_HOME/ENERGY_COMMUNITY (last accessed on 22/02/2014).
10. It was launched in 2004, soon after the ENP contents was unveiled, at a Ministerial Conference held in Baku that saw the participation of the European Commission and the Black Sea and the Caspian Littoral States and their neighbours, namely Azerbaijan, Armenia, Bulgaria, Georgia, Iran (observer), Kazakhstan, Kyrgyz Republic, Moldova, Romania, Russian Federation (observer), Tajikistan, Turkey, Ukraine and Uzbekistan.

11. www.consilium.europa.eu/uedocs/cms_data/docs/pressdata/en/er/107589.pdf (last accessed on 16/09/13).
12. Eastern Partnership Roadmap, http://ec.europa.eu/world/enp/docs/2012_enp_pack/e_pship_roadmap_en.pdf (last accessed on 22/02/14, p.3). The proposal to introduce Deep and Comprehensive Free Trade Agreements (DCFTAs) triggered the immediate reaction of Russia, which is building up its own Eurasian Customs Union project with Belarus and Kazakhstan, with a view to extend it to other countries in Eastern Europe in the future. Very recently, the EC stressed the incompatibility of EaP partner countries' membership to both DCFTAs and the Eurasian Customs Union and invited Russia to stop exerting pressures on EaP members to deter them from signing up to DCFTAs. http://europa.eu/rapid/press-release_SPEECH-13-687_en.htm (last accessed on 22/02/14).
13. However, the countries covered by the Global Mediterranean Policy were a greatly varied group: Albania, Algeria, Cyprus, Egypt, Greece, Israel, Lebanon, Libya, Malta, Morocco, Portugal, Spain, Syria, Tunisia, Turkey and Yugoslavia. Too varied a group to implement any sort of coherent policy, it has been argued (Tsoukalis) but a first move to put a step in the door of greatly resources-endowed countries.
14. Besides the EU member states, 12 Mediterranean partners were invited to join: Algeria, Cyprus, Egypt, Israel, Lebanon, Jordan, Malta, Morocco, the Palestinian Territories, Syria, Tunisia and Turkey. Later on, the ENP was launched and the EuroMed Partnership ended up under its umbrella. Nine partners to the EuroMed, therefore, became ENP partners as well; Malta and Cyprus became EU member states, and Turkey began pre-accession negotiations
15. www.enpi-info.eu/mainmed.php?id=394&id_type=10 (last accessed on 17/09/2013).
16. See http://ec.europa.eu/world/enp/docs/2012_enp_pack/e_pship_roadmap_en.pdf (last accessed on 16/09/13).
17. See http://ec.europa.eu/world/enp/pdf/com_11_303_en.pdf (last accessed on 16/09/13).
18. MedReg, today a regional regulatory association, brings together energy regulators for electricity and gas of the following countries: Albania, Algeria, Bosnia-Herzegovina, Cyprus, Croatia, Egypt, France, Jordan, Greece, Israel, Italy, Malta, Morocco, Montenegro, the Palestinian Authority, Portugal, Slovenia, Spain, Tunisia and Turkey.
    As one can easily spot, all member regulators of MedReg are also regulators of countries that are part of the Energy Community.
19. MedReg Concept and Strategy Paper, www.europarl.europa.eu/meetdocs/2004_2009/documents/dv/200/200903/20090316Medreg_concept_EN.pdf (last accessed on 22/02/2014).
20. See www.medreg-regulators.org/About/Mission.aspx (last accessed on 22/02/14).
21. See www.medreg-regulators.org/About/ExternalRelations.aspx (last accessed on 22/02/14).
22. See www.medreg-regulators.org/About/Accountability.aspx (last accessed on 22/02/14).

References

Bohne, E., 2011. Conflicts between national regulatory cultures and EU energy regulations. *Utilities Policy* 19, 255–269.

Browning, C.S., and Christou, G., 2010. The constitutive power of outsiders: The European neighbourhood policy and the eastern dimension. *Political Geography* 29, 109–118.

Cardwell, P.J., 2011. EuroMed, European Neighbourhood Policy and the Union for the Mediterranean: Overlapping policy frames in the EU's governance of the Mediterranean. *JCMS: Journal of Common Market Studies* 49, 219–241. http://onlinelibrary.wiley.com/doi/10.1111/j.1468–5965.2010.02119.x/abstract.

Coen, D., and Thatcher, M., 2008. Network governance and multi-level delegation: European networks of regulatory agencies. *Journal of Public Policy* 28, 49–71.

European Commission, 2006. *Green Paper - A European Strategy for Sustainable, Competitive and Secure Energy.* COM(2006) 105 final. Brussels. http://europa.eu/documents/comm/green_papers/pdf/com2006_105_en.pdf (last accessed on 22/02/2014).

European Commission, 2007. *Communication from the Commission to the European Council and the European Parliament. An energy policy for Europe.* COM(2007) 1 final. Brussels. http://ec.europa.eu/energy/energy_policy/doc/01_energy_policy_for_europe_en.pdf (last accessed on 22/02/2014).

European Commission, 2011a. *Commission Staff Working Paper, Key facts and figures on the external dimension of the EU energy policy, accompanying document to the communication from the Commission to the European Parliament, the council, the European Economic and Social Committee and the Committee of the Regions on security of energy supply and international cooperation – "The EU Energy Policy: Engaging with Partners beyond Our Borders".* SEC(2011) 1022 final. Brussels. http://eur-lex.europa.eu/LexUriServ/LexUriServ.do?uri=SEC:2011:1022:FIN:EN:PDF (last accessed on 22/02/2014).

European Commission, 2011b. *Communication from the Commission to the European Parliament, the Council, the European Economic and Social Committee and the Committee of the Regions. On security of energy supply and international cooperation - "The EU Energy Policy: Engaging with Partners beyond Our Borders".* COM(2011) 539 final. Brussels. http://eur-lex.europa.eu/LexUriServ/LexUriServ.do?uri=COM:2011:0539:FIN:EN:PDF (last accessed on 22/02/2014).

European Commission, 2011c. *Joint Communication to the European Council, the European Parliament, the Council, the European Economic and Social Committee and the Committee of the Regions. A Partnership for Democracy and Shared Prosperity with the Southern Mediterranean.* COM(2011) 200 final. Brussels. http://eeas.europa.eu/euromed/docs/com2011_200_en.pdf.

European Commission, 2012. *Joint Communication to the European Parliament, the Council, the European Economic and Social Committee and the Committee of the Regions. Eastern Partnership: A roadmap to the autumn 2013 summit.* JOIN(2012) 13 final. Brussels. http://ec.europa.eu/world/enp/docs/2012_enp_pack/e_pship_roadmap_en.pdf

Gehring, T., and Krapohl, S., 2007. Supranational regulatory agencies between independence and control: the EMEA and the authorization of pharmaceuticals in the European Single Market. *Journal of European Public Policy* 14, 208–226.

Gilardi, F., 2002. Policy credibility and delegation to independent regulatory agencies: a comparative empirical analysis. *Journal of European Public Policy* 9, 873–893.

Gilardi, F., 2005. The formal independence of regulators: A comparison of 17 countries and 7 sectors. *Swiss Political Science Review* 11, 139–167.

Haukkala, H., 2011. The European Union as a regional normative hegemon: The case of European Neighbourhood Policy, in R. Whitman (ed.), *Normative Power Europe.* London: Palgrave-Macmillan, pp. 45–65.

Kanellakis, M., Martinopoulos, G., and Zachariadis, T., 2013. European energy policy – A review. *Energy Policy* 62, 1020–1030.

Karova, R., 2009. *Energy community for South East Europe: Rationale behind and implementation to date* (Working Paper). EUI/RSCAS 2009/12, Florence School of Regulation, Florence (Italy).

Keleman, D. R., 2002. The politics of "Eurocratic" structure and the new European agencies. *West European Politics* 25, 93–118.

Laffan, B., 1992. *Integration and Co-operation in Europe*. London: Routledge.

Laffan, B., 2001. The European Union polity: a union of regulative, normative and cognitive pillars. *Journal of European Public Policy* 8, 709–727.

Lavenex, S., and Schimmelfennig, F., 2009. EU rules beyond EU borders: theorizing external governance in European politics. *Journal of European Public Policy 16*, 791–812.

Lavenex, S., and Stulberg, A., 2007. "Connecting the Neighborhood: Energy and Environment", in K. Weber, M. E. Smith, and M. Baun (eds.) *Governing Europe's New Neighborhood: Partners or Periphery?* Manchester: Manchester University Press, pp. 134–156.

Maggetti, M., and Gilardi, F., 2011. The policy-making structure of European regulatory networks and the domestic adoption of standards. *Journal of European Public Policy* 18, 830–847.

Majone, G., 1994. The rise of the regulatory state in Europe. *West European Politics* 17, 77–101.

Majone, G., 1999. The regulatory state and its legitimacy problems. *West European Politics* 22, 1–24.

Majone, G., 2009a. *Europe as the Would-Be World Power: The EU at Fifty*. Cambridge: Cambridge University Press.

Majone, G., 2009b. "Liberalization, Re-Regulation, and Mutual Recognition: Lessons from Three Decades of EU Experience", www.gla.ac.uk/media/media_111516_en.pdf.

Manners, I., 2002. Normative power Europe: A contradiction in terms? *JCMS: Journal of Common Market Studies* 40, 235–258.

Molle, W., 1997. *The Economics of European Integration: Theory, Practice, Policy*. Farnham, UK: Ashgate Publishing, Ltd.

Nicolaides, P., 2004. The political economy of multi-tiered regulation in Europe. *JCMS: Journal of Common Market Studies* 42, 599–618.

Nicolaides, P., 2005. Regulation of liberalised markets: A new role for the state? (or how to induce competition among regulators?). in D. Geradin, R. Muñoz, and N. Petit, (eds.) *Regulation through Agencies in the EU-A New Paradigm of European Governance*. Cheltenham: Edward Elgar, pp. 23–46.

Pinder, J., 1968. Positive integration and negative integration: some problems of economic union in the EEC. *The World Today* 24(3), 88–110.

Pointvogl, A., 2009. Perceptions, realities, concession – What is driving the integration of European energy policies? *Energy Policy* 37, 5704–5716.

Prange-Gstöhl, H., 2009. Enlarging the EU's internal energy market: Why would third countries accept EU rule export? *Energy Policy* 37, 5296–5303.

Schimmelfennig, F., 2009. Europeanization beyond Europe, *Living Reviews in EU Governance* 4, 3. Accessed 16 April 2014 from www.livingreviews.org/lreg-2009-3.

Smith, K. E., 2005. The outsiders: the European neighbourhood policy. *International Affairs* 81, 757–773.

Thatcher, M., 2002. Regulation after delegation: independent regulatory agencies in Europe. *Journal of European Public Policy* 9, 954–972.

Thatcher, M., 2011. The creation of European regulatory agencies and its limits: a comparative analysis of European delegation. *Journal of European Public Policy* 18, 790–809.

Thatcher, M., and Coen, D., 2008. Reshaping European regulatory space: An evolutionary analysis. *West European Politics* 31, 806–836.

Thatcher, M., and Stone Sweet, A., 2002. Theory and practice of delegation to non-majoritarian institutions. *West European Politics* 25, 1–22.

Tinbergen, J., 1954. *International Economic Integration.* Amsterdam: Elsevier.

Weinmann, J., 2007. *Agglomerative Magnets and Informal Regulatory Networks: Electricity Market Design Convergence in the USA and Continental Europe* (Working Paper). EUI/RSCAS 2007/15, Florence School of Regulation, Florence (Italy).

Youngs, R., 2009. *Energy Security: Europe's New Foreign Policy Challenge.* New York: Taylor & Francis.

Chapter 4

Will outsiders apply EU rules, and why?

Ana-Maria Boromisa[1]

This chapter examines to which extent exporting EU policies and institutions in the energy sector can be effective in balancing commercial, political and social interest in neighbouring regions.

The neighbouring regions that are in focus of this chapter are candidates (Macedonia, Serbia, Montenegro, Turkey) and potential candidates from South-East Europe (Bosnia and Herzegovina, Albania, Kosovo), Eastern neighbours (Ukraine, Moldova) and Mediterranean neighbours (Algeria, Egypt, Israel, Jordan, Lebanon, Morocco, the Palestinian Authority).

The starting hypothesis is that strategic and geopolitical importance of the South-East European (SEE), Eastern neighbours and Mediterranean countries[2] in providing access to energy sources and diversifying routes could, with time, serve as a basis for defining and promoting their regional interests.

First, a brief overview of evolution of external European energy policy towards enlargement countries and neighbours is provided. It is followed by presentation of development of the Energy Community and Medreg.

Next, planned energy infrastructure projects in the Energy Community and Medreg regions are presented. Roles and positions of selected stakeholders within and outside Energy Community and Medreg regions are outlined. Based on the preceding, the relevance of these projects for the countries concerned and the EU member states is evaluated, and conclusions and recommendations are formulated.

External European energy policy, enlargement and neighbourhood policy

The enlargement policy was considered as the most effective EU's foreign policy tool. It is about to reach its limits. On the contrary, the EU's energy policy (and its external dimension) is in an early stage of implementation. In 2007, when "big bang" enlargement was finalized, the Commission presented a package of measures to establish an energy policy that would tackle climate change while boosting energy security and competitiveness.

An action plan followed, which emphasized the need for an external energy policy towards main energy partners (e.g. Russia, Norway, the US, China, India and the Organization of the Petroleum Exporting Countries [OPEC]) and neighbouring regions.

The Treaty of Lisbon, which entered into force on 1 December 2009, gave energy policy a new legal basis. The Union became obliged to ensure security of energy supply, promote the interconnection of energy networks and improve energy efficiency and energy saving.

The strategy for competitive, sustainable and secure energy followed in 2010 with the aim to change the way in which Europe produces and consumes energy. The strategy set five priorities, including strengthening the external dimension of the EU energy market.[3] This primarily relates to encouraging the participation of neighbouring countries in the internal market and promotion of low-carbon energy in the world (e.g. through launching a major cooperation with Africa on sustainable energy).

Based on the need for better coordination of EU and member states' activities to ensure consistency and coherence in the EU's external relations with key producer, transit, and consumer countries the European Council in 2011 invited the Commission to submit communication on security of supply and international cooperation and to continue its efforts to facilitate the development of strategic corridors for the transport of large volumes of gas such as the Southern Corridor.[4]

In its communication (European Commission, 2011) a new approach to infrastructure policy which was confirmed by the EU Council conclusions (Council, 2011) and the European Parliament (European Parliament, 2011) resolution that followed, with aim to optimize network development at European level for the period up to 2020.

The European Council of 4 February 2011 underlined the need to modernize and expand Europe's energy infrastructure and to interconnect networks across borders. This was considered necessary to make solidarity between member states operational, to provide for alternative supply or transit routes and sources of energy and to develop renewable energy sources in competition with traditional sources. By 2015 all member states should be interconnected just with European gas and electricity networks.

The European Council also concluded that energy security should be fully reflected in the EU's neighbourhood policy. The energy partnerships should be established on the initiative of the EU and cover a wide range of issues. These issues include regulatory approaches on all subjects of common interest, such as energy security, safe and sustainable low carbon technologies, energy efficiency. Encouraging neighbouring countries to embrace the EU's relevant internal energy market rules, notably by extending and deepening the Energy Community Treaty and promoting regional cooperation initiatives, was recognized as a useful tool to achieve the EU's interests. The need to develop measures in the context of the Energy Strategy 2020 to ensure a level playing field for EU power producers vis-à-vis producers outside the European Economic Area was also recognized (European Council, 2011). In this context, the EU develops its external energy policy towards Mediterranean countries. Cooperation is based on the neighbourhood policy. Energy dimension promotes use of renewable energy sources and inclusion in the internal EU energy market, and the Medreg is one of the programmes aimed at integrating the Maghreb and Mashrek energy markets.

Current and potential candidates are also included in such effort in the framework of the enlargement policy, based on conditionality. The membership perspective for current candidates and potential candidates is rather distant. It does not provide sufficient incentives for alignment with the EU rules. Thus, the EU has offered model for energy market integration, which requires alignment with EU rules in areas of energy, environment, trans-European networks and competition policy. The motivation for alignment with EU rules in these areas is primarily economic. The same approach is extended to Eastern neighbours in the framework of the Energy Community.

The Energy Community and Medreg are models of European economic integration. They provide for participation in the internal market, and regional institutions have power to decide which parts of acquis will be transposed and when. These are elaborated on further in the next sections.

The Energy Community

The Energy Community is an international organization established in 2006 for a 10-year period. Its members are the European Union, on one hand, and eight parties (Albania, Bosnia and Herzegovina, Serbia, Montenegro, Macedonia, Kosovo, Moldova and Ukraine), on the other. It is based on the long-term strategy for energy in the South-East Europe developed by the European Commission, together with the partner countries and with help from the international financial institutions (notably the European Investment Bank [EIB], the European Bank for Reconstruction and Development [EBRD] and the World Bank) in 2001. It provides for energy market integration, that is an extension of the internal EU energy market.

The strategy has been endorsed in the framework of the Stability Pact.[5] Its implementation was fostered by the so-called Athens Process for regional energy cooperation. The Athens Process was formally launched in 2002 by the signature of the Memorandum of Understanding on the Regional Electricity Market in South-East Europe and Its Integration into the European Union Internal Electricity Market (the Athens Memorandum of 15 November 2002). By signing the Athens Memorandum, adhering parties, primarily countries of the South-East Europe, committed to implementation of the Community provisions related to creation of an internal electricity market in the area (see Table 4.1, column 1).

Table 4.1 The EU's formal reasons for launching the Energy Community and the Medreg

(1) *Athens I*	*(2)* *Athens II*	*(3)* *Treaty*	*(4)* *Status (2014)*
Adhering parties SEE, Turkey, Greece, Italy	*Participants* SEE, Turkey, EC	*Parties* SEE (Albania, Bosnia and Herzegovina, Bulgaria, Croatia, Kosovo, Macedonia Montenegro, Romania, Serbia), EC	*Parties* *SEE* (Albania, Bosnia and Herzegovina, Kosovo, Macedonia Montenegro, Serbia), Moldova, Ukraine, EU
Non-participating sponsors EC, Stability Pact	*Political participants* Greece, Italy, Austria	*Participants* (Open to EU member states) Austria, Cyprus, Czech Republic, Germany, Greece, Hungary, Italy, Slovakia, Slovenia, and United Kingdom	*Participants* Austria, Bulgaria, Croatia, Cyprus, Czech Republic, France, Germany, Greece, Hungary, Italy, the Netherlands, Poland, Romania, Slovakia, Slovenia, UK
	Sponsor Stability Pact		*Candidates* Georgia
Observers Austria, Hungary, Italy, Slovenia, Moldova	*Observers* Hungary, Moldova, Slovenia	*Observers* Moldova, Norway, Turkey and Ukraine	*Observers* Norway, Turkey, Armenia

The Athens Process was extended in 2003 to include the gas market, which was declared by the second Memorandum of Understanding. Under the memoranda, the SEE countries committed themselves to introducing EU rules in gas and electricity sectors, with some delay comparing with the relevant EU directives.

Signatories of both Athens memoranda are the same, but their roles changed during the process. Adhering parties of the first memorandum became participants in the second. Non-participating sponsors split into participants (the European Commission) and sponsor (the Stability Pact). "Old EU member states" (Italy, Austria, Greece) that were adhering parties to the first memorandum became political participants to the second. "New EU member states" (Hungary, Slovenia) kept their observer status, and so did Moldova.

Participants to the second Athens Memorandum became formal parties of the Energy Community Treaty (Turkey being the exception) signed on 25 October 2005 (Table 4.1, column 3). The political significance of the Stability Pact, which was sponsor under Athens memoranda was mentioned in the treaty preamble. During the process of ratification, the Energy Community functioned as association established under Austrian law. When the treaty entered into force (in 2006) the Energy Community (previously known as the Energy Community of South-East Europe and the European Energy Community) became an international organization. Its parties were the European Community, on one hand, and nine adhering parties (Bulgaria, Romania, Croatia, Albania, Bosnia and Herzegovina, Serbia, Montenegro, Macedonia and Kosovo), on the other hand. It has headquarters in Vienna and aims to establish an integrated market in gas and electricity, based on common interest and solidarity.

In 2007 Bulgaria and Romania acceded to the EU; thus, they ceased being parties and became participants. The same happened with Croatia in 2013.

Participation in the Energy Community is open to the EU member states, which means that they may be represented and discuss in the Energy Community institutions without voting. This privilege is currently used by 16 member states (Table 4.1, column 4).

Moldova and Ukraine acceded to the Energy Community in 2010. With their accession the Energy Community became an organization including enlargement and neighbouring countries. Considering the timeframe of the Energy Community (10 years from the date of entry into force, i.e. 1 July 2006) and the distant membership perspective of current candidates, the Energy Community is an organization of "excluded" countries.

Eastern enlargement of the Energy Community indicates the role of the excluded region in formulating and implementing external EU energy policy. The Energy Community is primarily tool for exporting the rules, but it is also an international organization with decision-making body, the Ministerial Council. Other institutions of the Energy Community are the Permanent High Level Group, the Regulatory Board, the Electricity Forum, the Gas Forum, the Oil Forum (the Fora) and the Secretariat.

The Ministerial Council consists of the ministers of energy of the member states and two representatives of the EU. It meets once a year and makes strategic decisions (such as on extension of the relevant acquis, membership), gives directions or formally adopts secondary legislation. Its presidency rotates on a yearly basis.

The Permanent High Level Group consists of representatives of the ministers (one for each party) and two representatives of the European Commission. It meets on the initiative of the Commission or the country holding the presidency. Its tasks are to prepare the Ministerial Council and to ensure the follow-up of the decisions.

The Regulatory Board consists of representatives of the energy regulators of each party and officials from the European Commission. It meets in Athens and considers issues of regulatory cooperation.

The Electricity Forum, the Gas Forum and the Oil Forum are composed of representatives of all interested stakeholders, including industry, regulators, industry representative groups and consumers. The Fora are chaired by the European Commission. The Fora advise the Energy Community, provided that they adopt conclusions by consensus.

The Secretariat provides administrative support to other institutions, reviews implementation and coordinates donors' activities. The largest single donor in the Balkans is the European Community, and the Commission contributed 94.9% to the Energy Community budget in the first five years.[6]

Currently, the Energy Community provides for application of the EU acquis in 35 European jurisdictions (28 EU member states, 5 signatories of the Energy Community from SEE and 2 Eastern neighbours – Ukraine and Moldova). In addition to electricity, oil and gas legislation, the Energy Community provides for alignment with the acquis on environment, competition, energy efficiency and renewables.

The Energy Community is more than an extension of the EU acquis. Its institutions (e.g. the Ministerial Council) can make "strategic decisions", so it might evolve into organization that formulates and promotes regional interests, thus strengthening its negotiation position towards main investors or suppliers.

Medreg

The Medreg is the Association of Mediterranean Regulators for Electricity and Gas, established in 2007 with seat in Milan. It gathers energy regulators (ministries or independent regulators) from 21 member countries, including nine EU member states, three members of the Energy Community (EnCT) and nine Mediterranean Partners (see Table 4.2).

Its main objective is to promote a transparent, stable and harmonized regulatory framework in the region (Medreg, 2006). Twelve member countries apply harmonized regulatory framework (i.e. EU rules) which implies that the goal is to extend the EU rules in that region as well. The Mediterranean countries are not (yet) legally obliged to implement EU energy acquis, but gradual alignment with EU rules is promoted by Medreg activities.

The General Assembly is Medreg's decision-making body, responsible for defining Medreg's strategy and approving the action plan, operational guidelines, budget and position papers elaborated by the groups and task forces. The Steering Committee prepares the work of the General Assembly and supervises the implementation of the action plan. It is composed of high-level representatives of Medreg members and meets every six months on a principle of rotation.

The Secretariat develops and implements Medreg strategy. It is in charge of the management and development of the Association's activities and fosters information exchange among its members through a Communication Officers network.

The groups of the Medreg (gas, electricity, environment, renewable energy sources and energy efficiency) correspond to the key areas of energy acquis included in the Energy Community,[7] whereas the institutional group promotes coordination among Medreg members. In particular, the

Table 4.2 Medreg members

Country	*National Regulatory Authority*
Albania (EnCT)	Energy Regulatory Entity (ERE)
Algeria	Electricity and Gas Regulation Commission (CREG)
	Regulatory Authority of Hydrocarbons (ARH)
Bosnia and Herzegovina (EnCT)	State Electricity Regulatory Commission (SERC)
Croatia(EU)	Croatian Energy Regulatory Agency (HERA)
Cyprus (EU)	Cyprus Energy Regulatory Authority (CERA)
Egypt	Egyptian Electric Utility and Consumer Protection Regulatory Agency (EGYPTERA)
France (EU)	Commission of the Regulation of Energy (CRE)
Greece (EU)	Regulatory Authority for Energy (RAE)
Israel	Public Utility Authority (PUA)
	Natural Gas Authority (NGA)
Italy (EU)	Regulatory Authority for Electricity and Gas (AEEG)
Jordan	Electricity Regulatory Commission (ERC)
	Ministry of Energy and Mineral Resources (MEMR)
Lebanon	Ministry of Energy and Water (MEW)
Malta (EU)	Malta Resources Authority (MRA)
Montenegro (EnCT)	Regulatory Agency for Energy (REGAGEN)
Morocco	Ministry of Energy, Mines, Water & Environment (MEMEE)
Palestinian Authority	Palestinian Energy Authority (PEA)
Portugal (EU)	Energy Services Regulatory Authority (ERSE)
Slovenia (EU)	Energy Agency (AGEN-RS)
Spain (EU)	National Energy Commission (CNE)
Tunisia	Ministry of Industry and Technology (MIT)
Turkey	Energy Market Regulatory Authority (EMRA)

Note: Bold are countries that are not members of the EU nor the Energy Community, that is, are not legally obliged to implement EU energy acquis.

electricity group is in charge of assessing the status and possible developments of electricity markets and regulatory frameworks in Medreg countries. It identifies and proposes basic requirements that will lead to the harmonization of energy regulation in the Medreg region, developing a competitive, strengthening and running regional electricity market in the region.

The gas group assesses the status of natural gas and liquefied natural gas (LNG) markets in Mediterranean countries, their relevant regulatory frameworks and possible evolutions. It elaborates on the Guidelines of Good Practice (GGP) and recommendations that could lead to the development of an integrated, competitive, secure and effective gas market in the region.

Environment, renewable energy sources and energy efficiency (RES group) focuses on the legislative and regulatory mechanisms used to promote renewable electricity generation, energy efficiency and RES deployment in the Mediterranean basin.

Specific task forces are established on

- integration of electricity markets of Maghreb Countries (Algeria, Morocco and Tunisia). Its activities are supported by the European Union.
- cooperation with the International Confederation of Energy Regulators (ICER TF) provides coordination among the Medreg representatives in the four ICER virtual working groups.
- consumer issues (CUS TF) works on enhancing the protection of household consumers in the Medreg countries. The work of the task force includes surveys and recommendations on energy billing, consumer protection and information. The task force also identifies and promotes best practices in consumer protection, including vulnerable customers and quality of service for electricity and gas from the point of view of final consumers.
- infrastructure. This task force was created in 2011 as internal observatory. It provides a regulatory analysis of the future investments in the area for the medium to long term, it builds a permanent dialogue with energy-sector stakeholders and institutions. At the time this task force was established, the EBRD announced that it would further expand its mandate towards this area, having the capital capacity to invest eventually up to around 2.5 billion euros per year.

As the next step, establishment of a Mediterranean Energy Community is foreseen by 2020; that is Medreg should evolve into international organization with legally binding rules.

Projects of common interest

The European Union has promoted and supported both the Energy Community and the Medreg within its external policies, which implies that both organizations are part of strategies in the interest of the EU. Formal reasons for spearheading the processes in both cases were threefold (see Table 4.3).

The Energy Community should (1) improve and sustain economic development in South-East Europe, (2) diversify supply of gas and electricity and

Table 4.3 The EU's formal reasons for launching the Energy Community and the Medreg

	The Energy Community	*Medreg*	*Benefits for the region*	*Benefits for the EU*
1.	Improve and sustain economic development in South-East Europe	Innovation	Innovation is tool for improving and sustaining economic development in region, technology transfer	Technology transfer, related to market access
2	Diversify supply of gas and electricity	Market access	Increased security of supply	Security of supply – new routes for conventional sources (EnCT) and new sources – renewables (Medreg)
3.	Achieve lasting peace and stability in the region and improving the security of private citizens in the SEE	Security	General security	General security

(3) help to achieve lasting peace and stability in the region and improve the security of private citizens in the SEE (European Commission, 2005).

The Medreg is aimed to promote (1) general security, (2) innovation and (3) market access.

General security in the Medreg framework corresponds to the first and third reasons for initiating the Energy Community (peace and stability and improving security of private citizens). Innovation should help to improve and sustain economic development in Medreg region, whereas market access is relevant from the perspective of diversifying supply.

The reasons are formally in the best interest of the region. They are also in the best interest of the EU. Both initiatives support EU energy policy goals, that is competitiveness, sustainability and security of supply.

The differences in EU approach towards two regions are primarily caused by differences among them. The Mediterranean region is significant for its huge potential for exploiting renewable sources of energy (primarily solar), whereas South-East Europe and its Eastern neighbours are relevant for the EU from transit perspective and access to conventional energy sources.

Both initiatives should facilitate investments in trans-European energy infrastructure. Trans-European energy infrastructure and its interconnections and interoperability with the energy networks of third countries are pre-conditions for the effective operation of the internal energy market and play a crucial role in ensuring the security and diversification of supply.

The EU identifies projects of common interest which contribute the most to the Union's energy and climate policy objectives. In order to be included into the EU list of projects of common interest, the projects shall be necessary for at least one of the energy infrastructure priority corridors and areas, its potential overall benefits outweigh the costs, including in the longer term; and should have a significant cross-border impact.

Projects with status of common interest might benefit from accelerated planning approvals, as well as potential funding from the EU.

During 2005 to 2011, about 300 projects had received funding under the trans-European energy network (TEN-E) budget, €323 million in total (European Commission, 2012).

The latest regulation on TEN-E projects (European Parliament and the Council, 2013b) identifies 12 strategic trans-European energy corridors and areas as priorities and provides for the identification of projects of common interest. These are four electricity corridors, four gas corridors, one oil corridor and three priority thematic areas.

For the Energy Community and Medreg countries, relevant TEN-E priorities relate to two electricity corridors (north–south electricity interconnections in Western Europe, north–south electricity interconnections in Central Eastern and South-East Europe) two gas corridors (north–south gas interconnections in Central Eastern and South-East Europe, Southern Gas Corridor) oil corridor and all priority thematic areas: creating smart networks, constructing electricity highways and cross-border CO_2 networks.

North–south electricity interconnections in Western Europe ("NSI West Electricity") are interconnections between EU member states with the Mediterranean area including the Iberian peninsula. They should enable integration of electricity from renewable energy sources and reinforce internal grid infrastructures to foster market integration.

North–south electricity interconnections in Central-Eastern and South-East Europe ("NSI East Electricity") are interconnections and internal lines in north–south and east–west directions to complete the internal market and integrate generation from renewable energy sources.

North–south gas interconnections in Central-Eastern and South-East Europe ("NSI East Gas") connect between and in the Baltic Sea region, the Adriatic and Aegean Seas, the Eastern Mediterranean Sea and the Black Sea, and creates basis for enhancing diversification and security of gas supply.

Southern Gas Corridor (SGC) is an infrastructure for the transmission of gas from the Caspian Basin, Central Asia, the Middle East and the Eastern Mediterranean Basin to the Union to enhance diversification of gas supply. Basically the SGC is a diplomatic term referring to the various projects to bring gas to the EU from Azerbaijan. Oil supply connections in Central Eastern Europe enable interoperability of the oil pipeline network in Central Eastern Europe, increase security of supply and reduce environmental risks.

Adoption of smart-grid technologies across the Union enables efficient integration of the behaviour and actions of all users connected to the electricity network, in particular the generation of large amounts of electricity from renewable or distributed energy sources and demand response by consumers.

Electricity highways provide for accommodating increasing renewable generation in the east and south of Europe and North Africa; and connecting these new generation hubs with major storage capacities and consumption centres.

Cross-border carbon dioxide networks are transport infrastructure between member states and with neighbouring third countries in view of the deployment of carbon dioxide capture and storage.

Based on the priority thematic areas and criteria for defining projects of common interest, numerous project promoters submitted their projects to be considered. In October 2013 the European Commission adopted a list of some 250 key energy infrastructure projects (European Commission, 2013). These "projects of common interest" (PCI) will benefit from accelerated licensing procedures and improved regulatory conditions and may have access to financial support from the Connecting Europe Facility, under which a 5.85-billion-euro budget has been allocated to trans-European energy infrastructure for the period 2014–2020 (European Parliament and the Council, 2013b). This will only make up a small part of EU's planned 200 billion euros of infrastructure investment in the period until 2020.

The money will include project bonds, grants and loan guarantees. The grants will be awarded to a select group of "common interest" projects which will benefit from a fast-track permit granting procedure.

Projects eligible for EU funding could then receive between 50% to 80% of their funding from the EU.

As mentioned before, some of the projects of common EU interest are connected with neighbouring regions. Selected projects that might be of relevance for the parties of the Energy Community and Medreg are presented in the following: seven gas projects (EU section of South Stream, the Nabucco pipeline, Nabucco West, the Trans-Adriatic Pipeline, the Trans-Anatolia Pipeline, the Ionian Adriatic Pipeline and the liquefied natural gas terminal), one oil project (Adria pipeline) and three electricity projects (the EuroAsia Interconnector, the Egypt-Greece Interconnector and TuNur). The list of project considered as projects of common interests of the EU is then compared with list of project of common interest of the Energy Community. The identified similarities and gaps are used to outline various interests of the EU member states and "outsiders".

South Stream was initiated in 2007 as part of Gazprom's export strategy. The pipeline was originally planned to carry 31 billion cubic metres of gas annually from Beregovaya in Russia, through Black Sea to Bulgaria[8], continue to run from Bulgaria through Serbia, Hungary and Slovenia to Italy, with branches to Croatia and Bosnia and Herzegovina. Russia had signed intergovernmental agreements in 2008 with Bulgaria, Hungary, Greece and Serbia, in 2009 with Slovenia and in 2012 with Croatia and Austria. Work on South

Stream began in December 2012, and it should be completed by 2018. The pipeline will be capable of supplying 63 billion cubic metres of gas per year.

South Stream provides for diversification of transport routes, but does not give access to any new sources of energy. The South Stream did not qualify for receiving the designation of a "Project of Common Interest" (European Commission, 2013). The South Stream project is positioned as a rival to Nabucco. Nabucco pipeline is planned to supply gas from the Caspian region to Europe and bypass Russia. The preparations began in 2002, and the pipeline route runs from Turkey (Erzurum) to Austria (Baumgarten an der March), via Bulgaria, Romania and Hungary (the pipeline length is approximately 3,300 km), with a maximum transport capacity of 31 billion cubic metres annually.

From a technical standpoint, the project never progressed very far: there is still no supplier. Iran, Turkmenistan, Egypt and Iraq all pulled out, and Azerbaijan finally rejected the idea.

In 2011, the pipeline was reduced in length to 1,300 kilometres. The eastern section, which was to have run from Azerbaijan across Georgia and Turkey to the Bulgarian border, was abandoned.

Nabucco West, which was to have carried gas from the Bulgarian/Turkish border via Bulgaria, Romania and Hungary to Austria, was the only remaining part of the original project. Nabucco West lost out to the Trans-Adriatic Pipeline in the "Southern Corridor" contest to move Azeri gas to Europe from the second phase of the Shah Deniz project. The project did not remain on the list of projects of common interest.

The Trans-Adriatic Pipeline (TAP) is contracted to carry 10 billion cubic metres of gas per year (bcm/y) to Europe. TAP will run from the Turkish–Greek border, through Albania and across the Adriatic Sea where it will connect to Italy's pipeline network. This route is 900 kilometres long (500 km shorter than that proposed by Nabucco West).

Although TAP is included into priority Southern Gas Corridor TAP is a defeat for the EU. Nabucco was a joint European project, whereas Turkey and Azerbaijan are behind TAP and Trans Anatolia Natural Gas Pipeline (TANAP).

TANAP will bring gas from Azerbaijan (and other possible neighbouring countries) through Turkey to Europe. A memorandum of understanding was signed between the governments of Turkey and Azerbaijan in 2011. TANAP will have a yearly capacity of 16 billion cubic metres, about 6 billion cubic metres of which is expected to go to the Turkish market. Construction is expected to start by the end of 2014 and it is expected to begin operations by 2018.

The TANAP and TAP pipelines will reduce Europe's dependence on Russian supplies of gas. However, the capacity of 10 billion cubic metres of gas per year is only around one-third of the amount Nabucco was to have carried. This equates to just 1% of Europe's total demand.

The increased independence of Turkey from Russian gas was one of the main goals of the southern route. Turkey is one of the largest importers of Russian gas, but at the same time is a key political partner of the North Atlantic Treaty Organization and the EU in Eurasia and the Caspian region. The route will now not be built under the direction of the EU, but will primarily under the control of Turkey and Azerbaijan. The laying of the TANAP and TAP pipelines will increase the geopolitical importance of these two countries as EU energy partners.

However, the economic crisis, above all in Greece, could place the completion of the TAP project in doubt. If constructed, TAP would be a sign of growing geopolitical role of the TAP countries (Turkey, Greece, Albania).

The Ionian Adriatic Pipeline (IAP) aims to deliver gas to northern Albania, Montenegro, Bosnia and Herzegovina, and Croatia. Two versions are considered: one with capacity of 5 billion cubic metres of gas per year (0.9 billion euros) and the second with a total capacity of 7 billion cubic metres per year, of which 4.5 bcm/annum in Croatia and Montenegro (1 billion euros). The IAP will connect the existing and the planned Croatian gas transmission system, via Montenegro and Albania with the TAP.

The TAP and the IAP have been included as projects of Energy Community interest and of the European interest.

As regards liquefied natural gas facilities on island of Krk in Croatia (LNG Krk), on the EU's list of projects of common interests is the Cluster Krk LNG Regasification Vessel and evacuation pipelines towards Hungary, Slovenia and Italy. It includes LNG Regasification vessel in Krk (HR), gas pipeline joining Zlobin, Bosiljevo, Sisak, Kozarac and Slobodnica (HR), the LNG evacuation pipeline from Omišalj to Jelšane (SI) or the gas pipeline to Italy. The Energy Community list includes LNG terminal and one of the pipelines which is also of the EU's interest: Zlobin–Bosiljevo–Sisak–Kozarc–Slobodnica.

The Adria (Janaf) Pipeline is considered as projects of common interest and the project of the Energy Community Interest (European Commission, 2013; Ministerial Council, 2013). It is based on increasing the existing capacities of Janaf oil pipeline and connecting it to the Southern Druzhba pipeline. The Druzhba Adria Oil Pipeline would enable the export of Russian oil via Omišalj. Inspection, evaluation, rehabilitation, upgrading and reconstruction of the existing Janaf oil pipeline in Croatia has been submitted for consideration as a project of Energy Community interest.[9]

In the electricity sector, projects that might be relevant for Medreg, EnCT and EU are mostly electricity highways, relevant to accommodate transfer of the renewable energy, such as the EuroAsian Interconnector, Egypt Greece Interconnection or TuNur.[10]

The EuroAsia Interconnector should connect Israel with Cyprus and Greece by underwater cable. Its planned capacity is 2,000 MW and its total length around 1,000 kilometre.

The Egypt–Greece Interconnector concerns the installation of new wind farms in Egypt (Suez area) with a total capacity of 3,000 MW and the transmission of energy from Egypt to Greece through submarine high-voltage direct current line (HVDC), with a total capacity 3,000 MW.

TuNur is a project connecting 2GW concentrating solar power tower plant in Southern Tunisia with storage, a 2 GW HVDC submarine cable between Tunisia and Italy, which might supply 9.5 TWh/y (terawatt-hour/year) of base load power. However, only the EuroAsia Interconnector is recognized by the Commission as project of common interest.

From foreign to internal policies?

The Energy Community member states have very limited energy sources. The EU activities towards the Energy Community reflect needs of the EU to connect with Caspian, North Africa and Middle East gas reserves. That is, this region is relevant for the EU from transit perspective. In addition to transport networks, the Energy Community promotes other regional energy infrastructure, such as generation and storage capacities. Generation capacities should serve as anchor loads and justify the large investment required by international gas transmission systems.

With such reasoning, the European Commission financed the Regional Balkans Infrastructure Study (REBIS-GIS), managed by the World Bank in the eve of creation of the Energy Community. The study was finished in 2005. It forecasted the least cost means to meet demand in the 2005–2020 period. The main objective was to assist the European Community, international financial institutions and donors in identifying an indicative list of priority projects for investments in power generation and related electricity transmission infrastructure. REBIS-GIS concluded that 11.5 GW of capacity should be rehabilitated and 13.5 GW of new capacity added to the system. Study outlined the scale of investments required in the SEE energy sector by 2020 – 16 billion euros, out of which 340 million will be spent on interconnections. It demonstrated that there are significant benefits in considering investments on a regional basis, that is implementing common expansion planning and operating practices could save up to 6.7 billion euros (constant 2005 euros) (PricewaterhouseCoopers LLP, Atkins International and MWH, 2004). However, the investments were not realized.

Thus, the investment needs are increasing. The key challenges for the Energy Community parties relate to the age of major facilities and to strategic decisions regarding energy supply routes. Individual countries have limited impact on strategic decisions related to supply routes. Thus, coordination of negotiating positions within the region with respect to the main suppliers and markets and need to attract investors were underlying reasons for the preparation of the Energy Community Strategy.

Opposed to the regional investment study from 2005, in which a top-down approach was used, the strategy provides synthesized collection of the existing policy goals and plans within the Energy Community including the national targets and objectives.

It is estimated that between 2012 and 2021, investments between 39 billion euros (minimal investment costs) and 73 billion euros (low emission/sustainability scenarios) are necessary to meet energy demand. Proposed projects include building of additional 20 GW generation capacity. Significant share of this capacity (8.1 GW) should be built in Ukraine.

In the gas sector, proposed projects include Ionian Adriatic Pipeline, LNG terminal and number of interconnections.

Based on the strategy a new list a of priority projects was adopted by the Ministerial Council in October 2013. Eligible project categories are power generation, electricity transmission, gas transmission, gas storage and oil. The categories are comparable to those used for EU's projects of common interest, but are not the same. The projects identified as projects of Energy Community interests are expected to attract more investors.

These projects have to be backed by more than one contracting party. Any project promoter, within or outside the Energy Community, was entitled to submit candidate projects to be assessed as project of Energy Community's Interest with the conditions that the project is located in at least one contracting party and that it will impact at least two contracting parties or a contracting party and an EU member state.

Also, planned are accelerated and coordinated permit granting and licensing procedures, coordinated regulatory authorizations and coordinated tariff methodologies adoption, support from relevant European Union funds such as the Western Balkans Investment Framework (WBIF) and pre-accession funds, degree of fulfilment of criteria for International Financial Institution's financing, increased visibility for investors. National regulatory authorities will also be involved in the process.

Lists of the projects submitted to be considered as projects of common interest (RCC 2013a, RCC 2013b, European Commission 2013b) show that only a few projects are on the list of the EU and the Energy Community, both. These include TAP and TANAP. Both projects are designed outside the EU and the Energy Community. Turkey and Azerbaijan are the main promoters.

Based on the preceding, it can be claimed that regional ownership, as exclusive possession or control of process is not likely to be reached. The reasons are regional and international.

Regional reasons include lack of regional trust and ownership. The Energy Community borders and members are changing: Bulgaria, Romania and Croatia became EU members. They ceased to be parties and became participants in the Energy Community. Ukraine and Moldova joined the community. Thus, the "region" changed its borders. The benefits of the membership are not recognized in the region.[11] The parties of the Energy Community are interested in participating in the implementation of the Energy Strategy in order to meet EU conditionality and accordingly to be able to apply for funding, rather than as a tool for furthering regional cooperation and hence improving energy security.

Thus, the potential of the Energy Community to develop into important regional organization is limited, related with its significance for the governments of the region.

This can be illustrated with the positions of Croatia and Turkey, which changed significantly during the process.

Croatia was slow to join the first memorandum,[12] which was explained by formal reasons. The EU sent several strong messages underlying Croatia's commitment to regional cooperation within the framework of the Stabilisation and Association Agreement (SAA). The SAA explicitly provided for tightening of cooperation in the adoption and planning of energy policy, including the modernization of the infrastructure, the security of supply and the reinforcement of efficiency. Accomplishment of this cooperation in the Athens Process was condition for further rapprochement with the EU. The European Commission in its *SAA Implementation Report* identified that Croatia had a potentially key role in the energy linkage of the region, particularly from the point of view of the development of regional energy links (physical infrastructure and market) (European Commission, 2002).[13] Faced with strong conditionality form the EU, Croatia committed to the first Athens Memorandum in February 2003, joined the community in 2006 and left in 2013.

Turkey was actively involved in the Athens Process, but has not become a member of the Energy Community. The EU and Turkey confirmed their interest to continue work for resolving the outstanding issues towards accession of Turkey to the Energy Community (e.g. Ministerial Council, 2006). Turkey's reservations related to transition period for alignment with the EU directive 2001/80/EC (Large Combustion Plants). Turkey asked for transition period until 2020, and the European Commission was willing to accept 2017 deadline.

It is interesting, however, that the Energy Community Treaty is signed in 2006 for a definite period: 10 years. It expires before either of proposed deadlines.

The EU would like to "help Turkey to make full use of its potential to become a major energy transit hub and in particular promote its rapid integration into the Energy Community Treaty" (Ministerial Council, 2006),[14] whereas Turkey, from its side, aims at addressing the alignment with the acquis on energy, environment and competition within the framework of the accession process (European Commission, 2006) and not in the framework of the Energy Community.

Involvement of external factors, such as Azerbaijan or Turkey as project promoters and international financial institutions will persist in long-term.

Such form of international intervention as a form of guarantee will persist, but region should become an important shareholder.

The EU activities towards the Medreg reflect needs of the EU to achieve renewable targets, and increase security of supply. This requires stable and predictable regulatory rules. The purpose of Medreg is to create such rules. Design of the Medreg initiative coincided with creation of the DESERTEC project aimed to install vast solar power plants across some 16,800 square kilometres of the Saharan desert and transferring it through high voltage direct current to consumption centres. Promoters of the project claimed that the project will be able to generate 100 GW by 2050 at a cost of 400 billion euros. The MENA region would be able to meet its needs for power with renewable energy, while exporting its excess power would provide 15% to 20% of Europe's electricity needs and create an export industry with an annual volume of more than 60 billion euros.

Critics of DESERTEC question the viability of the project. Doubts multiplied when founding shareholder Siemens pulled out of the venture in 2011. Since the inception of the project in 2009, there were no concrete ventures because of complicated negotiations over expensive projects by many partners with conflicting interests.

A less ambitious project, the Mediterranean Solar Plan (MSP) was developed under the scope of the "Barcelona Process: Union for the Mediterranean (UfM)". Its main objective is to develop 20 GW of renewable electricity capacity on the south shore of the Mediterranean, as well as the necessary infrastructures for the electricity interconnection with Europe. The plan also envisages saving, energy efficiency and technology transfer.

A key element for the development of this plan is the establishment of a suitable, new regulatory framework to promote the deployment of renewable energies and the facilitation of exchange of electricity (e.g. Medreg).

MSP would enable the Mediterranean countries to use their high potential in resources on renewable energy generation for energy self-sufficiency and trading. Its implementation would create social advantages of capacity building and, accordingly, local, qualified employment creation. For EU countries, the MSP provides opportunity of achieving renewable targets defined in 2009/28/CE Directive. For 2009–2010 the commission allocated 5 million euros for Mediterranean Solar Plan.

Evolution of the Medreg in a separate international organization rather than accession to the European Community suggests that sets of rules applied in two regions will not be the same. Institutions of each organization should decide upon which parts of the acquis will be legally binding and implemented. This might foster regional ownership.

In conclusion, while both organizations were initiated (and have seats) outside the region whose interests they should promote, the institutional structure enables tailoring the rules according to regional needs.

Conclusion

The Energy Community and Medreg are instruments of the EU's external energy policy. The EU policy is motivated primarily by energy security concerns and promoted by (market) integration perspective. These two initiatives are designed to enable export of EU's regulatory framework and thus facilitate investment. They build on previously applied policy instruments, such as enlargement and regional approach, including principle of conditionality.

With distant (or non-existent) membership perspective, the political motivation is not sufficient for alignment with the EU rules. Thus, incentives should primarily be economic and financial aid and access to finance conditional on implementation of the EU rules and projects of European interests.

The expected lifetime of such projects (which enables an appropriate return on investment) and their relevance for security of supply indicates that the EU and international financial institutions have long-term interests in the region. It can be expected that, provided projects are realized, their presence in the region will persist.

Energy industries in Energy Community and Medreg are not likely to create regional economic interest. Thus, global energy industries and financial institutions shall play a significant role in formulating national energy policies and regional policies. Such development makes it difficult to formulate regional ownership. Lack of regional interests and long-term EU interest indicate that regional ownership, as exclusive possession or control of process is not likely to be reached.

Still, the Energy Community and Medreg both formally promote regional interests. These interests, however, are not formulated within the region. Medreg and the Energy Community are launched and administered from outside: Medreg from Milan and the Energy Community from Vienna, where secretariats have seats.

Regional cooperation might have an impact on defining priorities and conditionality applied, but requires coordination of positions towards important regional issues. Capacity to formulate and implement truly regional interests is missing. Coordinating national and regional policies requires strengthening of the process of policy formulation, primarily enhancing the role of parliaments. National parliaments should be involved in decision making on the regional level and formulation of criteria for the identification of strategic projects and critical infrastructure at national and regional levels, including the relevant requirements regarding environmental sustainability and public participation.

Notes

1. The findings, interpretations, and conclusions expressed in this chapter are those of the author and do not reflect the positions of the institution.

2. In this chapter, the term *South-East Europe* is used for for Westen Balkans and Turkey. Western Balkans includes Montenegro, Serbia, Macedonia, Albania, Bosina and Herzegovina and Kosovo. Croatia was considered as Western Balkans country prior to its accession to the EU, that is until 1 July 2013. Eastern neighbours relate to Ukraine and Moldova, and Mediterranean countries include Algeria, Egypt, Israle, Jordan, Lebanon, Morocco, Palestinan Authority and Tunisia.
3. The priorities are (1) limiting energy use in Europe, (2) building a pan-European integrated energy market, (3) empowering consumers and achieving the highest level of safety and security, (4) extending Europe's leadership in the development of energy technology and innovation and (5) strengthening the external dimension of the EU energy market.
4. An example of such efforts is a visit of the commission's president Manuel Barroso to Azerbaijean and Turkmenistan in 2011, when visa facilitation for Azeri nationals was promised in return for bringing 10 billion cubic metres of gas each year to Europe.
5. The Stability Pact for South-East Europe was launched in 1999 as conflict prevention strategy of the international community. It provided a framework to stimulate regional cooperation and expedite integration into European and Euro-Atlantic structures. Over time it evolved in a forum of regional and international actors which identifies common problems and devises shared strategies to tackle them. As substantial progress on the ground was achieved, the internationally led approach driving the Pact at the beginning started to become obsolete. This was the main motive in launching the transformation of the Stability Pact into regionally led successor organization, the Regional Co-operation Council (RCC). The RCC was officially launched in 2008.
6. Annual budet of the secreatriat is about 2 million euros (2007).
7. The Energy Community also deals with oil sector, but this area has been gradually included in the acquis of the Energy Community since 2008 (Energy Community, 2012)
8. The EU consultation documents indicate two additional potential routes from the Russian Black Sea coast through (1) Bulgaria, Serbia and Hungary to Austria or (2) Bulgaria, Greece and the Ionic Sea to Southern Italy.
9. See www.energy-community.org/portal/page/portal/ENC_HOME/AREAS_OF_WORK/Regional_Energy_Strategy/PECIs.
10. A complete list of submitted projects is available at http://ec.europa.eu/energy/infrastructure/consultations/doc/pci_list_electricity.pdf.
11. For example, integration of Ukraine and Moldova into the Energy Community has been found to entail asymetrical costs and benefits for the EU and Ukraine and Moldova as well as entailing both coss and benefits for all parties (for more, see Petersen, 2012)
12. See page 13 of the memorandum of understanding, available at www.seerecon.org/infrastructure/documents/mou-rem-see.pdf.
13. European Commission (2002) 2001 SAP Report.
14. See *An External Policy to Serve Europe's Energy Interests*, paper from Commission/SG/HR for the European Council, 2006, available at www.consilium.europa.eu/uedocs/cmsUpload/st09971.en06.pdf.

References

Bertelsmann Foundation (2002). *Integrating the Balkans. Regional Ownership and European Responsibilities*, Discussion Paper, Balkan Forum, Berlin, July 15–16.

Energy Community (2012). Decision of the Ministerial Council of the Energy Community, Budva, 18 October 2012, www.energy-community.org/pls/portal/docs/1766178.PDF
European Commission (2002). The Stabilisation and Association Process for South Eastern Europe – First Annual Report. Brussels, 3.4.2002. COM (2002) 163 final, available at http://eur-lex.europa.eu/smartapi/cgi/sga_doc?smartapi!celexplus!prod!DocNumber&lg=en&type_doc=COMfinal&an_doc=2002&nu_doc=163
European Commission (2005). MEMO 05/397, 25.10.2005.
European Commission (2006). Turkey 2006 Progress Report. Brussels. 8.11.2006. SEC(2006)1390. Available at http://ec.europa.eu/enlargement/pdf/key_documents/2006/nov/tr_sec_1390_en.pdf
European Commission (2011a). Commission Staff Working Paper. Results of the public consultation on the external dimension of the EU energy policy. SEC (2011) 1023 final, 7.9.2011.
European Commission (2012). Projects financed under TEN-E programme 2005–2011. Available at http://ec.europa.eu/energy/infrastructure/tent_e/doc/2012_ten_e_financed_projects_1995_2011.pdf
European Commission (2013a). Commission Delegated Regulation (EU) No 1391/2013 of 14 October 2013 amending Regulation (EU) No 347/2013 of the European Parliament and of the Council on guidelines for trans-European energy infrastructure as regards the Union list of projects of common interest available from http://eur-lex.europa.eu/LexUriServ/LexUriServ.do?uri=CELEX:32013R1391:EN:NOT
European Commission (2013b). List of projects of common interest/clusters of PCIs. http://ec.europa.eu/energy/infrastructure/pci/doc/2013_pci_projects_country.pdf
European Council (2011). Council conclusions of 28 February 2011, 3150th General Affairs Council meeting, Brussels, 28 February 2011, www.consilium.europa.eu/uedocs/cms_data/docs/pressdata/EN/genaff/128255.pdf
European Parliament (2011). European Parliament Resolution of 5 July 2011 on Energy Infrastructure Priorities for 2020 and Beyond. OJ C 33 E, 5.2.2013, p. 46.
European Parliament and the Council (2013a). Regulation (EU) No 347/2013 of the European Parliament and of the Council of 17 April 2013 on guidelines for trans-European energy infrastructure and repealing Decision No 1364/2006/EC and amending Regulations (EC) No 713/2009, (EC) No 714/2009 and (EC) No 715/2009. Official Journal L 115, 25/04/2013 P. 0039 – 0075. Also available at http://eur-lex.europa.eu/LexUriServ/LexUriServ.do?uri=OJ:L:2013:115:0039:01:EN:HTML
European Parliament and the Council (2013b). Regulation (EU) No 1316/2013 of the European Parliament and of the Council of 11 December 2013 establishing the Connecting Europe Facility, amending Regulation (EU) No 913/2010 and repealing Regulations (EC) No 680/2007 and (EC) No 67/2010 . Official Journal L 348, 20.12.2013, pp. 129–171. Also available at http://eur-lex.europa.eu/LexUriServ/LexUriServ.do?uri=OJ:L:2013:348:0129:01:EN:HTML
Medreg (2006). *Concept and Strategy Paper*, available at www.medreg-regulators.org
Ministerial Council (2006). Conclusions of the Ministerial Council Meeting, 17 November 2006, Skopje.
Ministerial Council (2013). *Projects of Energy Community Interest.* Conclusions of the Ministerial Council Meeting 24 October 2013. Available at www.energy-community.org/portal/page/portal/ENC_HOME/AREAS_OF_WORK/Investments/PECIs

Petersen, A. H. (2012). *The Integration of Ukraine and Moldova into the Energy Community – a Liberal Intergovernmental Analysis*, http://rudar.ruc.dk/bitstream/1800/8813/1/Speciale%20RUB%20-%20Anne%20Hald%20Petersen.pdf

PricewaterhouseCoopers LLP, Atkins International and MWH (2004). Report on the South East Generation Investment Study (GIS). Issued to the European Commission for the purposes of European Commission CARDS programme for Balkan region Contract NO 52276. Available at http://siteresources.worldbank.org/INTECAREGTOPPOWER/Home/20551044/Volume%201%20-%20Exec%20sum_final.pdf

RCC (2013a). List of projects submitted to be considered as potential Projects of Common Interest in energy infrastructure – Electricity, available at http://ec.europa.eu/energy/infrastructure/consultations/doc/pci_list_electricity.pdf

RCC (2013b). List of projects submitted to be considered as potential Projects of Common Interest in energy infrastructure – Gas, available at http://ec.europa.eu/energy/infrastructure/consultations/doc/pci_list_gas.pdf

Chapter 5

A common strategy for closer EU–MENA cooperation in renewable energy

Miriam Bardolet[1]

Renewable energy sources are increasingly prominent in the energy mix of European member states as well as of Middle Eastern and North African (MENA) countries. Since the adoption of the binding renewable energy targets in 2009 (20% final energy consumption from renewable energy sources by 2020), renewables have experienced a significant growth in the European Union (EU), having reached a share of 12.7% of the EU's final energy consumption.[2] Recently MENA countries have followed a similar trend and have adopted ambitious renewable energy targets amounting to a total of approximately 50GW by 2020.[3]

Several studies have shown that both regions would significantly benefit from increasingly interconnected power systems strongly based in renewables.[4] MENA countries would benefit from further interconnections and renewable energy development for meeting their rising power demand, expected to increase at a rate of 5% to 9% per annum. Furthermore, renewables would significantly contribute to attracting investment and to local job creation.[5]

Regarding the context in the EU, an integrated market with MENA would enable a most cost-efficient transition towards a decarbonized economy, with potential savings of up to €33 billion per year,[6] and would improve security of supply. Figure 5.1 in the color section shows the options for renewables and grid infrastructure in MENA countries and around the Mediterranean until 2050.

The implementation of such an integrated EU–MENA power system requires strong cooperation efforts. Regional cooperation should lead to progressive regulatory convergence as well as to the establishment of common institutions. An approximation of the regulatory frameworks is crucial in order to create a level playing field for investments and to ensure the operability of the system. Common institutions will be necessary for the implementation, monitoring and enforcement of this framework.

In terms of regional integration, the starting points at the two sides of the Mediterranean are very different. Whilst during the last decades the EU has made important efforts in this direction, in the MENA region, cooperation is more case-specific and less far-reaching. Therefore, in order to achieve region-wide integration, cooperation efforts should not only address the interaction between EU and MENA but also within the MENA region. The debate on energy cooperation in the EU–MENA region has so far been divided into two regions, the Mediterranean and the MENA. In order to achieve an EU–MENA integrated power system, it will be essential to ensure consistency between these two discussions and to start including a Middle Eastern perspective in the Mediterranean debate.

The beginning of this transition is already at an advanced stage. Currently many initiatives are actively involved in fostering cooperation in EU–MENA with a particular interest in renewables. The objective of this chapter is to provide an overview of these institutions, to outline possible milestones to be achieved in the coming decades and to identify areas where interaction is needed.

Given the complexity of this framework, cooperation should be tackled from different angles. For the purpose of this chapter, cooperation efforts are grouped in three categories: political institutions, legal instruments and sector-specific organizations.

At a political level, the Union for the Mediterranean, the European Neighborhood Policy and the League of Arab States have all identified renewable energy as an axis for cooperation in their respective scopes of activity. All these processes have significantly contributed to create the current momentum for energy cooperation. Political achievements have mainly consisted in the adoption of declarations and communications as well as in the allocation of funds for specific projects. This is already a very good starting point. Nevertheless, more needs to be done in order to achieve concrete results.

Political institutions should strive to agree on a common long-term perspective for renewable energy development in the region to provide predictability for investments as well guidance and coordination for the different entities involved in the process. Given the important investments required, strong public–private cooperation will be a key success factor. Therefore, political stakeholders should aim at creating the right market conditions and investment environment for the private sector to actively take part in this transition. Pending the implementation of the necessary market conditions, political engagement will be of utmost importance to develop initial projects and new interconnections linking the EU and the MENA countries. First projects are important to bring in a tangible dimension and concrete success examples to the overall cooperation strategy.

Cooperation should lead to the creation of a common regulatory framework that ensures the compatibility of legal systems and offers a solid legal basis throughout this transition. Therefore, policy goals and general principles adopted by political institutions must be complemented with a set of common rules. The construction of this regulatory framework will be a gradual process. It is expected that cooperation starts at the project-specific level through the implementation of concrete projects involving different countries. The experience gained through these projects would provide key learning for the development of general provisions at a later stage. *Cooperation mechanisms* outlined in the European Renewable Energy Directive offer a good legal basis for the implementation of a series of projects among several governments at both sides of the Mediterranean. Why have these mechanisms not yet been implemented? Interference with national policies, legal uncertainty or lack of interconnection infrastructure are some of the reasons that could have led to this situation.

In parallel to the implementation of first projects, a general legal framework should start to develop with the adoption of non-binding principles. Such principles should lead, in the coming decade, to adoption of binding minimum standards. Currently there are different proposals that point in this direction. The *Arab Renewable Energy Framework* currently promoted by the League of Arab States and by the Regional Centre for Renewable Energy and Energy Efficiency (RCREEE), as well as the Mediterranean Renewable Energy Framework, suggested by the EU Commission, could become legal instruments that can play a key role in the construction of a common legal framework for renewables across the EU–MENA. These instruments could be preceded by the adoption of the Energy Charter Treaty that provides a level playing field for energy trade and investment.

The gradual process of market integration will also entail that groups of countries sharing particular synergies will integrate first. Sub-regional integration is an essential interim step for the creation of an EU–MENA-wide system. The Algiers Declaration concerning the implementation of an integrated electricity market in the Maghreb region, Integration des Marchés Maghrébins d'Electricité (IMME) and the Gulf Cooperation Council Interconnection Authority (GCCIA) are excellent examples of how sub-regional integration can take place.

The integration process should aim for a pragmatic progression that reflects the day-to-day activities of the entities operating in the power sector across the region such as regulators, transmission system operators (TSOs) and industrial players. Sector-specific organizations group key actors in a particular field with the aim of developing a common understanding through the sharing of experiences and know-how. These organizations will be crucial to provide a bottom-up approach to the cooperation strategy.

The Association of Mediterranean Regulators (Medreg), the Association of Mediterranean TSOs (MedTSO) and the RCREEE play an essential role in shaping the integration process by issuing good-practice recommendations and providing input for the adoption of legal instruments. As integration increases, these organizations are expected to gain weight in their respective fields of activity, particularly to monitor and, if applicable, enforce the implementation of common rules.

Industrial stakeholders have also joined forces to promote energy cooperation in the region. The Observatoire Méditerranéen de l Energie (OME) and the Arab Union of Electricity (AUE) were already founded in the 1980s. In the recent years, other industrial initiatives have followed, namely Dii and Medgrid, which promote regional integration in the region with a particular focus in renewables and interconnection infrastructure respectively. Involving industry platforms as stakeholders in the process will be important to ensure that private sector needs are duly addressed, thereby enabling the large investments that are needed.

Ultimately, an integrated power system will require a comprehensive and robust legal framework as well as common and well-established institutions applicable throughout the EU–MENA region. This final goal is referred to in this chapter as the Mediterranean and Middle Eastern Renewable Energy Partnership. This partnership would encompass political, institutional and legal aspects. The Energy Community, which established a single energy market between the EU and Eastern and South-Eastern European countries, provides important lessons learned for a potential market integration process in the EU–MENA region. The European Commission has pointed out in several occasions at a "Mediterranean Energy Community"[7] as a long-term objective. Nevertheless, the specific characteristics of the MENA region will require a different setup from the current Energy Community. The Mediterranean and Middle Eastern Renewable Energy Partnership should be seen as a natural progression of a cooperation strategy and should be based on commonly agreed-on principles rather than as the extension of EU law.

The key to success of this transition processes is to coordinate and align all the entities involved in order to promote effectiveness, and to aim for a pragmatic progression of international cooperation. The long-term vision for renewables will have to be complemented with short-term achievements. For this purpose the goal must be to create win-win situations that combine the immediate needs and priorities of the actors concerned.

Figure 5.2 provides an overview of the main actors involved in this process, their possible milestones and the potential for interaction.

Political engagement for the development of renewables in EU–MENA

Currently, several political institutions with different goals and geographical scopes promote cooperation in renewable energy as a means to bring about economic and social development in the EU–MENA regions. The Union for the Mediterranean is pursuing the Mediterranean Solar Plan, the European Neighbourhood Policy has placed renewables as a key topic of the bilateral dialogue between the EU and Southern Mediterranean countries, and the League of Arab States is currently fostering the adoption of common

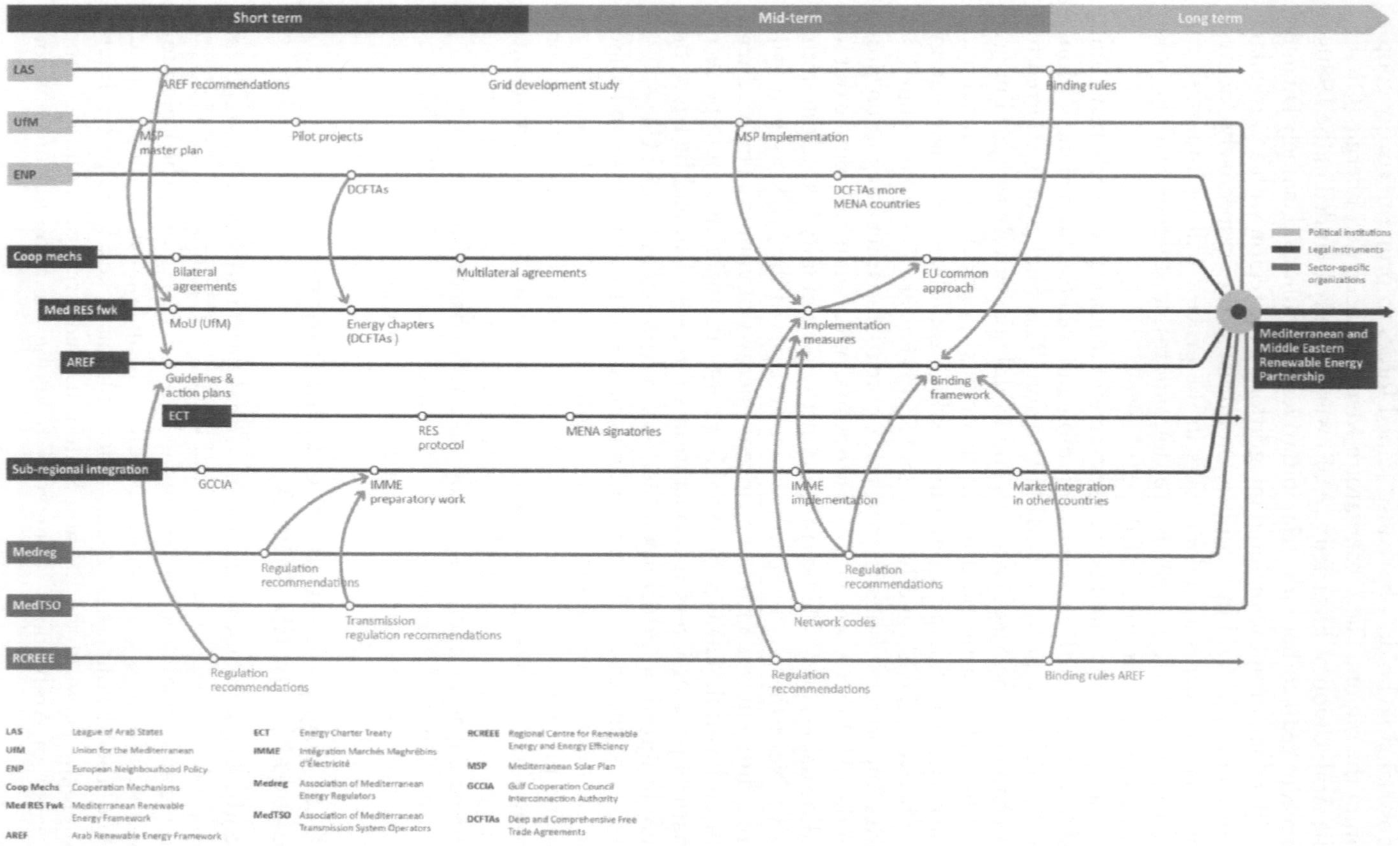

Figure 5.2 Cooperation framework for renewables in EUMENA

Source: Dii GmbH, Desert Power: Getting Started (2013).

principles for renewable energy as well as further integration of Arab countries' transmission grids.

Thanks to these initiatives, today the starting phase of this cooperation process is already at an advanced stage. Many political declarations have been issued supporting the mutual benefits of renewables, key organizations are already in place and funds have been dedicated to the implementation of policies and specific projects, such as the first concentrated solar power (CSP) plant in Morocco.

These international trends have also had an influence on national policies. In the recent years, most MENA countries have adopted ambitious renewable energy targets, and some of them, particularly Morocco and Algeria, have expressed their interest in the development of renewable energy projects for export to the EU.

However, and despite the significant efforts made so far, a clear political leadership is still missing. This entails one of the most important challenges for a timely and efficient progression of regional cooperation. In order to overcome this challenge, it will be paramount to coordinate all political efforts and to aim for a shared commitment of all the relevant political institutions. Such strong coordination will require, on one hand, the joint development of key principles for cooperation in renewable energy and, on the other, the agreement on a common long-term objective.

Regarding the joint development of key principles, the Union for the Mediterranean (UfM) has recently made important progress through the elaboration of the *Mediterranean Solar Plan – Master Plan*. This document is the result of the discussions among UfM member countries, organized in several working groups, during the past three years. The most relevant contribution of this master plan is the development of a common understanding of basic aspects for the development of renewables such as policy and regulation, financial support tools, transmission infrastructure and industrial development. The Master Plan should now be endorsed by all UfM countries in order to fully develop its potential. If implemented successfully, the Master Plan can make a substantial contribution to the creation of an integrated and sustainable electricity market.

General principles, such as those of the Master Plan, must be complemented with more precise provisions. At this stage, concrete measures can be better negotiated bilaterally. The bilateral relations between the EU and the Southern Mediterranean countries are channeled through the European Neighbourhood Policy. Renewable energy has been consistently strengthened as key element of the European Neighbourhood Policy to bring about prosperity and sustainable development in the region. In 2011, the EU Joint Communication "A Partnership for Democracy and Shared prosperity with the Southern Mediterranean"[8] stated that "there is clear potential for building an EU Mediterranean partnership in the production and management of renewable energy." As a consequence, the EU should strive to include more precise provisions on renewables in its bilateral instruments. The upcoming deep and comprehensive free trade agreements (DCFTAs), to be negotiated initially with Morocco, Tunisia, Jordan and Egypt offer a good occasion for this purpose. In order to facilitate the materialization of renewable energy development, the European Neighbourhood Policy should continue to provide financial support to renewable energy and transmission infrastructure projects.[9]

As mentioned earlier, the development of a shared long-term view will be critical for renewable energy cooperation. A common goal will provide the necessary guidance to all entities involved in this process, making them progress in the same direction. Additionally, a shared long-term objective would provide the private sector with a credible perspective which is necessary to engage in capital intensive investments such as renewable energy projects and transmission infrastructure.

The UfM provides a suitable platform for the adoption of a political declaration regarding a shared long-term goal given that it brings together all the relevant countries on equal footing. This long-term objective of intensified regional cooperation has been referred to by the European Commission as a "Mediterranean Energy Community". The content and form of this community is yet to be defined.

The political declaration on a "Mediterranean Energy Community" should start the process to shape this concept and to elaborate a road map for its implementation. It is important that this political process keep a pragmatic approach at all times. Therefore, despite being an ambitious endeavor, unnecessary complexity and intricate architectures should be avoided, and concrete implementation measures should be focused on.

The UfM and the European Neighbourhood Policy, discussed earlier, focus in the relations between the EU and Mediterranean countries. In order for an integrated system to benefit from the excellent renewable energy resources of the Middle East, that these countries are included in the political debate as soon as possible is essential. The League of Arab States, a political organization that brings together 22 MENA countries, is a key player in promoting south–south cooperation and could act as a catalyst for bringing the Middle Eastern perspective to the ongoing Mediterranean energy dialogue.

The League of Arab States[10] aims to promote cooperation in several areas, including renewable energy and energy efficiency. In 2010 the Arab Ministerial Council for Electricity adopted the *Arab End Use Electricity Efficiency Improvement Guidelines*. These guidelines were promoted by the League of Arab States together with RCREEE and the EU-funded project MED-ENEC.[11] Based on the EU's experience, the guidelines have led to the development of National Energy Efficiency Action Plans by most MENA countries. Currently these two entities are developing the *Arab Renewable Energy Framework*, which is addressed in detail in the following section.

The creation of an integrated power system for EU–MENA countries based on renewables will entail significant investments. Therefore, a strong cooperation between public and private will be vital. In this regard, the political process should entail a close dialogue with industrial partners, namely through industrial organizations. Political efforts should aim at establishing a framework to promote investments and to create the adequate market conditions to attract private capital. An investment friendly environment would enable private actors to take an active role in driving this transition. The work of the League of Arab States as well as the new free-trade agreements to be negotiated by the EU will be of utmost importance for the creation of a favourable environment for the development of renewable energy projects.

Building a common legal framework

A common set of rules laying down minimum standards will be necessary to capture the benefits of an integrated electricity market for the EU–MENA region. Common minimum standards regarding power-sector regulation will be necessary to create a level playing field for energy investments and to foster cross-border trade. Regarding the promotion of renewables, common rules should be adopted to create a common traceability scheme to certify the renewable origin of electricity. Last but not least, a common framework for transmission activities will have to be in place for the operation and development of these infrastructures.

Currently, the frameworks for power sector regulation in the EU and in the MENA region are very different. Whilst EU regulation is largely based on liberalization, unbundling and third-party access, the power sector in the MENA region is characterized by strong state domination, through vertically integrated monopolies. Energy policy and energy cooperation perspectives vary also significantly among MENA countries, namely deepening on their role as energy importer or exporter. Nevertheless, the development of renewable energy is a point of encounter between energy policies on both sides of the Mediterranean.

The disparity among existing regimes will be an important hurdle for the development of common norms. The need to find a balanced solution reinforces the importance of the political processes as well as of sector-specific organizations, such as Medreg or MedTSO, which will be vital to create an ongoing dialogue and find an adequate compromise.

Given the complexity of EU–MENA cooperation in renewables, the legal framework will evolve over time as part of a gradual process. A fully developed framework is only expected to be in place in the long term. The dynamics of this process will entail an evolution of legal instruments in several fronts:

- from project-specific cooperation to a general framework;
- from the adoption of general principles to binding provisions; and
- from sub-regional to region-wide integration.

Regulatory convergence is expected to start with cooperation among several countries on a project-by-project basis. The practical experience gained through these projects would lead to the adoption of a generalized framework at a later stage. Initial projects will play a key role in the overall cooperation strategy given that they offer a tangible dimension to this process and are a short-term success story.

Currently cooperation mechanisms of the Renewable Energy Directive provide a legal basis for project-specific cooperation between countries on both sides of the Mediterranean. Under "Joint Projects with Third Countries" (Article 9), one or more member states may agree with a third country on the development of new renewable energy projects in the third country's territory. The electricity generated by those projects which is physically imported to the EU may be counted in the 2020 targets of participating member states.

Despite several attempts for implementing joint projects with southern Mediterranean countries, none is yet in place. The European Commission has recently issued a guidance document aimed at rendering these mechanisms more operational.[12] This paper entails an important improvement for the implementation of these projects. It includes model agreements for Article 9 projects and identifies the concerns of member states for using these mechanisms such as potential interference with national support schemes or perceived technical complexity. Additionally, the EU Commission encourages member states to enter into multilateral agreements for Article 9 implementation, which would have the potential of creating a pipeline of projects.

However, the main challenges for Article 9 implementation relate to the structure of this mechanism and thus should be tackled in the course of the revision of the Renewable Energy Directive. This is particularly the case of the physical import requirement. Missing interconnection infrastructure makes imports unfeasible (except between Morocco and Spain), and in the particular case of Spain, due to over capacity and lowering power demand, exports are currently not market oriented. Additionally the 2020 time horizon does not give sufficient visibility for investments. Cooperation mechanisms should therefore be adapted to a more market-friendly design. Power flows should follow market signals rather than stem from a legal obligation. Should the physical requirement remain, renewable energy imports would be better dealt with at EU level given that from a system perspective renewable energy imports would benefit the whole EU region instead of specific member states. Moreover, an EU approach would promote efficiency in financial support.[13]

At the same time that initial projects are implemented, a common framework of general character should take shape. Given the large number of countries involved and the heterogeneity of their legal systems, it is essential that common rules are sufficiently flexible to adapt to the needs of each country. For this purpose, the framework should evolve from general non-binding principles (soft law) adopted in the short term, into a binding set of rules (hard law) in the medium term.

Today, legal uncertainty and instability constitute significant barriers to renewable energy investments in MENA countries. Multilateral legal instruments, particularly focused in the energy sector, would significantly mitigate this risk given that they provide for minimum legal standards and international dispute settlement mechanisms.

The Energy Charter Treaty (ECT) can mark an important step forward in this direction. The ECT is a multilateral legally binding treaty that tackles the most important aspects for energy trade, transit and investment by reference to World Trade Organization (WTO) rules. To date, the ECT has been signed by 54 states, namely covering the EU and European countries and former Soviet countries. Most MENA countries are already observers, but only Morocco and Jordan have signed the Energy Charter Declaration. Given its international character, the ECT would not only contribute to trade and investment between EU–MENA but also among MENA countries as well as with other ECT members.

Since the ECT entered into force in 1998, the energy landscape has undergone important changes partly due to renewable energy development and environmental concerns. Consequently, the Energy Charter secretariat is already conducting analysis and consultations on the promotion of low-carbon investments. In the coming years, the ECT should strive to strengthen the role of renewable energy. The adoption of an association agreement targeting the scope of the ECT to renewable energy or to electricity would be particularly appropriate. Such an association agreement would allow for countries to join the ECT only for electricity-related activities which might be of interest for some MENA countries.[14]

The wide scope of the cooperation framework, both in content and in geography, will require additional legal instruments. The *Mediterranean Renewable Energy Framework*, suggested by the European Commission, and the *Arab Renewable Energy Framework*, promoted by the League of Arab States and RCREEE, are very suitable mechanisms to initiate regulatory convergence with a particular focus on renewables.

In the recent communication "Renewable Energy: A major player in the EU electricity market", the European Commission outlined the possible content of a "legal framework for trade and investment in renewable energy in the Mediterranean".[15] The main objective of this framework is to facilitate the development of renewables in this region by establishing a favorable regulatory environment that covers all the relevant topics on a legally binding basis. This legal instrument would build on the general principles agreed upon by Mediterranean countries as part of the *Master Plan of the Mediterranean Solar Plan* and on the energy chapters of upcoming DCFTAs.[16] The European experience in the promotion of renewables and power markets integration will be indeed an important reference;

nevertheless, an automatic extension of EU law should be avoided. The provisions of this framework should be developed jointly with southern Mediterranean countries in order to adapt to the current context in these countries. The involvement of stakeholders such as Medreg and MedTSO will be crucial to this end.

Concerning cooperation within MENA countries, the League of Arab States and RCREEE are currently working on the promotion of an *Arab Renewable Energy Framework*, a set of recommendations aimed at establishing a common approach to the regulation of renewables in the region. This remarkable initiative could be a major advancement towards the approximation of renewable energy policies among MENA countries.

The topics covered by the *Arab Renewable Energy Framework* are very comprehensive. Recommendations refer to renewable energy targets, administrative procedures as well as access to and operation of the grid. Additionally it would provide for the development of renewable energy action plans, reporting obligations and the setting up of an information sharing platform. The provisions of the *Arab Renewable Energy Framework* are inspired in EU regulation (in particular the Renewable Energy Directive), which is an important factor to foster the compatibility of the regulatory frameworks of these two regions. In order for this framework to have full effects, it will need to be agreed to by the Arab Ministerial Council of Electricity. Building on the experience gained through these guidelines in the medium term the *Arab Renewable Energy Framework* should lead to the adoption of a binding framework regarding minimum standards for the development of renewable energy in the region.

Given the different starting points in terms of energy sector regulation across the EU–MENA region, regional integration is not expected to occur simultaneously. Groups of countries with similar characteristics and mutual interests will integrate first. Therefore, sub-regional integration is a crucial interim step towards region-wide integration across the EU and MENA countries.

Today several MENA countries have already started this process. Since 2001 Gulf countries have in place the GCCIA,[17] an entity in charge of the common Gulf Cooperation Council (GCC) grid. It aims at fostering power exchange in order to gain efficiency in power markets, provide security of supply and system stability. The GCCIA has provided its members with substantial practical experience regarding power system integration at a technical, institutional and regulatory level.

Maghreb countries have also started a process for the integration of their electricity markets, the IMME. By means of the Algiers Declaration, in 2010, Algeria, Morocco and Tunisia agreed to undertake the appropriate regulatory measures to create an integrated electricity market by 2015. These measures are to align with EU standards in order to progressively be part of the EU's internal electricity market. Together with the declaration, the previously mentioned governments agreed to an Action Plan 2010–2015 for the implementation of this initiative. The action plan develops the main commitments of the declaration and translates them into specific milestones for the integration of electricity markets.

The EU is pro-actively supporting the Maghreb sub-regional integration process with an emphasis on renewable energy development. In its recent joint communication "Supporting Closer Cooperation and Regional Integration in the Maghreb: Algeria, Libya, Mauritania, Morocco and Tunisia,"[18] the European Commission proposed concrete measures in this direction such as the elaboration of common network codes and the support for energy sector reforms.

A bottom-up approach to cooperation in renewables

A key success factor for this cooperation strategy will be to keep at all times a pragmatic approach towards this transition. In order to include a practical dimension to the overall cooperation framework, it is essential to involve the entities that operate in the energy sector on a daily basis. Regulators, transmission system operators and industrial players are considered the most relevant actors to bring in a hands-on perspective to market integration.

Sector-specific organizations bring together the entities mentioned earlier. These organizations aim to achieve a common understanding among their respective members which stems from the sharing of experiences and know-how, thus providing a bottom-up form of cooperation. This interaction normally results in the issuance of best practice recommendations and specific studies, along with dedicated activities to support implementation.

Medreg was created in 2007, and today brings together 23 regulatory entities across the Mediterranean[19] with the aim of promoting regional market integration by means of regulatory convergence.[20] In the short and medium terms, Medreg will be a key actor for the establishment of well-functioning independent regulators in the region as well as for assisting Mediterranean countries in undertaking power sector reforms. Medreg's input in the adoption of common rules, particularly regarding electricity market reform and transmission regulation, will be essential.

RCREEE,[21] based in Cairo, presents certain parallels with Medreg given that it also brings together energy regulators but it has a different geographic scope – the MENA region – and it focuses on renewables. As of today RCREEE includes 13 Arab countries[22] with a view to include all the 22 member states of the League of Arab States in the future, with the aim of fostering the development of renewable energy and energy efficiency. Together with the League of Arab States, RCREEE published templates for Land Use and Power Purchase Agreements as well as Energy Efficiency Guidelines and is currently engaged in the development of an Arab *Renewable Energy*

Framework, described earlier. RCREEE has also been entrusted with monitoring competences and is actively involved in supporting its members with the implementation of the instruments it develops.

Regarding cooperation in transmission issues, MedTSO[23] has been recently incorporated to group together the TSOs operating around the Mediterranean. MedTSO's main objective is to promote the coordination of infrastructure development plans and grid operation.[24] Infrastructure development is a must to create an integrated power system. Therefore, MedTSO's activities are very promising as a means to accelerate this development.

Certain working areas of the previously mentioned organizations are overlapping. In order to ensure effectiveness in the overall process, strong coordination and interaction amongst them is necessary. Over time, as market integration is strengthened, these institutions are expected to play an increasingly relevant role in their respective field of activity. For instance, the creation of a supranational body with regulatory competences should be considered.

Last but not least, the industrial sector has also joined forces for promoting energy cooperation. Since the late 1980s, the OME brings together energy companies operating in the Mediterranean region to foster cooperation in energy field, and the AUE groups 19 Arab countries with the aim of providing a coordinated approach to the development of the electricity sector by strengthening cooperation.

Another important forum for the coordination of electricity activities in the southern Mediterranean region is the Comité Maghrébin de l'Electricité (COMELEC). This organization, established in 1989, is an initiative of Algeria, Libya, Morocco, Mauritania and Tunisia aimed at fostering electrical interconnection within the region. Given that it had similar interests as the Union of the Electricity Industry EURELECTRIC for Europe, the Euro-Mediterranean Electricity Cooperation (MEDELEC) was created in 1992 to coordinate these initiatives with a focus on the Mediterranean. MEDELEC's most renowned initiative is the Mediterranean ring aimed at creating a fully interconnected system along the Mediterranean coast.

More recently, Dii and Medgrid were created as industrial consortiums with similar objectives. Dii was established in 2010 with the goal to accelerate the creation of a market for renewables in EU–MENA regions, and Medgrid was established in 2011 with the objective of facilitating the development of interconnection infrastructure in the Mediterranean.

All these initiatives have contributed substantially to the current energy cooperation debate through the development of innovative studies and state-of-the art methodologies. Given the important investments that will be necessary to develop sufficient generation capacity and transmission infrastructure, the involvement of energy companies is crucial to adequately reflect the needs of the private sector.

A comprehensive framework for an integrated EU–MENA power system

The gradual process of regulatory convergence and market compatibility should ultimately lead to the implementation of an integrated EU–MENA power market. This will require the adoption of a comprehensive framework that establishes common principles and regulations as well as common institutions. For the purpose of this chapter, this framework is referred to as the Mediterranean and Middle Eastern Renewable Energy Partnership.

The experience of the existing Energy Community between the EU and South-East and East European countries offers very important lessons learned regarding the challenges that could be faced when implementing an integrated market across the EU–MENA region. In this regard, the European Commission has pointed in several occasions its interest in a "Mediterranean Energy Community" as long-term policy objective.

The Energy Community was adopted in 2005 through the signature of its founding treaty that provides for the creation of a single energy market based on regulatory convergence through the extension of EU Law; that is it requires Contracting Parties to implement the *acquis communautaire*. The scope of the Energy Community is very comprehensive, and beyond the adoption of the relevant EU directives, it also establishes a set of institutions, as well as a decision-making process and dispute settlement forum.

The specific characteristics of the MENA region would require a different set-up from the current energy community. The scenario in South-East European countries at the time of entering into the energy community was very different from that in MENA. Among others, these countries came from a previously unified power system and considered the energy community as a tool for future accession to the EU. The key requirements for a future partnership between the EU and MENA that can be drawn from the energy community process are, among others,

- binding commitments based on common principles which are jointly developed are crucial for the effectiveness of the framework.
- providing for a step-wise approach and allowing for a progressive implementation of applicable regulation is necessary to adapt to the needs of different countries
- a secretariat with sufficient monitoring, enforcement and consultation competences should be in place for rendering the framework operational. An active participation of all the countries involved will be a key success factor for its implementation. Enforcement and dispute settlement process should be clearly designed and should entail limited political involvement.

Conclusion

A sustainable and integrated EU–MENA power system would trigger significant benefits for all the countries in the region. On one hand, building such a system entails important political, regulatory and institutional challenges. On the other hand, from an energy perspective, there is hardly a choice but to understand the Mediterranean as a hub rather than as a border in the long term.

The initial phase for an intensified regional cooperation is already at an advanced stage. Political processes for renewables and grid integration around the Mediterranean and in the Middle East are in progress and key organizations such as Medreg, MedTSO and RCREEE are in place.

In the years to come, shared efforts should focus on ensuring a pragmatic and efficient progression of cooperation. For this purpose, it will be paramount that the actors involved agree on a long-term perspective facilitating coordination and interaction. The experience of the Energy Community can provide very valuable learning regarding the process of power markets integration. Nevertheless the specific characteristics of the MENA region will require a different set-up based on the joint development of common principles rather than in the extension of EU Law.

The *Mediterranean Solar Plan – Master Plan* developed under the umbrella of the UfM, is a good example of a joint exercise for the construction of a common understanding based on the dialogue of the different countries involved. In the medium term, general principles, such as those of the Master Plan, should develop into binding legal instruments that provide a solid basis for market integration.

The construction of a common legal framework will entail a long and gradual process. The current disparity of regulatory regimes in the EU and MENA countries will certainly entail an important challenge. Nevertheless, certain aspects of energy policy are common throughout the EU and MENA countries. The vast majority of countries in the region have developed ambitious renewable energy targets, and there is a consensus that these can be better achieved under a cooperative approach. The cooperation process should focus on these overlapping areas in order to find a balanced solution to bring the currently heterogeneous regimes closer together.

Regional cooperation should at all times be guided by a practical approach and avoid unnecessary complexity and intricate institutional structures. For this purpose, political institutions should engage in a close dialogue with the organizations that gather key stakeholders in the energy sector, namely regulators, TSOs and industrial players. A bottom-up approach will be vital for an organic development and implementation of common policies.

Given the important investments required to construct this integrated and sustainable power system, strong public–private cooperation will be a success factor. Therefore, cooperation efforts should aim at creating the right market conditions and investment environment for the private sector to actively take part in this transition. Pending the implementation of the necessary market conditions, political engagement will be of utmost importance to develop initial projects and new interconnections linking the EU and MENA countries, which will provide a tangible dimension and concrete success examples to the overall cooperation strategy.

Notes

1. The findings, interpretations, and conclusions expressed in this chapter are those of the author and do not reflect the positions of the institution.
2. European Commission, *Renewable Energy Progress Report*, COM (2013) 175 final (27 March 2013), Brussels, available at http://ec.europa.eu/energy/renewables/reports/doc/com_2013_0175_res_en.pdf.
3. Dii GmbH, *Desert Power: Getting Started, The Manual for Renewable Electricity in MENA*, Munich (June 2013); and Dii GmbH, *Desert Power 2050: Perspectives on a Sustainable Power System for EUMENA*, Munich (June 2012).
4. Ibid.
5. Dii GmbH, *Economic Impacts of Desert Power*, Munich (June 2013).
6. See note 2.
7. European Commission Joint Communication, *A New Response to a Changing Neighborhood*, Brussels COM (2011) 303 (23 May 2011).
8. Joint Communication: *A Partnership for Democracy and Shared Prosperity with the Southern Mediterranean*, Brussels COM(2011)200 final (8 March 2011). See also Joint Communication, *A New Response to a Changing Neighbourhood,* Brussels COM (2011)303 (25 May 2011).
9. In recent years the Neighbourhood Investment Facility has already actively provided financial assistance to specific renewable energy projects in MENA countries, for example CSP plant in Morocco, and has supported neighboring countries in the implementation of ambitious energy reforms, such as the Energy Sector Policy Support Programme in Egypt.
10. More information on the League of Arab States can be found at www.lasportal.org.
11. More information on MED-ENEC can be found at www.med-enec.com/.
12. Commission Staff working document "Guidance on the Use of Renewable Energy Cooperation Mechanisms", accompanying the commission's communication, "Delivering the Internal Electricity Market and Making the Most of Public Intervention", Brussels, (2013) draft.
13. For a detailed analysis on the potential savings that could stem from an EU-wide approach to renewable energy imports from MENA countries, see *Desert Power: Getting Started*, chapter 3.5: Quantification of RE Support Needs. Munich: Dii, 2013. www.dii-eumena.com/fileadmin/Daten/Desert_Power/Desert%20Power%20Getting%20Started-Full%20Report%20English-Screen.pdf
14. Association agreements are foreseen in part VII of the Energy Charter Treaty.
15. Commission Staff Working Document – SWD, Brussels (2012) 164 final (6 June 2012), accompanying the Commission's Communication *Renewable Energy: A Major Player in the European Energy Market*, Brussels, COM (2012) 271 final (6 June 2012).
16. DCFTAs will be an extension of current Euro-Mediterranean Association Agreements aimed at a progressive economic integration of these countries into a EU single market. DCFTAs will reach beyond removing tariffs and will cover a full range of regulatory areas of mutual interest.

17. More information on GCCIA can be found at www.gccia.com.sa/.
18. Joint Communication *Supporting Closer Cooperation and Regional Integration in the Maghreb: Algeria, Libya, Mauritania, Morocco and Tunisia*, Brussels, JOIN(2012) 36 final (17 December 2012).
19. Medreg gathers 23 energy regulators from the following countries: Albania, Algeria, Bosnia-Herzegovina, Croatia, Cyprus, Egypt, France, Greece, Israel, Italy, Jordan, Malta, Montenegro, Morocco, Palestinian Authority, Portugal, Slovenia, Spain, Tunisia and Turkey.
20. Medreg has developed very interesting studies for market integration, such as "The Master Plan towards a Regional Market", Milan (June 2012), available at www.medreg-regulators.org/portal/page/portal/MEDREG_HOME/PUBLICATIONS/Electricity.
21. More information on RCEEE can be found at www.rcreee.org/.
22. Member states of RCREEE, to date, are Algeria, Bahrain, Egypt, Iraq, Jordan, Lebanon, Libya, Morocco, Palestine, Iran, Sudan, Syria, Tunisia and Yemen.
23. More information about MedTSO can be found at www.terna.it/default/home_en/the_company/about_terna/Terna_in_Europe_and_the_Mediterranean/medtso_en.aspx.
24. The founding members and future associates of MedTSO are Albania – OST; Algeria – SONELGAZ, GRTE, OS; Cyprus – Cyprus Transmission System Operator; Egypt – EETC; France – RTE; Greece – ADMIE; Israel – IEC; Italy – TERNA; Jordan – NEPCO; Libya – GECOL; Montenegro – CGES; Morocco – ONE; Portugal – REN; Slovenia – ELEKTRO-SLOVENIJA; Spain – REE; Tunisia – STEG; and Turkey – TEIAS.

Chapter 6

Regional cooperation in the Western Balkans

The eighth regional energy market in Europe

Milica Uvalic[1]

Regional cooperation in the Western Balkan (WB) region has acquired a fundamental importance over the past decade, in the aftermath of the 1999 conflict in Kosovo. After the end of the conflict, an important initiative – the Stability Pact for South-Eastern Europe (SEE), adopted in Cologne on 10 June 1999 – was launched by the European Union (EU) and the wider international community with the aim of organizing international donors' financial assistance and of promoting regional cooperation among the WB countries. In order to transfer "policy ownership" to the SEE region, the Stability Pact for SEE has in the meantime been transformed into the Regional Cooperation Council based in Sarajevo, maintaining its initial objectives of promoting cooperation among the countries in the WB region as one of the most important instruments for bringing more lasting peace and stability in this part of Europe. The new EU policy for the Western Balkans, embedded in the stabilization and association process, has also strongly promoted regional cooperation among the WB countries (see Cerovic and Uvalic, eds., 2010). Since 1997, in addition to the Copenhagen criteria, regional cooperation has been one of the official criteria for the WB countries' integration and accession to the EU. Initiatives on regional cooperation in the WB region have involved many specific fields that come within three broad areas – politics, economics and security. The various initiatives have included trade liberalization, improving the investment climate, promoting SMEs, fighting corruption and many others. One of the most recent initiatives is the creation of a regional energy market in South-East Europe.

The present chapter discusses some aspects of regional cooperation in SEE, focusing particularly on the energy sector. For the purpose of this chapter, SEE will include primarily the seven transition countries in the Western Balkans – Albania, Bosnia and Herzegovina, Croatia, Kosovo, Macedonia, Montenegro and Serbia, although Bulgaria and Romania are also occasionally considered. The structure of the chapter is as follows. First, we discuss the question of economic integration in the Western Balkan region, by looking at some of the main arguments that have been proposed in favour of regional cooperation among the WB countries. We then consider more specifically the situation in the energy sector of the WB countries in order to assess the differences and degree of complementarities. Because all WB countries have undertaken important reforms of the energy sector over the past decade, the results achieved so far are also briefly examined. Finally, European Union policies towards the WB region aimed at stimulating regional cooperation in the energy sector are discussed in order to underline some reasons why the creation of a more integrated regional energy market could indeed be beneficial. The chapter ends with some concluding remarks.

Benefits of regional cooperation

Regional cooperation in SEE has been on the international and national agendas for over a decade. Although regional cooperation in the broader SEE region – in addition to the WBs, Bulgaria and Romania – has been promoted by the EU since the second half of the 1990s, until recently it has not led to significant results. On the contrary, peace and stability in SEE have been seriously undermined precisely by the *lack* of regional cooperation, which has contributed to several armed conflicts in the region, including the war in Kosovo/Yugoslavia in March–June 1999 (Uvalic, 2001).

A number of arguments have been proposed in favour of regional cooperation in the WBs. Closer cooperation in various areas would in itself be beneficial, but a global regional strategy could bring additional spill-over effects – more cooperation in one area could clearly facilitate and reinforce closer cooperation in others. Within such a comprehensive regional strategy for the SEE region are a number of *economic arguments* why closer cooperation among these countries would be highly beneficial. These arguments are even more important considering that, in comparison with the more advanced economies in Central-Eastern Europe (CEE) that today are EU member states, the WB countries are in a worse economic situation – regarding growth recovery, FDI, trade integration and the EU integration process (see Uvalic, 2012a). It is sufficient to mention that by 2013, three countries have still not reached their pre-1989 level of gross domestic product (GDP), with Serbia being in the worst situation (see Uvalic, 2010, 2013).

The WB countries are small and still relatively underdeveloped, in spite of rapid growth during the 2001–2008 period. For years, they have been isolated because of high tariff and non-tariff barriers, exclusion from key EU programs sustaining transition, closed borders due to the EU visa regime and, in some cases, sanctions and embargos. During the 1990s, isolation and underdevelopment determined low purchasing power and impeded more intensive regional trade, stimulating smuggling, illegal trade and the development of underground activities. Regional economic cooperation could help the WB region pull out of this vicious circle of backwardness through its contribution to faster economic growth and development.

The main economic arguments in favor of regional cooperation fall within four main areas: trade, investment, regional dimension of problems and EU integration (Uvalic, 2001). These arguments are closely interrelated. Intensifying trade and other forms of economic links among SEE countries could contribute to faster economic growth, whereas faster development is crucial for political and economic stability in the region; only if the WB countries become economically and politically stable are they likely to attract more private capital from abroad; and only if conditions for more sustainable growth and development are fulfilled can these countries integrate with the rest of Europe.

1. ***Trade:*** One of main ways of strengthening economic links among the WB countries is through increased mutual trade. Since 2001, this objective has been pursued through trade liberalization among the WB countries. The launch of the trade liberalization initiative in 2001 provided for the signing of bilateral free-trade agreements among the SEE countries after June 2001, which has been further institutionalized through the transformation of bilateral trade agreements into the Central European Free Trade Agreement – CEFTA 2006. There is no doubt that these initiatives have given a strong impulse to increasing regional trade and that the increase in regional trade has had a beneficial effect on growth of the WB economies (see Uvalic, 2006, 2009). There could be potentials for expanding trade among the WB countries further, primarily some of the successor states of former Yugoslavia. Today, in the light of the strong effects of the global economic crisis on the WB economies, in which exports to the EU have fallen or have stagnated, sluggish export performance on EU markets could partly be compensated by increasing exports to the other countries in the WB region.
2. ***Investment:*** Regional cooperation can further contribute to improving the investment climate in the WB region. In the 1990s, the level of national savings and investment in the WB countries was very low, because of wars, recessions, underdevelopment, general lack of capital and loss of confidence of the population in local banks; we should recall that four countries (Albania, Bulgaria, Macedonia and the Federal Republic (FR) of Yugoslavia) have been involved in pyramid schemes which have deprived the population of their life savings. Moreover, until 2001, Foreign Direct Investment (FDI) inflows into the WB region have been limited due to high political risk, major interest of foreign partners in other transition economies, and smallness of markets. In the 1990s, the cumulative FDI into the WB countries has been extremely low compared to the net inflows into other transition economies (see Uvalic, 2012a; Estrin and Uvalic, 2013). Such a low level of FDI is undoubtedly a regional problem, as illustrated by the case of Macedonia, a country that has not been involved in the military conflicts in the

1990s yet has also received a very low amount of FDI. Although the situation has greatly improved over the past decade, FDI in the WBs has been relatively limited in comparison to the CEE countries (see Estrin and Uvalic, 2013). Capital inflows from abroad remain of fundamental importance for the WB economies in the future. A further increase in FDI could indeed be stimulated by the creation of a more integrated regional energy market.

3. ***Common regional problems:*** Many economic problems in the WB region today are of a regional character. After the 1999 military conflict in Kosovo, a number of common problems have emerged which necessarily have to be resolved by all the WB countries jointly. These common problems of the WB countries concern cooperation regarding transport, the Danube, migration, energy, environment or industrial policies. These problems have to be addressed in their regional dimension, requiring close collaboration of all countries. Regional cross-national projects are needed for rebuilding railways, highways, other transportation routes, the communications and the energy network. Economic development in the WBs depends to a great extent on efficient regional infrastructure networks, because the reintegration of regional infrastructure can lead to important economies of scale. All cross-national projects of regional importance will necessarily require strong public policy, because energy infrastructure, bridges or the river Danube need to be treated as public goods.
4. ***EU integration:*** Regional cooperation in the WB region could ensure smoother integration of these countries into the EU, not simply because it is one of the official criteria the WB countries must fulfill. During the past decade, the process of EU integration has proceeded at such a slow pace, that only Croatia signed the Accession Treaty and thus was able to join the EU in July 2013. Because the remaining WB countries will not become EU member states soon, they should, in the meantime, strengthen relations with neighboring countries in the region. Regional cooperation can facilitate mutual intra-regional trade, large investment projects and major inflows of FDI, which in turn could contribute to faster economic growth of the WB countries. This would reduce the huge income gap which presently exists with respect to the earlier incomers to the EU.

Some of the most important economic arguments for promoting regional cooperation in the WBs have been stressed. However, even if one rejects these economic arguments, there are other extremely important non-economic motives for intensifying regional cooperation. Despite increased political stability over the past decade, there are problems of status, borders, refugees or return of property, that have still not be definitely resolved. Given the delicate problems that still need to be tackled in the WB region, there are strong political reasons to promote closer regional cooperation. The need to establish more permanent conditions for security, stability, peace and development in the Balkan region should be a major stimulus for closer cooperation among these countries in all fields.

The energy sector in the Western Balkans

The Western Balkans are located between hydrocarbon-rich regions (including Russia, the Caspian basin and the Middle East) and key energy-consuming regions of Western and Central Europe. The WB region therefore has the potential of playing an important role in the transit of energy sources, including hydrocarbon resources and oil and gas. Presently, gas markets in the Western Balkans are small, but do have the potential for growth (see Organisation for Economic Co-operation and Development [OECD]/International Energy Agency [IEA], 2008). These are some of the reasons why the WB region has become increasingly important for the EU, in addition to the more general reasons related to the EU integration process.

The WB region has experienced dramatic political and economic changes over the past twenty years, which have also greatly influenced energy supply of the individual countries. From the early 1990s onwards, the WB countries have been negatively affected by a series of important events. The transition-related policies and economic reforms which started in 1989–1990, the disintegration of Socialist Federal Republic (SFR) of Yugoslavia, the military conflicts that accompanied the break-up of the Yugoslav federation and economic sanctions imposed against some of the countries, have all had devastating effects on the economies in the region (see Uvalic, 2010, 2012a). One of the most pronounced effects of the early 1990s was a very strong drop in GDP and even steeper reduction in industrial production, because many industrial plants had to be closed. The deep recession of the early 1990s that in some WB countries lasted until 1996 led to a remarkable drop in domestic consumption and living standards, including domestic consumption of electricity by households. Although there has been substantial economic recovery during the past decade, some of the effects of the 1990s have not been completely eliminated.

Another important consequence of the events of the early 1990s was the fragmentation of the energy infrastructure. Until late 1991, the WB region consisted of only two countries, Albania and SFR Yugoslavia. The electricity network of SFR Yugoslavia until 1992 was interconnected with the Union for the Coordination of Transmission of Electricity (UCTE), namely the Western European grid, that was separated in 1992 when the country disintegrated. Croatia and the Federation of Bosnia (the southern and western part of Bosnia and Herzegovina) remained connected to the UCTE Zone 1, whereas Republika Srpska (the northern and eastern part of Bosnia and Herzegovina), Serbia and Macedonia (together with Bulgaria, Romania and

Greece) made up the South-East European UCTE Zone 2. During the military conflicts in the 1990s, energy infrastructure in Bosnia and Herzegovina, Kosovo and Serbia was severely damaged. During the North Atlantic Treaty Organization (NATO) intervention in 1999, two of Serbia's main oil refineries, in Novi Sad and Pancevo, were demolished.

After the normalization of the political situation in the region in 2000, as part of the EU and wider international community's reconstruction assistance to the SEE, there were major efforts to improve the interconnections and electricity transmission system operators in the WB region. This led to the reconnection, in 2004, of the two sub-regional networks (the UCTE Zones 1 and 2) and their re-synchronization with UCTE (OECD/IEA, 2008, p. 18). Since then, the UCTE norms and standards are again fully applicable in the WB region, with the total interconnection capacity of 5,800 MW in 2007 (see Karova, 2010). Only Albania's electricity system, which for long was underdeveloped and whose transmission system does not satisfy all criteria for parallel operation with UCTE, is not a member of this network.

The current economic situation in the individual WB countries is different regarding the achieved degree of macroeconomic stability, level of economic development, social indicators and economic reforms required by the transition to a market economy (see Uvalic, 2012a). Serbia is the largest country in terms of territory and population (a third of total population in the Western Balkans lives in Serbia), whereas Montenegro is the smallest (less than 3% of the Western Balkans' population). Croatia is the second most populated country and is also the most developed WB country, with 45% of the region's total gross domestic product and at 60% (2009) of the EU-27 average GDP per capita. The WB economies have had high growth rates over the 2001–2008 period, on average close or greater than 5%, and have attracted increasing FDI, but the recent global financial and economic crisis has brought a halt to fast growth. Given that the WB countries have become highly integrated with the EU economy through trade, FDI, financial flows and banking systems, the effects of the global crisis were felt very heavily through various transmission mechanisms. After late 2008, the WB countries experienced a substantial reduction in capital inflows (FDI, credits, remittances) and a strong decline in export demand on EU markets, which led to a serious deterioration in the most important macroeconomic indicators.

The heterogeneity of the WB countries regarding size, geography and economic characteristics is equally present when it comes to energy. There are wide differences regarding total primary energy supply (TPES), the energy mix, the volume of national energy production and dependence on imported energy (see Table 6.1). Most WB countries are not able to cover their essential energy needs through domestic production but import from other countries. The size of the WB countries markets, in terms of final energy consumption, varies, but most of the markets are very small.

Table 6.1 Energy supply in the Western Balkan countries, 2005 (Mtoe)

	TPES	*Domestic production*	*Imports*	*Exports*	*Net Imports*	*Import dependency*	*Total final consumption*
Albania	2.4	1.2	1.2	0.0	1.2	51%	2.1
Bosnia and Herzegovina	5.0	3.3	1.9	0.3	1.6	32%	3.0
Croatia	8.9	3.8	7.8	2.6	5.2	58%	7.1
FYR Macedonia	2.7	1.5	1.6	0.3	1.3	45%	1.7
Montenegro	1.0	0.6	0.5	0.1	0.4	40%	0.8
Serbia	16.7	11.4	6.4	1.1	5.3	32%	9.7
Kosovo	2.0	1.2	0.9	0.1	0.8	40%	1.0
Total	38.7	23.0	*	*	*	*	25.4

Source: OECD/IEA (2008, p. 16).

Note: Import dependency is calculated as net imports/TPES. Data on Serbia are based on the official submission of the Ministry of Mining and Energy of Serbia. For the purpose of this survey, data directly from the administrations in Montenegro and Kosovo were used.

*Not summed up to avoid double counting due to intra-regional trade.

Given the different size of countries, there are wide differences in the supply of energy across the WB countries. Serbia, as the largest country, has the dominant share in total primary energy supply, as much as 43% of TPES of the seven WB countries, followed by Croatia (23%), Bosnia and Herzegovina (13%), Macedonia (7%), Albania (6%) and Kosovo (5%). However, domestic production represents less than 60% of total primary energy supply in the WB region, so the region as a whole relies on imports of substantial amounts of energy. The smallest market in energy terms is Kosovo with a final electricity consumption of 3.2 TWh (2005), closely followed by Albania and Montenegro, whereas the largest is Serbia with 25.6 TWh (2005; Poyry and Nord Pool Consulting, 2011).

There is also great diversity among the WB countries regarding the energy mix. The principle energy sources in the region, which are also those mainly utilized in the WB countries, consist of coal (mostly lignite) and hydropower, followed by oil and natural gas. Serbia has the largest installed generation capacity and depends mostly on lignite-fired thermal power plants, but its overall fuel efficiency and utilization rates remain low. Bosnia and Herzegovina and Macedonia also rely heavily on lignite-powered generating capacity, whereas Kosovo gets almost all of its domestic electricity generation from thermal plants (lignite). Albania, by contrast, derives almost all (98%) of its electricity generation from hydropower by its three large hydropower plants which provide more than 85% of total generation; however, only 30% of available sources are actually utilized, so there is major potential for future expansion of mini-hydropower plants. Croatia, Bosnia and Herzegovina, and FR Yugoslavia (Serbia and Montenegro) also rely on hydropower, which represented 53%, 47% and 31%, respectively, of total energy sources for electricity generation in 2004.

Presently, gas markets in the Western Balkans are small but do have the potential for growth. Oil and gas production is limited and is mostly located in Albania, Croatia and Serbia. Until now, only Croatia, and to a much lesser extent Serbia, are important consumers of natural gas. Albania, Montenegro and Kosovo are not gasified. A more diversified mixture of energy sources would certainly contribute to energy efficiency, including more reliance on renewable sources, such as hydropower.

The situation in the single countries of the wider SEE region is presented in Figure 6.1 in the color section, which shows the sources of domestic electricity production (thus not taking into account imports). The figures confirm that the most frequently used sources for electricity generation in the WB countries are conventional thermal energy and hydropower. Other renewable sources, besides hydropower, have played a very limited role so far in the WB region.

The WB region is dependent on imported primary energy, not only oil and natural gas but also electricity. The only countries in the SEE region that are exporters of energy are Bosnia and Herzegovina and Bulgaria, but the SEE region as a whole is one of the major net importers of energy in Europe. Croatia depends most on imports, followed by Albania (see Table 6.1). The energy infrastructure is in poor condition. The thermal power plants, built in the 1960s and the 1970s using Eastern European technology, were not modernized during the 1980s and the 1990s, which has increased the region's dependency on energy imported from outside.

Albania is a net importer (2.1 TWh) with large imports not only from Greece, but also from Montenegro and to some extent Kosovo. Croatia is also a net importer (6.6 TWh) with large imports from Serbia, Hungary and Bosnia and Herzegovina; however, it is also a net exporter to Slovenia. Kosovo is a net importer (0.6 TWh) with large imports from Serbia, but it also exports to Montenegro and Macedonia (transit trade). Montenegro is a net importer (1.8 TWh), importing mainly from Bosnia and Herzegovina and to a lesser extent from Kosovo. Montenegro has a relatively balanced trade with Serbia and is a net exporter to Albania. Macedonia is a net importer (2.7 TWh), importing mainly from/through Kosovo and Bulgaria, but it is also a net exporter to Greece. Serbia has a relatively balanced energy balance. Only Bosnia and Herzegovina is a net exporter to the WB region, exporting mainly to Croatia and Montenegro, but it is also a net importer from Serbia (Poyry and Nord Pool Consulting, 2011).

Regarding regional trade, the most frequently traded energy products among the WB countries are electricity, oil products and fuel wood. However, there is lack of reliable data on actual trade of single energy sources. Estimates indicate that intra-regional energy trade accounts for about 13%

of the region's energy needs (or 4.8 Mtoe), of which the most traded are oil products (52% of total), electricity (41%) and fuel wood (7%). Regarding electricity trade, in the mid-2000s it amounted to 15.8 TWh, or 23% of final electricity consumption. Serbia accounts for 56% of total exports to other WB countries, followed by Bosnia and Herzegovina with 29%. On the side of intra-regional imports, Croatia is the biggest importer, accounting for 42% of total imports from the other countries in the region. Macedonia is the only country that does not export to the other WB countries, because it exports primarily to Greece.

Energy sources for electricity generation in five WB countries, as reported on the database of the European Network of Transmission System Operators for Electricity, are presented in Table 6.2 (data are not available for Albania and Kosovo). Only two countries in 2010 had primary energy production higher than their domestic consumption – Bosnia and Herzegovina and Serbia, although only somewhat less. Montenegro has a relatively balanced situation (its electricity consumption is only slightly higher than its production). The remaining two countries, Croatia and Macedonia, have a substantial deficit in production, thus having to cover their energy consumption needs by imports from other countries.

Lack of reliable electricity supply is generally a serious problem in the region. A report by the OECD and the IEA (2008) on the WBs mentions the erratic electricity consumption pattern of the poorer parts of the population as a factor for concern. Fuel wood is used for heating needs by the poor, but during the heating season, electric heaters are often used when fuel wood demand spikes. This exacerbates seasonal and weather-related peaks in electricity demand, which can cause blackouts or require rationing. The utilities are forced to maintain considerable reserve requirements, which then reduce potentials for exports and revenues. Low tariffs and insufficient payment discipline also limit the revenues.

Another problem shared by all WB countries is the high inefficiency of the electricity network, which causes substantial losses of energy (as

Table 6.2 Primary energy sources for electricity generation and consumption (in GWh), 2010

	Hydropower	*Fossil fuels*	*Other renewables*	*Total production*	*Total consumption*	*Deficit/ surplus*
Bosnia and Herzegovina	7,870	7,684	0	15,554	11,725	3,829
Croatia	8,313	4,801	135	13,251	17,594	–4,343
Macedonia	2,316	4,282	0	6,598	8,328	–1,730
Montenegro	2,738	1,267	0	4,005	4,044	–39
Serbia	12,453	28,508	0	40,961	39,525	1,436

Source: Compiled based on data provided by the European Network of Transmission System Operators for Electricity (www.entsoe.eu).

much as 30% in Albania). The WB economies generally have high energy intensities, considerably higher than the average OECD level. This is due to the degraded state of energy infrastructure, high energy losses in transformation, transmission and distribution and inefficiency in the end-use sector. Croatia has relatively high energy efficiency, but according to some estimates, it still has an energy savings potential of around 25% of total primary energy supply. The region as a whole could save 13.4 TWh annually by bringing losses down to the level of Croatia. The WB countries also have high carbon intensities compared to OECD averages. Serbia has the highest level of carbon intensity (1.2), whereas Albania the lowest (0.3) thanks to its intensive usage of hydropower resources (Poyry and Nord Pool Consulting, 2011).

In 2005, network losses amounted to, on average, 22% of the region's final electricity consumption (or 14.5 TWh), as compared to the OECD average of only 8% (see Table 6.3). Network losses were particularly high in Albania (36%) and Kosovo (37%), whereas they were the lowest in Croatia (14%). Losses caused by technical problems of transmission and distribution amount to 25% of total electricity consumption. Providing households with alternative fuels to replace electric heating would further reduce peak demand, thus helping to reduce the stress presently placed on the electricity systems in the WB region. The low level of electricity prices also causes substantial losses to the electricity sector in terms of reduced revenues that could provide for maintenance and investment.

Table 6.3 Electricity demand and network losses in the Western Balkan region, 2005

	Peak demand	*Final electricity consumption* (FEC)*	*Share of residential in FEC*	*Network losses***	
	GW	TWh	%	%***	TWh
Albania	1.3	3.6	75	36	1.30
Bosnia and Herzegovina	1.7	7.7	52	28	2.16
Croatia	3.0	14.4	44	14	2.02
FYR Macedonia	1.4	6.2	48	22	1.36
Montenegro	0.8	3.8	30	29	1.10
Serbia	6.9	25.6	55	21	5.38
Kosovo	1.3	3.2	73	37	1.18
Total/average %	**16.4**	**64.5**	**52**	**22.5**	**14.50**

Source: OECD/IEA (2008, p. 37).

*Calculated as gross production + import – export – energy used in electricity production.

**Transmission and distribution technical losses, as well as commercial losses.

***Percentage of final electricity consumption.

Today, households are responsible for the highest share of electricity consumption in the large majority of countries – 50% on average for the whole WB region – but in Albania and Kosovo, the share is as high as 75%, extremely high compared to 29% in European members of OECD. Among the main reasons for such a very high consumption of electricity by households are low electricity prices and low payment discipline on the part of the users. Electricity prices remain relatively low, greatly contributing to wasteful practices of high electricity consumption. Electricity prices are most frequently lower for households than for non-residential consumers (see Uvalic, 2012b). The low electricity prices translate into very high peaks in electricity consumption that overload the networks, while low payments reduce revenues across the electricity sector, having a very negative impact on maintenance and investment (OECD/IEA, 2008, p. 36). Despite relatively low prices, high energy consumption puts a significant pressure on household budgets. It is estimated that 16% of the people in the region are exposed to energy poverty. It is reasonable to believe that the current high energy intensities/low energy efficiency are likely to be affected by changes in prices that market opening would bring.

The environmental impact of the present use of energy is very different across the WB countries (see Table 6.4), precisely because the energy sources are so different. Despite the situation today, we can assume that all WB countries will also try to limit polluting emissions in line with EU regulations regarding sustainable environment. This could be done through an increase in the use of natural gas coming mainly from Russia, but increasing imports of natural gas would also worsen the situation regarding the volatility of

Table 6.4 Main energy and environmental indicators in the WB region, 2005

	TPES/GDP in toe per USD thousand (PPP)	*Electricity* consumption TWh*	*Electricity intensity in kWh/GDP (PPP)*	*CO_2 in Mt*	*Carbon intensity in CO_2/GDP (PPP)*
Albania	0.16	3.7	0.25	4.6	0.31
Bosnia and Herzegovina	0.19	9.1	0.35	15.9	0.62
Croatia	0.17	15.4	0.3	20.8	0.4
FYR Macedonia	0.21	6.9	0.53	8.3	0.64
Montenegro	0.26	3.8	1	2.5	0.66
Serbia	0.41	29.1	0.72	50.4	1.24
Kosovo	0.42	3.2	0.67	3.9	0.81

Source: OECD/IEA (2008, p. 24).

Note: Data on Serbia are based on the official submission of the Ministry of Mining and Energy of Serbia. For the purpose of this survey, data directly from the administrations in Montenegro and Kosovo were used.

*Production + import – exports – transmission/distribution losses.

prices, given that the price of gas, being anchored to the price of petroleum, would imply increased uncertainty regarding the price of production (and sales) of electricity. If the prices of electricity were to become more volatile, households would pay the cost and the higher burden, not the big enterprises (see Hooper and Medvedev, 2008).

Despite the discussed problems, the recent liberalization of the electricity market in the WB states has stimulated various large European energy groups to invest in the Western Balkans (Enel, Eni, Cez Group) in order to get market shares and important production capacity, a trend that is likely to continue in the future.

Energy sector reforms

All WB countries have undertaken substantial economic reforms, as part of their overall efforts to introduce a market economy. Particularly after 2001, there was an acceleration of economic reforms throughout the region, leading to greater convergence of WB countries with the institutional model present in CEE economies (see Uvalic, 2012a).

The energy sector in the WB countries is also in a phase of transition, from a monopoly in production, distribution and transmission of electricity to a liberalized market. Under the Energy Community Treaty (ECT) signed in 2005 (discussed further in the next section), the WB countries are expected to adopt the EU *acquis communautaire* in energy, competition and environment (see www.energy-community.org). Under the ECT, the WB countries have made commitments towards implementing a common energy policy: to gradually liberalize power markets, to restructure energy companies, to maintain cost recovery tariffs, to adopt tariff methodologies and technical codes for network access, to enforce payments, to introduce social safety nets and to set up independent regulators to scrutinize third party network access (Deitz, Lindsay, and Wright, 2009, p. 8). The WB countries are expected to adopt the EU electricity reform model, although some doubts have been raised whether this policy choice is the best for the Western Balkans (see Pollitt, 2009).

In the field of energy, two main groups of reforms in the WB countries, similarly to the experience in other countries, have been (1) the restructuring of state-owned energy companies represented by monopolies, through the unbundling of their activities by separating generation from transmission and distribution, along with the adoption of a new regulatory framework that should enhance energy efficiency, and (2) the creation of more open, liberalized and competitive energy markets that require the reduction or elimination of subsidies and the gradual increase in prices that have remained below effective cost levels.

All WB countries have started the process of restructuring state-owned energy enterprises. By now, all WB countries have primary legislation for the energy sector, as well as a regulatory authority but only two countries – Croatia and Serbia – have legislation for the gas sector. However, the liberalization process is only in its initial phase in most WB countries.

The European Bank for Reconstruction and Development (EBRD) evaluates progress in economic reforms in various specific areas and the remaining gaps in its client countries. Infrastructure is given specific emphasis, whereas one of the main sub-areas is energy. The EBRD has recently also created a specific blog focusing on reforms in the area of renewable energy sources (www.ebrdrenewables.org). The recent progress in the most relevant reforms for the energy sector in the WB countries is presented in Table 6.5.

The EBRD transition indicator scores which assess results achieved in the various areas of economic reform go from 1, which implies no or limited progress, to 4+, which indicates that the situation is comparable to that of an advanced market economy. In evaluating reforms in the energy sector, various elements are taken into account: the degree of restructuring through the separation of activities, the percentage of market determined prices, the existence of subsidies, participation of the private sector, tariff reform, legal framework, competition and liberalization and establishment of an independent regulator.

The EBRD assessment (see Table 6.5) suggests that only Croatia has a relatively satisfactory situation, having been evaluated with a high 4– score for reforms regarding natural resources, with a 3– regarding efforts to promote sustainable energy sources, and with a 3 regarding electricity sector reforms. Albania is also evaluated rather well, having obtained a relatively high score, 3, in all three areas, probably due to its very high reliance on hydropower. Macedonia and Montenegro seem to have applied most reforms regarding, respectively, the electricity sector and natural resources (score 3), but are lagging behind in the other two areas. The remaining two countries, Bosnia and Herzegovina and Serbia, have been assessed with the low score of 2 (with the addition of a plus or a minus) in all three areas, indicating that most reforms are in their initial stage.

Table 6.5 Energy sector reform ratings in the Western Balkan countries, 2011

	Natural resources	*Sustainable energy*	*Electric power*
Albania	3–	3+	3
Bosnia and Herzegovina	2	2	2+
Croatia	4–	3–	3
Macedonia	2+	2+	3
Montenegro	3+	2	2+
Serbia	2	2	2+

Source: Compiled based on data in EBRD (2011, p. 9). Data for Kosovo were not available.

The EU initiatives: Towards the eighth regional energy market

The EU initiative to create the eighth energy market in Europe was launched in 2002, when a memorandum of understanding was signed by nine countries from the region, with the Stability Pact and the European Commission acting as sponsors. The clear enlargement perspective of the SEE countries was one of the main incentives for reforms of the national electricity markets and for initiatives in this area (Karova, 2010). Cooperation was further expanded to the gas sector through a second memorandum of understanding signed in December 2003. In 2004, the South East European Regulators Board for Electricity and Gas was established by the Ministerial Council. These memoranda represented only political intentions and did not impose any legal commitments on the parties involved (Karova, 2011).

A legally binding agreement exists only since 2005, when the Energy Community Treaty (EnCT) was signed in 2005 by the EU and the SEE countries – the Western Balkans in addition to Bulgaria and Romania – with the aim of creating a new regional energy market in SEE, that would be connected to the EU energy market. After Bulgaria and Romania entered the EU in 2007, their legal status changed to participants. As of April 2008, there were fourteen participants of the Energy Community and four countries with the status of observers (Georgia, Norway, Turkey and Ukraine). Moldova joined the Energy Community Treaty in 2010.

The EnCT aims at regulating the relations between the countries' signatories in order to create a common legal and regulatory framework for the energy markets. The EnCT also envisages the coordination of mutual assistance in case of serious disturbance to the energy networks or external disruptions and possibly the achievement of a common external energy trade policy. The EnCT should encompass the principles and policies of the EC, taking into consideration the specificities of all parties, by ensuring that the countries signatories adopt the *acquis communitaire* in areas such as energy, environment, competition and renewables (Karova, 2010).

The other obligations taken by the SEE countries particularly in the sphere of electricity are the following: (a) establishing common rules for the functioning of the national electricity markets and establishing mechanisms for crisis situations (safeguard measures); (b) establishing the Regional Energy Market itself – prohibition of taxes and quantitative restrictions for import and export of electricity, and common rules for trade with third countries; and (c) opening of the electricity markets – of all non-household markets by 2008, and of household markets by 2015. The EnCT was concluded for a period of 10 years from the date of entry into force, but the Ministerial Council may decide to extent its duration (Karova, 2011, p. 81).

In 2006, the European Regulators Group for Electricity and Gas (ERGEG), with support of the European Commission, launched two regional initiatives: the Electricity Regional Initiative (ERI) and the Gas Regional Initiative

(GRI). These initiatives are considered an interim step in moving from national electricity and gas markets towards an internal regional market.

There is no doubt that regional integration has the potential to improve the performance of the electricity markets in the WB region. The harmonization of the economic and technical infrastructure, combined with deepening intraregional integration through investment in transmission lines and efficient congestion management are the most important issues. Here are some of the benefits expected from regional integration of energy markets in the WBs (Karova, 2010, 2011):

- the integration of markets ought to decrease price volatility, which in turn would benefit consumers and allow for higher political acceptance of deregulated electricity markets;
- more optimal use of existing capacity between several countries could lead to an increase in trade of electricity between them;
- price signals will be increasingly used as a benchmark for the players active on the market, which would be beneficial for both consumers and potential investors;
- integration would lead to a reduction in the potential market power of generators, by expanding the size of the market and consequently reducing their share of total generation capacity;
- reliability, that is security of supply, could also be increased by physical and energy systems integration;
- fuel diversification would be facilitated by the creation of a large regional market;
- benefits regarding the environment and health are also expected, because integration would decrease the risk of construction of generation and transmission facilities that degrade the environment and expose the local population to long-term health risks; and
- size and configuration of the regional market would be dictated by technical and economic factors rather than political factors as was the case until recently.

There are also obstacles which might prevent effective regional integration of electricity markets in the WBs, among which the following are the most important (Karova, 2010):

- political reluctance to share control over a politically sensitive sector of the economy, because state-owned utilities have been used as an instrument of industrial policy, by providing cheap electricity to certain industries;
- opposition from entrenched interests, as suggested by the public choice theory: parties with an interest in maintaining the status quo will resist integration of local markets into regional ones; and

- strategic behavior by governments negotiating the terms of integration, since each government may prefer institutional arrangements that are favorable to their domestic consumers and producers, rather than those with regional benefits.

In balancing the arguments in favour and against regional integration of energy markets in the Western Balkans, there are strong reasons why the former seem far more important. In part, this derives from the main characteristics of the energy sector in the WB countries. The energy markets in the WB countries are small, their economies are energy intensive, energy prices are generally below economic costs, tariff structures are inappropriate, while more intensive energy trade is prevented by poor infrastructure. At the same time, the WB countries today share the same regional energy infrastructure, which has created a high level of interdependence within the region. Some examples of such interdependencies are that all countries participate in extensive daily and seasonal exchanges of electricity, or that Serbian oil refineries rely on deliveries through the Croatian pipeline network. Given that the electricity generation systems in the Western Balkans are based on complementary energy sources, the establishment of a regional energy market would provide a much larger market, with the possibility of substituting thermal and hydropower in peak and off-peak periods. Given the specific features of various energy sources, the risks deriving from the use of the different sources could also be better shared by the individual WB countries within such a regional framework. For some sources, such as oil and gas, prices are bound to fluctuate due to changing quotations on international markets, so strengthening the common and shared infrastructure could provide concrete benefits, particularly for those countries that have intense daily and seasonal trading in electricity. On the contrary, hydropower supply depends primarily on weather conditions, so countries like Albania and Bosnia and Herzegovina are the most exposed to this kind of risks. The political significance of regional energy markets in the WB should also not be underestimated; this is why the SEE Energy Community has been compared to the European Coal and Steel Community, which paved the way for the creation of the European Economic Community.

Concluding remarks

There are a number of reasons why the WB countries ought to coordinate their energy sector strategies further. The great diversity among the WB countries regarding the availability of energy sources – some countries are particularly rich in hydropower, other in traditional fossil fuels – suggests there are strong motives for the further development of an integrated energy market in the region. Such a market could bring a number of economic benefits, including a reduction in the volatility of electricity prices, better connection of networks and reduction of losses, which would be an important advantage for all the WB countries. For the small and underdeveloped WB economies, regional economic cooperation, also in the energy sector, could be an answer to some of the present challenges. Energy issues are bound to play an increasingly important role also within the South-East Europe 2020 Strategy, adopted in November 2013 by the respective governments at the Regional Cooperation Council in Sarajevo.

The recent EU initiative to create a regional energy market in the WBs is a good example of regional cooperation that can be seen more as a form of "open regionalism" achieved from within the region, with regionally owned institutions and solutions for solving the regional problems (Karova, 2010), contrary to many other initiatives launched from outside the region that have prevailed during the last fifteen years. The development of the SEE Regional Energy Market has been impressive during its short period of existence, suggesting that the energy community does represent a positive example of a regional initiative in the region.

Note

1. The findings, interpretations and conclusions expressed in this chapter are those of the author and do not reflect the positions of the institution.

References

Cerovic, B. and M. Uvalic (eds) (2010), *Western Balkans' Accession to the European Union: Political and Economic Challenges*, Belgrade: Faculty of Economics Publishing Centre.

Deitz, L., S. Lindsay, and K. Wright (2009), "South East Europe's Electricity Sector: Attractions, Obstacles and Challenges of Europeanisation", *Utilities Policy* 17: 4–12.

Energy Community (www.energy-community.org)

Estrin, S. and M. Uvalic (2013), "Foreign Direct Investment into Transition Economies: Are the Balkans Different?", LSE Europe in Question Discussion Paper Series, LEQS Paper No. 65/2013, July.

European Bank for Reconstruction and Development (2011), *Transition Report 2011. Crisis and Transition: The People's Perspective*, London: EBRD.

European Network of Transmission System Operators for Electricity (www.entsoe.eu)

Hooper E. and A. Medvedev (2008), "Electrifying Integration: Electricity Production and the South-East Europe Regional Energy Market", Electricity Policy Research Group Working Paper 0803, Cambridge Working Paper in Economics 0804, Cambridge.

Karova, R. (2010), "The Energy Community – the Eight Regional Electricity Markets in Europe", in B. Cerovic and M. Uvalic (eds.) (2010), *Western Balkans' Accession to the European Union: Political and Economic Challenges,* Belgrade, Faculty of Economics of the University of Belgrade Publishing Centre, pp. 313–342.

Karova, R. (2011), "Regional Electricity Markets in Europe: Focus on the Energy Community", *Utilities Policy,* 19: 80–86.

Organisation for Economic Co-operation and Development/International Energy Agency (2008), *Energy in the Western Balkans. The Path to Reform and Reconstruction*, Paris: OECD/IEA in cooperation with the UNDP.

Pollitt, M. (2009), "Evaluating the Evidence on Electricity Reform: Lessons for the South East Europe (SEE) Market", *Utilities Policy*, 17: 13–23.

Poyry and Nord Pool Consulting AS (2011), *South East Europe Wholesale Market Opening Final Report,* December, available at www.ecrb.eu/portal/page/portal/ECRB_HOME/ECRB_DOCUMENTS/STUDIES/SEE%20WMO%20Final%20report%20version%202.02.pdf

Uvalic, M. (2001), "Regional Co-Operation in South-Eastern Europe", *Southeast Europe and Black Sea Studies,* 1 (1): 64–84.

Uvalic, M. (2006), "Trade in Southeast Europe: Recent Trends and Some Policy Implications", *European Journal of Comparative Economics,* 3 (2): 171–195.

Uvalic, M (2009), "Regionalism in Southeast Europe", in P. Della Posta, M. Uvalic and A. Verdun (eds.), *Globalization, Development and Integration: A European Perspective,* Basingstoke: Palgrave Macmillan, pp. 176–194.

Uvalic, M. (2010), *Serbia's Transition: Towards a Better Future,* Basingstoke: Palgrave, Macmillan.

Uvalic, M. (2012a), "Transition in Southeast Europe: Understanding Economic Development and Institutional Change", in G. Roland (ed.), *Transition in Eastern Europe. The Long-Term,* Basingstoke: Palgrave-Macmillan, pp. 364–399.

Uvalic, M. (2012b), "The Regional Electricity Market in the Western Balkans", in M. Uvalic (ed.), *Electricity Markets and Reforms in Europe,* Milano: Franco Angeli.

Uvalic, M. (2013), "The Economic Development of the Western Balkans since Thessaloniki", in Eviola Prifti (ed.), *The European Future of the Western Balkans. Thessaloniki 10 (2003–2013)*, Paris: European Union Institute for Security Studies, pp. 73–82.

Chapter 7

Parliamentary support to Euro-Mediterranean energy initiatives

Sergio Piazzi[1]

Energy represents a fundamental sector in ensuring the prosperity and the stability of the Mediterranean region. For this reason, energy-related issues have always been very high on the agenda of the Parliamentary Assembly of the Mediterranean (PAM[2]), whose members analyzed, over the years, several crucial aspects of the energy sector, in order to contribute in the process of regional and international cooperation in this field.

Energy consumption in the Mediterranean countries has more than doubled in recent decades. Despite the unfolding new era in the energy sector, the world is increasingly dominated by concerns regarding unimpeded access to energy sources and energy demand, as well as the need to strike a balance between development, environment, food security and access to a secure supply of energy.

This chapter is a summary of the activities carried out by the PAM on energy-related issues, underlining the importance and the key role of parliamentarians, in their capacity of policymakers, in support of cooperation in legislative matters at the regional level.

PAM action in support of regional cooperation in the energy sector

The PAM has tackled energy-related issues through the work of its 2nd Standing Committee, dedicated to Economic, Social and Environmental issues. This focus resulted in a series of reports and resolutions adopted over the years, at each PAM Plenary Session.

In 2008, PAM focused on the need to support investment in the Mediterranean and developing educational programs. This can be achieved through the adoption of policies and legislation to encourage the private sector to invest in renewable energies. The resolution adopted on the occasion[3] of the 2008 Plenary Session, underlined the importance of the EU financial aid initiatives to integrate renewable energy in the European energy markets and the significance of reaching a Mediterranean treaty for organizing, implementing, transferring and promoting renewable energy technologies in the

Mediterranean. The development of the work of regional, national and local energy structures in the region and the concept of adapting national policies to climate change are aimed at developing plans to prevent expected risks in the Mediterranean, particularly on the islands and in agricultural areas in the Mediterranean region.

In 2009, PAM tackled the issue of energy security and cooperation strategy between the two sides of the Mediterranean. The main focus of the standing committee dealt with climate change, sustainable development, non-traditional mechanisms to provide funds for energy enterprises and efficient energy use and nuclear energy cooperation.

In 2010, the Assembly centred its work on "Energy and the Global Financial Crisis: Opportunities for Sustainable Development in the Mediterranean Region". The recommendation in the resolution, unanimously adopted on the occasion of the PAM 5th Plenary session (Rabat, Morocco), included the need for the development of regional trade in energy, of improving regional integration and the strengthening in the implementation of the Mediterranean Solar Plan (MSP).

In 2011 PAM focused on "civil nuclear energy". The resolution adopted in Palermo, at the PAM 6th Plenary Session, stressed the need, for governments of PAM member states, to conduct urgent assessments of existing nuclear installations on their territory and to proceed with the immediate closure of facilities which do not meet the minimum levels of safety. The resolution urges all institutions to inform in a transparent way the citizens on the current risks and to hold periodic debates with the agencies responsible for nuclear safety at the national level, in order to ensure a continuous monitoring of government decision on nuclear investments, as well as revisions of safety standards. Particular attention was dedicated to the importance of submitting emergency plans in case of nuclear accidents, as well as the compilation of an annual report on the status of their installations and on the crisis management policies proposed in case of accidents, especially the ones related to seismic or volcanic risks.

In May 2010 PAM launched the "PAM Panel on Trade and Investments in the Mediterranean[4] (Lisbon, Portugal, 27–28 May), at a special meeting hosted by the Portuguese parliament, following a request, made by several national delegations, for a stronger commitment from PAM on economic cooperation and integration in the region. The aim of the PAM panel is to gather, around the same table, parliamentarians and representatives of the economic and financial sector (primarily investments promotion agencies, chambers of commerce, financial institutions, governmental bodies and agencies, business and consumers' associations and international organizations), in order to assess, promote and support measures needed to contribute to economic progress and sustainable development, including through legislative action. The PAM panel acts as a comprehensive and multidisciplinary group with the purpose of addressing common challenges, identifying obstacles and proposing concrete actions for the promotion and support of trade and investment of this region, with, amongst others, objective to

- stimulate trade in the Mediterranean area and support the establishment of a Free Mediterranean Trade Zone,
- sustain small and medium enterprise (SME) activities,
- facilitate investments and access to finance and
- promote job creation.

The steering committee of the PAM panel identified energy as a priority area, and, since then, several meetings were organized by PAM on specific topics related to Energy. On 17 November 2011 PAM organized, in cooperation with MEDREG, the conference "Mediterranean Energy Highways: An Urgent Need". The conference, held in Naples, Italy, was divided in four sessions focusing respectively on

- the need to promote cross-border interconnections in the context of diversity and respect for National priorities and geographic particularities. In this regard, cooperation at the regional level among electricity transmission system operators (TSOs) through the newly founded association Med-TSO is strongly supported by PAM and MEDREG, confirming the need for a strategic dialogue between grid operators and regulatory authorities to create a Mediterranean Energy Free Trade Area.
- the participation of different stakeholders from both the public and private sectors of the gas energy community. The need to support global energy security emerged. Considering that the EU's dependence on gas imports would increase to more than 80% by 2030, better distribution of energy in Europe becomes a priority concern.
- The financial perspectives and the importance to define priority areas to finance sustainable, competitive and secure projects were highlighted. At the regional level, the World Bank confirmed it was able to offer assistance in energy integration and with restructuring and improving the system of the governance for the energy sector.
- the creation of an internal energy market, in which the Mediterranean region will play an increasingly important role as an exporter of Energy. There was strong consensus and support for PAM's decision to organize a conference with the chairs of national energy parliamentary committees from the PAM member states to discuss the future architecture of a shared Mediterranean energy policy.

What emerged from the conference was that political, economic and social progress in the Mediterranean basin also requires improved availability of and accessibility to energy systems. Electricity and gas interconnections represent the most concrete and enabling tools for international energy cooperation in terms of trade and energy exchanges among the countries of the region. It follows that collaboration and cooperation at the institutional level should be strengthened and supported in order to achieve the establishment of a Mediterranean Energy Free Trade Area to identify shared principles and objectives. In this context, MEDREG has a key role to play to facilitate investments in infrastructures and to support greater efficiency and integration of energy markets, based on safe, cost-effective and environmentally sustainable energy systems in the Mediterranean basin. Increased cooperation between the shores of the Mediterranean could be reached by considering a governance solution where all Mediterranean countries would be able to make a substantial contribution and play a significant role.

In the international conference "Energy Challenges in the Euro Mediterranean Region", held in Ouarzazate, Morocco, on 14–15 September 2012, PAM adopted a set of recommendations, among which included

a. setting up of the necessary legislative and regulatory frameworks including fiscal measures, to promote investments in the renewable energy sector;
b. creation of a favourable environment among the PAM countries, conducive to the realization of the Mediterranean Solar Plan; and
c. establishment of a "Mediterranean Community of Energy", to ensure coordination among the key actors on renewable energy, the reinforcement of national and regional transmission grids, as well as the promotion of investments and clear legal frameworks.

Setting up a core group (executive members of the Energy Group) was also agreed to, in order to prepare a road map and a series of activities to implement the conclusions reached in Ouarzazate.

The main issues raised during the meeting of the executive members of the Energy Group of the PAM panel on trade and investments in the Mediterranean (Rabat, Morocco, 4 May 2013) were the following:

1. Regional cooperation and integration as a priority for energy systems in the Mediterranean. With the aim of facilitating the creation of a Mediterranean integrated system, there was agreement on the necessity of the harmonization of a shared set of rules in terms of investments and energy transit and trade, in order to attract investments uniformly in all the countries of the region. It was also underlined that investments are needed at the regional level, and for this reason, all countries needed common rules. In this regard, the framework of the Energy Charter represents an example of convergence towards a shared set of rules.

2. Regulatory activities are fundamental in the energy sector. For this reason, Regulatory Authorities should be established in all the Mediterranean countries. Furthermore, it was noted that, after MEDREG and Med-TSO, the creation of an Association of Mediterranean Gas operators would be beneficial for the region, in terms of network, price and quality.
3. It was noted that the Joint Communication of the European Commission of 17 December 2012 identifies energy as one of the most sensitive and crucial issues for the Mediterranean region. In this context, and in the framework of regional cooperation, it was agreed that PAM should engage in a constructive dialogue with the European Parliament and the European Commission on energy issues. The content of the joint communication matches the priorities of the Panel, and the documents represents a good opportunity for PAM to open a cooperation channel with the European Parliament, with the aim of gathering Mediterranean and European MPs to work together on these crucial matters. It was nevertheless noted that the document does not mention the importance of the gas sector, in the macro energy field.
4. The proliferation of uncoordinated activities on energy related issues. There was consensus that the Mediterranean region is moving towards fragmentation, with several initiatives having similar objectives but with different action plans. Overlapping and duplications are becoming common practice. Because of this, resources are not allocated in the most efficient way. Participants underlined the need of a regional neutral entity to be used as a starting point and possibly coordinate all the initiatives, and many of them indicated PAM as the ideal institution to serve with this function. The role of the Union for the Mediterranean needs to be clarified, because its initiatives were contradicting the plans undertaken by operators in the region.
5. The mechanism of subsidies should be redesigned in order to facilitate the integration of the energy market. It was also noted that, in this regard, countries should not prefer the least expensive energy, because very often this has an impact on other crucial factors, such as environment protection, quality of life and efficiency, so the specificities of energy must be taken into account.
6. Education, training and know-how transfer as a priority. Solar energy projects are very rich in terms of lessons learned, and these should be shared with those countries interested in implementing the same projects, such as Egypt or Jordan.
7. In connection with investments, it was noted that although their need is very high, this does not represent a problem due to the profitability of the projects, which offer the right margins to recover the investments. Furthermore, it was highlighted that, in order to receive funds for projects, countries should present concrete projects and a strong political will and commitment.

During the discussion, the importance of the "Energy for All" UN initiative[5] was underscored. PAM was encouraged to take all the necessary steps to become an active partner of the initiative, due to the relevance of involving members of parliaments – the policymakers – to the process. It was also mentioned that, in order to release the full potential of solar energy, and of all renewable energy, there is a need for launching not only centralized projects but also micro and decentralized initiatives, such as water heating. Emphasis was also given to the necessity of having a binding treaty to support the implementation of a master plan. There are numerous studies and reports on this issue, providing specific and useful recommendations. However, if the proposed measures are not brought forward strongly and are supported by a binding agreement, there is the risk of not achieving any concrete and significant result.

PAM cooperation with international institutions

The PAM fosters regional cooperation not only through its own activities but also by participating and contributing to initiatives promoted by partner institutions and other international players.

The first organization to be granted the status of observer to PAM was MEDREG, a key partner for PAM in the energy sector. The two institutions have established a fruitful cooperation platform, and the members of each institution are constantly aware of the activities of the other organization. This synergy contributes to the dissemination of information and to the process of raising awareness on energy-related issues in the Mediterranean basin. MEDREG regularly contributes to PAM's activities. A clear example is the address by PAM at the UN General Assembly high-level segment on energy in 2012, to which MEDREG contributed by providing technical expertise for the message that PAM delivered in New York. MEDREG is one of the executive members of the Sustainable Development Group of the PAM panel, and its participation provides PAM with high-level input in terms of agenda setting, substance and quality of the debates and the production of targeted recommendations.

PAM enjoys excellent relations with many of the most relevant institutions in the energy sector: Energy Charter, Med-TSO, MEDGRID, Moroccan Agency for Solar Energy (MASEN), the European Commission, the European Bank for Reconstruction and Development (ERBD), International Renewable Energy Agency (IRENA) and other regional and international parliamentary organizations, such as the Parliamentary Assembly of the Council of Europe, active on energy related issues, as well as the United Nation agencies and the initiative Energy 4 All, launched by the UN secretary-general and led by the former director general of United Nations International Development Organization (UNIDO), Kandeh Yumkella.

PAM and UNIDO have established a strong partnership, which resulted, for example, in the parliamentary debate on "The Role of Parliaments in Support of the Economic Process", led by PAM as part of the "International Conference on Productive Work for Youth in Tunisia and the MENA Region", co-organized by the Tunisian Ministry of Industry, the Tunisian Agency for Investments and UNIDO on 28–29 November 2012. The cooperation between PAM and UNIDO consist also in a constant contribution to each other initiative. PAM benefitted from the expertise of UNIDO representatives on many occasions in the framework of the activities of the PAM panel on trade and investments in the Mediterranean.

UNIDO is not the only UN agency partner of PAM's on energy-related issues. Cooperation has been successfully initiated with the United Nations Economic Commission for Europe and with the United Nations Economic Commission for Africa, especially on renewable energy.

In connection with the European Bank for Reconstruction and Development, a memorandum of understanding was signed between the two institutions in order to ensure a more coordinated effort in support of the development and prosperity of the Mediterranean region. The geographical extension of EBRD's mandate to include the countries of the southern and eastern Mediterranean has increased the importance of its role in addressing the challenges in the PAM region. PAM had in fact strongly lobbied for and supported the call to entrust this mandate to the EBRD, to establish a dedicated mechanism to finance infrastructure projects (also energy infrastructure) and facilitate access to credit by SMEs in the Mediterranean region.

The cooperation with the institutions mentioned earlier is developed through several approaches, depending on the mandate, scope and membership base of each partner, with the main aim of involving members of parliaments in the process, and creating the conditions for them to give their contribution, through their legislative actions, in addressing the challenges that the Mediterranean region is facing.

Conclusion

The current challenges are multi-faced and represent the result of complex trans-boundary processes. For this reason, solutions need to be planned, discussed and agreed at the international level, and implemented at the national level using a common approach and a shared target, for the benefit of all the people of the Mediterranean.

PAM will continue to be committed to contribute to the process of energy integration in the Mediterranean region, together with its partners. The future initiatives to support the Euro-Mediterranean energy community include, among others, a meeting of the presidents of the national parliamentary committees responsible to energy related issues.

Appendix

The Parliamentary Assembly of the Mediterranean: Origin and objectives

The Parliamentary Assembly of the Mediterranean (PAM) is an international organization established in 2005, following fifteen years of cooperation among states of the Mediterranean region. Cooperation focused on themes of common interest, such as political, socio-economic and environmental issues. The process, known as the Conference on Security and Cooperation in the Mediterranean (CSCM), and which began in 1990, developed under the auspices of the Inter-Parliamentary Union (IPU).

The main objective of PAM is to forge political, economic and social cooperation among the member states in order to find common solutions to the challenges facing the region and to create a space for peace and prosperity for the Mediterranean peoples.

PAM is the centre of excellence for regional parliamentary diplomacy and is a unique forum whose membership is open exclusively to Mediterranean countries, which are represented on equal footing. This is reflected in the composition of the bureau and the alternating presidency. The current president of PAM is Francesco Maria Amoruso (Italy).[6] Each national delegation has five members with equal voting and decision-making powers. Associate members and observers do not have voting rights.

PAM conducts the bulk of its work within three standing committees.[7] It may also set up *ad hoc* committees or special task forces to tackle a particular topic (i.e. Middle East, migration, free trade, terrorism, climate change, natural resources management, energy and gender issues, etc.). Although the reports and resolutions adopted by PAM are not legally binding *per se*, they are a "soft diplomacy" tool when dealing with parliaments, governments and civil society in the region.

PAM is also entrusted with the parliamentary dimension of the 5 + 5 dialogue.

The structure of the PAM secretariat

The secretariat assists the PAM president, the PAM bureau and all members in the execution of their mandate and is responsible for the follow-up on the decisions taken by the Assembly providing coordination, assistance and support to the work of the committees and all other bodies established under PAM.

The secretariat interacts with national delegations, as well as with regional and international bodies sharing an interest in the Mediterranean region. It stimulates the activities of the Assembly and coordinates the awarding of the PAM Prize, dedicated to individuals or institutions whose work is considered important for the Mediterranean region.

The secretary general, assisted by international and local staff, directs the Secretariat. Ambassador Sergio Piazzi is the current secretary general.[8]

PAM: relations with non-Mediterranean countries and international institutions

The membership criteria and the geographical location of its members position PAM as a central actor in the Mediterranean region.

The General Assembly of the United Nations granted the Observer Status to the Parliamentary Assembly of the Mediterranean, by Resolution A/RES/64/124, at its 64th Session on 16 December 2009, in recognition of its excellent work.

This further strengthens PAM's contribution to the region, in addition to that of other parliamentary platforms, states and other international bodies which have a direct interest, or a specific role, in the region.

Today's challenges, including migration, climate change and, most important, the resurgence of conflicts, require a collective effort, good will and confidence building measures.

An essential link with civil society, parliamentarians are key to any political decision-making process. In order to ensure coherence and coordination in decision processes, including social peace, it is necessary that parliamentarians fully participate in today's regional debates thereby contributing to the elaboration of policies that bear both an immediate impact and a long-term impact for future generations.

Likely, the political dialogue established among members of PAM, in particular with respect to Middle East issues, is of vital importance.

In furtherance of its objectives, PAM encourages the inter-governmental support of the main actors in the region as well as regional institutions.

PAM membership

The Member States of PAM: Albania, Algeria, Andorra, Bosnia-Herzegovina, Croatia, Cyprus, Egypt, France, Greece, Israel, Italy, Jordan, Lebanon, Libya, Malta, Mauritania, Monaco, Montenegro, Morocco, Palestine, Portugal, Serbia, Slovenia, Syria, The Former Yugoslav Republic of Macedonia, Tunisia and Turkey.

Associate Country: Romania

Partner States and Organizations: San Marino, Bulgaria, Georgia, the Russian Federation, the Holy See, the UN System, the IPU, the League of Arab States, the Arab Inter-Parliamentary Union, the Arab Parliament, the UN Alliance of Civilizations (UNAOC), UNESCO, CTBTO, the Parliamentary Assemblies of the Council of Europe (PACE), OSCE, NATO, the Anna Lindh Foundation (ALF),the Organization of the Black Sea Economic Cooperation, UfM, OIC, the Arab Maghreb Union and the Islamic Conference

(PUIC), the EBRD, the World Meteorological Organisation (WMO), the Association of the Mediterranean Regulators for Electricity and Gas (MEDREG), the Fondazione Mediterraneo and the Euro-Mediterranean University (EMUNI).

Headquarters

The Headquarters of the General Secretariat of PAM were established in Malta in November 2007, in recognition of Malta's strategic role and commitment in setting up PAM.

Notes

1. The findings, interpretations and conclusions expressed in this chapter are those of the author and do not reflect the positions of the institution.
2. The Annex reports informative documents on the Parliamentary Assembly of the Mediterranean. It provides useful background information on the mission of PAM as a tool of parliamentary diplomacy in the Mediterranean region.
3. The resolution is available for download at the following link: www.pam.int/welcome.asp?m=documents&a=cat&cid=55.
4. Members of the PAM panel are grouped by sectors and national delegations. Each sectoral group appoints a coordinator. The panel presidency is co-chaired by a parliamentarian and a representative of the economic sector. The two co-chairs of the PAM panel are currently the Honorable Nevena Stojanovic (Serbia) and EngineerAlessandro Ortis (former president of MEDREG and of the Italian Regulatory Authority for Electricity and Gas). The two co-chairs and the coordinators of each sectoral group, with a number of selected individuals and institutions, make up the steering committee of the PAM panel.
5. "Sustainable Energy for All" is an initiative launched by the United Nations secretary-general and is guided by his High Level Group that brings all key actors to the table to make sustainable energy for all a reality by 2030.
6. Senator Francesco Maria Amoruso is the secretary of the Foreign Affairs Committee of the Italian parliament. He assumed the PAM presidency on 13 October 2012, on the occasion of the PAM 7th Plenary Session.
7. 1st Standing Committee: Political and Security-related Cooperation and Regional Stability; 2nd Standing Committee: Economic, Social and Environmental Cooperation, Development and Partnership; 3rd Standing Committee: Dialogue among Civilisations and Human Rights.
8. Italian born, Ambassador Piazzi worked successively as a diplomat and UN senior official, mostly dealing with crisis management and EU and NATO affairs.

Chapter 8

Enhancing the rule of law in the Middle East and North Africa

A role for the Energy Charter Treaty

Ernesto Bonafé[1]

The European energy market is evolving towards a Euro-Mediterranean energy market. The Middle East and North Africa (MENA) are important for the EU in terms of security of gas and oil supplies, being responsible for about 25% of today's EU imports of gas and 6% of crude oil. As for the future, the East Mediterranean region has considerable potential to supply gas to the European Union (EU). The EU transition to a competitive low-carbon economy is based on efforts to reduce its greenhouse gas emissions by 80% by 2050. This has unlocked the potential for widening the EU–Mediterranean partnership to also include renewable energy sources.

International energy cooperation of this nature must be based on a sound regulatory framework in order to increase stability and transparency in the investment environment. The comprehensive EU *acquis*, which may be considered as a benchmark for regional integration of energy markets, has extended to the Western Balkans by the Energy Community. Building on this experience, stakeholders are now suggesting a convergence of South Mediterranean partners' energy policies with EU energy policy.

However, the situation in the Mediterranean countries differs from the one that led to the Energy Community Treaty, which was conceived to reunify an energy system which had disintegrated during the conflicts of the 1990s. In contrast to energy integration in the Western Balkans, regulatory convergence with the Mediterranean countries may have to follow a step-by-step approach. EU–Mediterranean integration is based on a differentiated and gradual approach as implemented by the European Neighbourhood Policy (ENP). The ENP is to be complemented by regional cooperation initiatives such as the Union for the Mediterranean (UfM).

The Energy Charter Treaty (ECT), of which the EU and its member states are already members, provides an international framework to enhance the rule of law on a global basis, including transition countries. The ECT is equipped to play a key role in the Euro-Mediterranean energy market. In that sense it should be noted that the countries of the Western Balkans adopted the ECT as the first basic legal framework for energy reform and later supplemented it with the Energy Community Treaty.

Existing Euro-Mediterranean regional associations provide policy guidance, coordinated action, guidelines, studies and recommendations for best practices for cross-border energy exchanges. The regulatory outcomes are to be approved and implemented on a voluntary basis. A consensus-based approach for energy regulation is suitable to reach compromises among countries with different energy sectors. However, those incipient practices need to be endorsed by a series of legally binding provisions. The Energy Charter Treaty contains a minimum set of rules that may contribute to enhance the rule of law in the Mediterranean energy market according to international legal standards.

This chapter looks into the ECT as a valuable international legal framework for the countries from MENA. The ECT would be a necessary complement to the bilateral and regional approaches that are currently developing.

Bilateral relations on energy between the EU and MENA countries

The ENP is proposed to ten southern Mediterranean neighbours: Algeria, Egypt, Israel, Jordan, Lebanon, Libya, Morocco, Palestine, Syria and Tunisia.[2] The goal of the ENP is to avoid the emergence of new dividing borders and achieve political association and economic integration between the EU and its neighbours based on common values: democracy, the rule of law, respect for human rights and social cohesion. According to the ENP "more for more" principle, partners that are embarking on more ambitious democratic reforms are offered increased market access (e.g. Deep and Comprehensive Free Trade Agreements – DCFTA), cross-border mobility partnerships and further financial support.

The ENP builds on partnership and cooperation agreements or association agreements between the EU and the partner in question. Furthermore, action plans set out political and economic reforms for every three to five years, reflecting the specific needs and commitments between the EU and each of its partner countries. The EU has recently concluded negotiations on action plans with Lebanon, Morocco, Palestine and Tunisia. The new action plan for Jordan entered into force in October 2012. Also, the EU started DCFTA negotiations with Morocco and announced the preparation of negotiations with Tunisia and Jordan.[3]

In the energy sector, efforts have focused on energy security, market reform and regional integration. The priorities are to develop infrastructure, improve energy efficiency and use more renewable energy sources. The idea of setting up a Mediterranean Energy Community, inspired by the Energy Community Treaty, has so far made little progress. In June 2013, the EU and Algeria signed a memorandum of understanding in the field of energy.[4]

The Neighbourhood Investment Facility (NIF), the financial instrument of the European Neighbourhood Policy, supports infrastructure projects in energy and other economic sectors. It brings together grant funding from the European Commission and the EU member states, and loans from national and European financial institutions such as the Agence Française de Développement (AFD), the Entwicklungsbank (KfW), the European Investment Bank (EIB) and the European Bank for Reconstruction and Development (EBRD). The NIF budget from the European Commission and from EU member states amounted to €745 million for the 2007–2013 period.

The bilateral approach of the ENP is enriched and complemented by regional cooperation initiatives. The UfM and the Mediterranean Solar Plan (MSP) have favoured the establishment of political, economic, regulatory and industrial regional institutions in the energy sector.[5]

The establishment of Euro-Mediterranean energy institutions

The Barcelona Declaration established in 1995 the Euro-Mediterranean Partnership. It set out the goals to create an area of peace, stability and shared economic prosperity, as well as the respect of democratic principles, human rights and mutual understanding and promotion between cultures and civilizations. The 2008 Paris Summit created the UfM and gave new political impulse to the economic and social cooperation between the EU and the Mediterranean. Today the UfM comprises the 28 member states of the EU, the European Commission and 15 Mediterranean countries;[6] the League of Arab States participates in all meetings. The UfM has a co-presidency, the EU has held the northern co-presidency since 2012 and Jordan currently assumes the southern co-presidency. The Secretariat of the UfM is located in Barcelona.

Energy is one of the six priority areas of the UfM. The 2008 Paris summit fixed the target to develop 20 GW of new installed renewable energy capacity in the South Mediterranean by 2020. This was accompanied by the launch of the MSP, which contributes to the achievement of a series of goals: to develop renewable energy to supply electricity to local markets and export part of it if possible, to improve energy efficiency, to create domestic industrial capacities and to foster regional market integration. As a contribution to the implementation of the MSP, the European Commission launched the project "Paving the Way for the MSP" to work on the legal, policy, financing, technological and regulatory conditions conducive to greater use of renewable energy and energy efficiency in the region. Likewise, the task of the UfM Secretariat was to develop a master plan for the MSP with the collaboration of all stakeholders. However, the MSP master plan could not be formally endorsed by the Euro-Mediterranean Ministerial Meeting on 11 December 2013, as initially planned, which revealed the difficulties of the process.

Several international and regional organizations participate in the reform of the energy sector in the region: the League of Arab States (LAS), the Regional Center for Renewable Energy and Energy Efficiency (RCREEE), the International Renewable Energy Agency (IRENA), the Parliamentary Assembly of the Mediterranean (PAM), the Observatoire Méditerranéen de l'Énergie (OME), the Arab Union of Electricity (AUE), the Euro-Mediterranean Electricity Cooperation (MEDELEC) and the Comité Maghrébin de l'Electricité (COMELEC) as well as industrial private initiatives created specifically in the context of the MSP, namely Dii (Desertec), Medgrid and Res4Med. In order to gain institutional efficiencies and avoid overlapping activities, streamlined communication between these institutions is desirable, as well as a clear delineation of tasks and competencies.

In this context, it is worth referring to two specialized technical bodies that gather national regulatory authorities and transmission networks. These bodies are the Association of Mediterranean Energy Regulators (MEDREG) and the Association of the Mediterranean Transmission Operators (Med-TSO) for electricity. The two entities are in a privileged position to play a crucial role in the integration of the Euro-Mediterranean regional energy system. In this sense, echoing their common mission, the two associations signed a cooperation protocol in Algiers in September 2013.[7] Their envisaged mandate is to play a similar role that the Council of European Energy Regulators (CEER)[8] and the European Network of Transmission System Operators (ENTSO) effectively play in the EU internal energy market. Of course, MEDREG and Med-TSO have to adapt their work to the specific needs of the Euro-Mediterranean region.

MEDREG started as a voluntary working group and became a permanent regional organization in 2007. Today, it plays a growing role in the context of the ENP.[9] MEDREG aims to improve the transparency, attractiveness and investment climate by forging a regulatory framework in the fields of electricity and gas. In 2012, the MEDREG working group on electricity demonstrated the need for an EU–MENA backbone grid using high-voltage direct-current power transmission to accommodate the increasing renewable energy in the overall electricity production in the region. Also, the working group on renewable energy designed a scenario to verify the effectiveness of Article 9 of Directive 2009/28/EC on the promotion of renewable energy sources as a flexibility mechanism allowing for joint projects between EU member states and third countries. Furthermore, two studies on gas dealt with the external dependence and security of supply in the region, and the improvement of national data transparency.[10]

Med-TSO was established in April 2012 as the association of the Mediterranean Transmission Networks for Electricity.[11] It should benefit from the work undertaken by the European Network of Transmission System Operators for Electricity (ENTSO-E) and, when necessary, adapt to the specific needs of the South Mediterranean. Med-TSO should aim to adopt common network codes on priority management capacities and interconnections, develop market procedures to manage international exchanges and define homogeneous criteria of cost-sharing and cost-allocation analysis. It should also work on the coordination of national network development plans and contribute, alongside other initiatives such as Medgrid, to the discussion of long-term North–South submarine electrical interconnections.[12] The coordinated tasks of Med-TSO and MEDREG are recognized as a concrete step to develop common technical rules and grid codes towards the development of a West-Mediterranean electricity area.[13] This would build on the Maghreb Electricity Market Integration initiative established in 2010 by the Algiers Declaration.

The need for an international framework: The ECT

The existence of vital North-South associations such as the UfM, MEDREG and Med-TSO suggest that adequate institutions are in place to move towards progressive reform of the regulatory frameworks and the integration of the energy systems in the Euro-Mediterranean region. Those institutions provide valuable political orientation and support, coordinated action, guidelines, studies and recommendations for best practices for cross-border exchanges. The regulatory outcomes, resulting from comprehensive consultations among all stakeholders, are to be approved and implemented on a voluntary basis. Although consensus may be praised as a suitable decision-making process among players from countries with substantial differences on their energy sectors, consensus alone is insufficient and needs to be supplemented and reinforced by an international legal system.

The ECT may contribute to filling that gap. The ECT provides a legal framework for energy relations as well as dispute-settlement mechanisms. Beyond the resolution of particular disputes, the ECT provisions have the effect of helping to strengthen the rule of law and transparency, thereby contributing to an improvement in the general investment and business climate. At the same time, the ECT explicitly recognizes the national sovereign rights over energy resources.

The ECT was designed to make possible energy collaboration in transition economies. In 1991, after the fall of the Berlin wall and the break-up of the Soviet Bloc, an Energy Charter Declaration was signed as a political and legal foundation for East–West energy cooperation. It was based on common objectives and principles such as development of open and efficient energy markets, the stimulus of private investments, non-discrimination among participants, respect for state sovereignty over natural resources and the recognition of the importance of environmentally and energy-efficient policies.

The Energy Charter Declaration was a political document that also emphasized the need for the establishment of an appropriate international legal framework for energy relations, leading to the signing in 1994 of the ECT, which then entered into force in 1998. The ECT aims "to promote long-term cooperation in the energy field, based on complementarities and mutual benefits" (Article 2) and it imposes the obligation "to promote access to international markets on commercial terms, and generally to develop an open and competitive market" in the energy sector (Article 3).[14]

The ECT contains a set of rights and obligations of "hard law" nature in the areas of investment protection, trade and transit, which are enforceable in legally binding arbitration mechanisms. The ECT is supplemented with the Protocol on Energy Efficiency and Related Environmental Aspects that entered into force simultaneously with the ECT in 1998. It also contains "soft law" provisions related to competition, technology transfer and access to capital. However, the confirmation of the principle of national sovereignty over energy resources means that the ECT does not prescribe the structure of the domestic energy sector or the ownership of energy companies and does not oblige member countries to open up their energy sector to foreign investors.

The ECT offers a legal framework and policy forum that is open to all countries along the energy chain: producers, consumers and transit states, as well as industrialized, transitioning and developing economies. The Energy Charter process expanded beyond its traditional borders and gained a global dimension. Asian and Central Asian states became contracting parties: Japan, Mongolia and the Central Asian states of Kazakhstan, Kyrgyzstan, Tajikistan, Turkmenistan and Uzbekistan. To date, the ECT has 54 members while 25 countries have observer status. It must be noted that the EU and its member states are contracting parties to the ECT, and with the expected accession of Montenegro, all northern Mediterranean countries will be contracting parties to the ECT.

The ECT created a ministerial Energy Charter Conference and a Secretariat to serve it. The Charter Conference is an international organization that has the function to keep under review and facilitate the implementation of the principles of the Energy Charter Declaration and the provisions of the ECT. The Charter Conference also decides on policy directions, adopts new instruments and protocols and approves the accession of new members.

The ECT sets an international-level playing field in the energy sector. Its principles provide for a long-term global model of energy cooperation within the framework of a market economy based on mutual assistance and the principle of non-discrimination. This was clearly referred to by EU Commissioner Günther Oettinger at the Energy Charter Conference in Warsaw in December 2012. The commissioner noted that the more countries who accede to the ECT, the more they would set the standard for international energy relations on a global basis.[15] In the same vein, the 2011 Commission Communication on the EU external energy policy expressed that "the Energy Charter Treaty should seek to extend membership towards North Africa and the Far East".[16]

The Energy Charter Conference has adopted a policy on consolidation, expansion and outreach aiming to promote the accession of countries from North Africa and the Middle East to the ECT. The policy objective is to enhance the rule of law in the areas of investment, trade and transit. Accession to the ECT by MENA countries would send out clear signals indicating the willingness to join an international energy community that enhances wider and deeper cooperation in the energy sector.

With regard to the current ECT membership status in the South Mediterranean, Jordan in 2007 and Morocco in 2012 have signed the Energy Charter Declaration, which is the first step towards accession to the Energy Charter Treaty. Other countries, widely across the region, particularly Algeria, Egypt, Oman, Palestine, Qatar, Saudi Arabia, Tunisia, United Arab Emirates and Yemen are observers by invitation of the Energy Charter Conference.

Investment promotion and protection under the ECT

An integrated electricity market all around the Mediterranean would be of benefit for all countries. The huge investments needed in generation and networks cannot be conceived without private capital. A regional electricity market is also more attractive for private capital than segmented national markets. Furthermore, a regional market would allow complementarities between national systems experiencing different load profiles, fuel mix and back-up capacity for renewable electricity production. Finally, a South Mediterranean market tied to the EU market would improve the overall stability and the security of electricity supply.

According to the analysis made by the OME, electricity demand in the South Mediterranean is expected to double by 2030, and therefore, massive generation investments are needed. Power capacity in 2009 amounted to 122 GW, whereas the planned new generation capacity by 2030 would represent between 160 and 200 GW. The installed renewable capacity would amount from 2 GW in 2009 to some 40 GW in 2030, or to 82 GW in a proactive green scenario.[17]

The NIF, in the context of the Mediterranean Solar Plan, provided the Ouarzazate solar power plant in Morocco, in 2011, with a grant of €30 million in addition to the €515 million loans from the European finance institutions, particularly the EIB, the AFD and the KfW.[18] In 2012 the NIF provided €15 million of support to reinforce Morocco's high-voltage electricity transmission network as well as €5 million of technical assistance to support wind and solar projects in the region.[19]

But simply providing development assistance will not be enough to bring about the necessary changes. Private investment flows remain insufficient compared to the energy needs of the region. More financing is needed to develop infrastructure, and the role of the private sector in mobilizing investments is crucial. The ECT is well equipped to attract private investments. For an open and private sector, confidence, predictability and a stable energy framework are fundamental. This is the important value added that the ECT can bring to the South Mediterranean countries.

The perceived degree of political risks in the host country considerably affects the decision of foreign companies whether to make an investment or not and what level of return it would imply. The lower the perceived risk, the more capital is likely to be invested and the more potential revenue the host country will attract. By reducing the political risks, the ECT seeks to improve investor confidence and to contribute to an increase in international investment flows.

The investment chapter is a cornerstone of the ECT. Its provisions aim to promote and protect foreign investments. To this end, the ECT grants a number of rights to foreign investors with regard to their investment in the host country. Foreign investors are protected against the most important political risks, such as discrimination, expropriation and nationalization, breach of individual investment contracts, damages due to war and similar events and unjustified restrictions on the transfer of funds. Investor rights are further protected by dispute-settlement provisions of the ECT, covering both interstate arbitration and investor–state dispute settlement.

A distinctive feature of the ECT is that it provides a set of rules that covers the entire energy chain, including not only investments in gas production and electricity generation but also the terms under which energy can be traded and transported across various national jurisdictions to international markets. The ECT's provisions on trade and transit are based on those of the World Trade Organization (WTO). WTO trade rules are thus extended to ECT contracting parties that are not yet members of the WTO.

The ECT addresses in more detail than the WTO the strategic issue of energy transit. In accordance with the principle of freedom of transit, ECT contracting parties undertake to facilitate the transit of energy through their territory. This includes the prohibition of discrimination in terms of origin, destination and ownership of energy as well as equal treatment of transit and transport facilities in terms of governmental measures. A general obligation to secure established flows of energy implies that transit countries must not interrupt or reduce existing transit flows, even if they have disputes with any other country concerning this transit, nor must they place obstacles in the way of new capacity being established.

ECT market-based rules do not mean that a particular model of energy market structure is imposed on national governments at the international level. The ECT fully respects the sovereign right of each of its signatory states to determine the system of property ownership of its national energy resources. Moreover, each state continues to hold the right to decide, inter alia, the geographical areas to be made available for exploration and development of its energy resources and to determine the rate at which such energy resources may be depleted or exploited. The explicit recognition of national sovereign rights over energy resources by the ECT means that, at all times, the balance between the protection of investors and the sovereignty of host states needs to be maintained.

Conclusion

The ECT benefits from a general recognition that it is a unique international legally binding treaty for cooperation in the energy sector. The Energy Charter Conference has adopted a strategy on expansion that contains the MENA region as one of the most important focus areas. The ECT legal provisions are a necessary complement to the political, regulatory, operational arrangements and industrial initiatives that have been set up in the Euro-Mediterranean region during the past years.

A regional electricity market is more attractive for private capital than fragmented national markets as it substantially improves business opportunities. Foreign direct investments in the energy sector are very important for strengthening national and regional economies. This area is comprehensively covered by the ECT and should be highlighted as one of the main advantages of acceding to the ECT. At the same time, enhancing the rule of law and the promotion of investments by the Energy Charter is consistent with the EU's objectives in the Mediterranean region and should be pursued in line with the ENP.

Furthermore, if countries are to integrate large amounts of electricity from renewable energy sources into the network, they need to share not only interconnections, but also common principles and rules. The creation of a Euro-Mediterranean energy market will likely result from the reinforcement of existing regional institutions such as the UfM (political), MEDREG (regulatory) and Med-TSO (operational). In this context, the ECT would contribute to fill the legal gap for the development of investments, trade and transit. Ultimately, the ECT would provide a legal bridge for energy cooperation between the EU, the MENA countries and West Africa.

Notes

1. The findings, interpretations and conclusions expressed in this chapter are those of the author and do not reflect the positions of the institution.
2. Six Eastern countries, which are all members of the ECT, namely Armenia, Azerbaijan, Belarus, Georgia, Moldova and Ukraine, are also part of the ENP.
3. Joint Communication "European Neighbourhood Policy: Working towards a Stronger Partnership", Brussels, (2013) 4 final (20.3.2013); and Joint Staff Working Document accompanying "European Neighbourhood Policy: Working towards a Stronger Partnership", SWD (2013) 86 final, Brussels, (20.3.2013).
4. European Commission – MEMO/13/665 (7.7.2013).
5. OME and Medgrid, *Towards an Interconnected Mediterranean Grid. Institutional Framework and Regulatory Perspectives*, Nanterre (2013).

6. From the north shore, Albania, Bosnia and Herzegovina, Monaco, Montenegro and Turkey and, from the south shore, Algeria, Egypt, Morocco, Tunisia, Israel, Jordan, Lebanon, Mauritania, Palestine and Syria.
7. Press release available at www.medreg-regulators.org.
8. The EU Third Energy Legislative Package of 2009 created the Agency for the Cooperation of Energy Regulators (ACER), which repealed the European Regulatory Group of Electricity and Gas (ERGEG) composed of the CEER and the European Commission.
9. MEDREG gathers regulatory energy authorities from Albania, Algeria, Bosnia and Herzegovina, Croatia, Cyprus, Egypt, France, Greece, Israel, Italy, Jordan, Lebanon, Malta, Montenegro, Morocco, Palestine, Portugal, Slovenia, Spain, Tunisia and Turkey.
10. MEDREG, *Annual Report 2012*. Available at www.medreg-regulators.org.
11. The funding members and future associates are electricity system operators from Albania, Algeria, Cyprus, Egypt, France, Greece, Israel, Italy, Jordan, Libya, Montenegro, Morocco, Portugal, Slovenia, Spain, Tunisia and Turkey.
12. See speech of Fabrizio Barbaso, Deputy Director-General, Directorate General for Energy, European Commission, on "Multilateral cooperation for the integration of the Mediterranean electricity systems", Rome, 4 May 2012. Available at http://ec.europa.eu/energy/international/doc/20120504_med-tso_conference.pdf.
13. Joint Communication, "Supporting closer cooperation and regional integration in the Maghreb: Algeria, Libya, Mauritania, Morocco and Tunisia", JOIN (2012) 36 final, Brussels, 17.12.2012.
14. See Y. Selivanova, "The Energy Charter and the International Energy Governance" in Y. Selivanova (ed.), *Regulation of Energy in International Trade Law. WTO, NAFTA and Energy Charter*, 2011, Kluwer Law International, pp. 373–406; and C. Bamberger and T. Wälde "The Energy Charter Treaty" in M. Roggenkamp (ed), *Energy Law in Europe – National, EU and International Regulation*, 2007, Oxford University Press, pp. 145–194. The ECT and related documents are available at www.encharter.org/fileadmin/user_upload/document/EN.pdf.
15. See speech of Günther Oettinger, EU Commissioner for Energy, at the 23rd Meeting of the Energy Charter Conference and Ministerial Conference, Warsaw, 27 November 2012. Available at http://ec.europa.eu/commission_2010–2014/oettinger/headlines/speeches/2012/11/doc/20121127_warsaw.pdf.
16. Commission Communication, "On Security of Energy Supply and International Cooperation – The EU Energy Policy: Engaging with Partners beyond Our Borders", COM (2011) 539 final, Brussels, (7.9.2011).
17. OME, Mediterranean Energy Perspectives, Nanterre (2011).
18. *NIF Neighbourhood Investment Facility, Operational Annual Report 2011*, Luxembourg: Publications Office of the European Union, (2012). Available at http://ec.europa.eu/europeaid/where/neighbourhood/regional-cooperation/irc/documents/annual_report_2011_nif_en.pdf
19. *NIF Neighbourhood Investment Facility, Operational Annual Report 2012*, Luxembourg: Publications Office of the European Union, (2013). Available at http://ec.europa.eu/europeaid/where/neighbourhood/regional-cooperation/irc/documents/annual_report_2012_nif_en.pdf.

Part II

Export of the EU model

Chapter 9

Enhanced sectoral multilateralism

Will the Energy Community lead the way to a new European Neighbourhood Policy?

Josefine Kuhlmann[1]

The European Neighbourhood Policy (ENP) and especially its cooperation in the field of energy are built on soft law instruments. It is held that using an additional instrument with legally binding commitments such as a sectoral agreement could facilitate the achievement of the ENP's objectives, especially, but not only in the sector covered. The Energy Community (EnC) provides such a legally binding framework and has already attracted attention amongst ENP countries. The following is the attempt at providing thoughts on EnC's ability as a very specific concept of integration to provide a model for the EU's external relations with ENP countries.

Enhanced sectoral multilateralism

The idea of enhanced (sectoral) multilateralism constitutes a variation of the concept of differentiated integration. In a European Union (EU) context, differentiated integration was first intensively discussed in the 1970s on the occasion of the upcoming enlargement to Greece, Spain and Portugal. Since this "Southern Enlargement", scholars have been trying to find a solution to overcome the heterogeneity within the Union that has been increasing with every new member state that joined. Despite serious doubts about ever-ongoing enlargements, effective alternatives have not been developed. Instead, concepts of flexible and differentiated integration were born, discussed, dismissed and partly integrated into EU Law.

In the context just described, "[e]nhancement denotes closer legal relations, going far beyond mere approximation of laws, implying the clear-cut obligation to apply acquis", which results in legal integration by the establishment of a common legal space where EU law is applied but without participation in EU decision making.[2] In this vein, the EU has repeatedly used the tool of establishing organizations of enhanced sectoral multilateralism to shape its external relations with third countries; most successfully with the countries connected to the EU within the sophisticated framework of the European Economic Area (EEA). The EnC as another form of enhanced sectoral multilateralism is strictly limited to (non-nuclear) energy and concluded with countries far less developed than the EEA partners or Switzerland.

The Energy Community as a concept of enhanced sectoral multilateralism

In a nutshell, the EnC's task is the creation of a legal and economic framework with regard to "Network Energy" defined in Article 2 (2) of the Treaty establishing the Energy Community (TEnC) as the electricity and gas sectors falling within the scope of Directives 2003/54/EC and 2003/55/EC[3] as well as the oil sector falling in the scope of Directive 2006/67/EC.[4] Originally, these two directives, together with several other acts of Union secondary legislation on energy, environment, and renewables were to be implemented by the Contracting Parties (Article 3 (a) and Title II TEnC). In the shadow of the third legislative package implementation at EU level, the EnC adopted the most important pieces of EU energy legislation to be implemented with a lag of a few years.[5]

The implementation of EU legislation shall attract investments in necessary energy infrastructure eventually leading to enhanced security of supply for the region and its neighbours (most importantly the EU itself), an improved environmental situation and competition in the single regulatory space developed (Article 2 TEnC). Another aim is to overcome entrenched state-level political problems, including issues of regulatory and physical market cooperation.[6]

The parties to the TEnC are, on one hand, the EU and the contracting parties Albania, Bosnia and Herzegovina, the Former Yugoslav Republic of Macedonia, Moldova, Montenegro, Serbia, Ukraine and the UN Interim Administration Mission in Kosovo, on the other. EU member states are represented by the EU and can only be participants pursuant to Article 95 TEnC. According to Article 96, neighbouring third countries may apply for observer status; currently Armenia, Georgia, Norway and Turkey enjoy this special relationship. The former observers Moldova and Ukraine became contracting parties on 1 May 2010 and 1 February 2011, respectively, shifting the focus from the (potential) member states of the Western Balkans to other neighbouring countries.

The EnC is the first and only international organization comprising the EU, (potential) EU candidates and ENP countries. This alone would not be so remarkable, were it not for its ambitious objective of the establishment of a common Internal Energy Market encompassing this selection of third countries. Although the contracting parties progress in implementing these pieces of EU legislation, the application and enforcement of fundamental principles inherent to the EU's Internal Energy Market are still lacking.[7] In the same vein, the EnC's progress as an organization is for the most part limited to keeping up with the introduction or the adaptations of EU legislation in the field of network energy. Thus, the EnC as an organization struggles to keep pace with developments at EU level, while the contracting parties face great challenges as regards the implementation of the ever-demanding rules.

From the EU's point of view, the most problematic issue as regards the EnC should be the high degree of heterogeneity under the rules of the current framework. Even if "homogeneity with the internal market" is considered as being of great importance,[8] it must not be forgotten that currently there is no homogeneity between the legal frameworks of the contracting parties and EU member states. As long as there is a time lag for implementing new legislation, heterogeneity will be inherent to the EnC. In addition, the TEnC stipulates solutions in terms of adoption, interpretation, implementation and enforcement of EU law, which are potentially harmful to homogeneity and thus opposed to the establishment of a real, common Internal Energy Market. At best, market rules will be approximated to EU legislation in some of the contracting parties and will applied correctly by able authorities. In the worst case, the rules will be similar to EU legislation in their wording but inapplicable in the national legal order in question and/or ignored by the relevant authorities.

There is much to be said about what works well and which issues could need improvement within the EnC framework. On reflection, however, it is most of all important that an organization fostering and requiring cooperation was introduced into the Western Balkans region. It is of equal importance that the contracting parties have to attend to energy issues in a broader context; their citizens will appreciate any improvements made to their (not necessarily immediate) benefit. For the EU it is an attempt capable of offering additional options for the EU's external relations policy. However, "[i]f the reform model cannot be transferred successfully to this set of countries [i.e. the contracting parties], it bodes ill for the further extension of the reform model to other countries in Asia or Africa."[9] Thus, if the EU wanted to extend the concept of the EnC to neighbouring countries, it would need a success in the Western Balkans beforehand. Considering recent developments at EnC level, this leads to the question of whether its enlargement to Moldova and Ukraine was the right move at the right time. It has to be kept in mind, however, that the pressure to enlarge the EnC to Moldova and Ukraine was most probably exerted on the EnC by the EU itself in order to undergird the special relations with both countries. So, in fact the major challenge for the EnC might be to strike a balance between fulfilling the EU's requests and developing at its own pace and into the direction predetermined by the TEnC's provisions and backed by the contracting parties.

Energy in the EU's Neighbourhood Policy

The European Neighbourhood Policy

The idea to extend the EnC to other countries surrounding the EU has been inherent to this organization. Piebalgs, the former energy commissioner, indicated that future contracting parties were those ENP countries willing to reform their energy sector and to apply Internal Energy Market rules. It is essential to stress that Piebalgs directed his offer to join the EnC to all ENP countries and not just to a selection of them.[10]

The ENP was established in 2003. It works with instruments not explicitly envisaged in primary law, with legally non-binding character but an aspiration to further develop existing external relations.[11] It was designed "to enhance relations with its neighbours on the basis of shared values". Its aim is the establishment of "a zone of prosperity and a friendly neighbourhood – a 'ring of friends' – with whom the EU enjoys close, peaceful and cooperative relations".[12] The countries forming this "ring of friends" are Algeria, Armenia, Azerbaijan, Belarus, Egypt, Georgia, Israel, Jordan, Lebanon, Libya, Moldova, Morocco, the Palestinian Authority, Syria, Tunisia and Ukraine. Placing Ukraine, Belarus, Moldova and the countries of the Southern Caucasus into one group with the Euro-Mediterranean countries draws a clear political dividing line between these Eastern European states and the (potential) EU candidates of the Western Balkans, which explicitly do not take part in the ENP "since they have a membership perspective".[13] In order to mitigate this "one size fits all" policy, several regional initiatives were launched.[14]

The main characteristics of the ENP are differentiation between the various regions and countries encompassed, joint ownership as regards definition of priorities, and its complementary character to already existing, mostly bilateral legal relations with the countries in question. The incentives offered by the EU are a "significant degree of integration" including "a stake [. . .] in the EU's Internal Market".[15] Traditional trade concessions are not that appealing in a world where not tariffs but non-tariff barriers are the real obstacles to trade.[16] In addition, to enhanced political dialogue, economic and social cooperation, trade, and cooperation in the field of justice and home affairs, the ENP also engages in the integration of transport, energy and telecommunications networks. The extent to which each of the policy areas would be shared with a specific neighbouring country depends on its progress and reforms towards commitment to the shared values.

The international agreements on which cooperation with the ENP countries is based were in most cases concluded already years before the introduction of the ENP; these are the so-called partnership and cooperation agreements (PCA) and the Euro-Mediterranean Association Agreements (EMA). The PCAs were designed as "entry-level" agreements for the former republics of the USSR; as their name suggests, they do not establish any form of association between the EU and the countries in question. The EMAs, on the other hand, are real association agreements concluded under Article 310 TEC (now Article 217 TFEU). Both types of agreements provide the overall organizational and institutional framework for each bilateral relationship, while priorities with regard to content are specified in action plans (APs), a legally non-binding ENP instrument. It is important to note that these APs do not know any graduation of goals or priorities; human rights issues are – according to these documents – no more important than trade or energy issues.

Energy and the ENP

Most of the ENP countries are crucial for the EU's supply with hydrocarbons, as producer, consumer, or transit countries.[17] Additionally to legal and political coordination, well-functioning cooperation in the field of energy most naturally should also consider the extension of networks together with mechanisms ensuring their interoperability with EU systems.

It is telling that in 2006, the European Commission offered and even urged all ENP countries to membership in the EnC,[18] not only those with "membership potential" due to their "Europeanness". Time showed, however, that the Energy Community seems to appeal only to those countries with at least the ability to fulfil the pre-conditions for membership application (Article 49 TEU-L): so far, only Moldova, Ukraine and Georgia have applied for membership in the EnC. Two of them, Moldova and Ukraine, have already acceded. Other countries which exert a dominant influence on the EU's external energy policy negotiated their own framework for cooperation in the field of energy, most importantly, the Russian Federation.

The extent to which energy is dealt with at the ENP level shall be assessed in order to ascertain whether the ENP framework concerning energy is compatible with the EnC's rules and vice versa. The pre-ENP agreements, that is the PCAs and the EMAs, which serve as the institutional basis for the bilateral relationships also after the introduction of the ENP, include energy articles with varying degrees of attention to detail depending on the country in question.

Energy in the PCAs

The PCAs are bilateral agreements concluded with the countries of Eastern Europe and the Southern Caucasus. The PCAs for Armenia,[19] Azerbaijan[20] and Georgia[21] include rather detailed provisions on energy as regards cooperation in various areas. They do not themselves speak of any approximation or convergence towards the EU's Internal Energy Market. Instead, their focus clearly lies with investments in infrastructure and development. This view is endorsed by recital 12 of these PCAs, which stresses the importance of "promoting investment [. . .], including in the energy sector, and in this context the importance attached by the Community and its Member States to equitable conditions for transit for export of energy products". Generally, their energy article seems to be more of a wish list of the EU fearing for its security of energy supply.

Energy in the EMAs

As disparate the countries taking part in the Euro-Mediterranean Partnership are, they are also diverse regarding the energy provisions in their bilateral agreements. To begin, the most dissimilar examples of the EMA with Israel[22] and the interim EMA with the Palestinian Authority[23] is compared. The main objective of the latter is to help acquire energy technologies and infrastructure (Article 48 PS EMA) as opposed to the former which includes the only energy article of all EMAs and PCAs worded as if two equal parties had drafted the text. It focuses on cooperation in development of renewable energy sources, reduction of pollution of the environment, and promotion of energy conservation (Article 51 IL EMA).

The other association agreements of relevance are the ones with Algeria,[24] Egypt,[25] Jordan,[26] Lebanon,[27] Morocco,[28] Syria[29] and Tunisia.[30] The EMAs with Egypt, Jordan, Lebanon, Morocco and Tunisia share a common core provision, namely that the EU and the country in question will cooperate in promoting renewable energy and saving of energy as well as in applied research relating to networks of databases linking the two parties' economic and social operators. The agreements also foresee cooperation in modernization and the development of energy networks and interconnections with the EU's networks.[31] As is the case with the PCAs, the level of differentiation between the energy provisions in the EMAs with those five countries is again rather low. Only the EMA with Algeria and the one with Syria (which has not yet entered into force) show a higher degree of adaptation to the countries' particularities in the energy sector.

Article 61 DZ EMA defines three aims of cooperation between the EU and Algeria in the energy and mining sector followed by a lengthy list of priority areas. They do, of course, include the ones mentioned previously and found in the other EMAs, but add several more. According to this article, cooperation in the field of energy between the EU and Algeria focuses on institutional, legislative and regulatory as well as technical and technological upgrading with the intention to develop partnerships and encourage private investment. In this respect, it may not be forgotten that Algeria ranks third behind Russia and Norway as exporter of natural gas to the European Union.[32] The priority areas of cooperation with Syria – as a (potentially) important transit country – are developing interconnections, improving the safety and security of infrastructure, technologically cooperating in the petroleum sector and opening of energy markets with an effective regulatory framework (Article 106 SY EMA). It is important to recall, however, that considering the current political situation the entry into force of the SY EMA has become very unlikely.[33]

Hence, looking at the majority of those EMAs one could get the impression that cooperation in the field of energy is only of secondary interest to the parties. The APs, however, tell another story.

Energy in the action plans

In those genuine ENP instruments energy plays a more prominent role than in the various bilateral agreements and there is far more differentiation between the priorities established with each ENP country. This comes as no surprise, because APs are to "be based on common principles but be differentiated, as appropriate, taking into account the specificities of each neighbour, its national reform processes and its relations with the EU".[34] On one hand, they seek to ensure fulfilment of the bilateral agreements, be they PCAs or EMAs, and on the other, they provide guidance on what exactly the commitments are and how they should be achieved.[35] Regarding content, the initial notion of what an AP should be and entail seems to have shifted from focusing on promoting and sharing common values such as democracy, rule of law or human rights[36] to an ever-increasing extent to cooperating in more "tangible" issue such as energy. In all the APs analyzed, some issue in relation to energy is always listed as one of the priority areas for action.

We concentrate on those energy aspects common to all or a majority of APs.[37] Indeed, one item is found in two-thirds of the relevant APs, namely (gradual) "legal and regulatory convergence towards the principles of the EU internal electricity and gas markets".[38] Most countries are to "establish a list of measures" in this respect; others should only explore possibilities for such a venture. In almost all cases, this task is connected to the development of time schedules and/or financing plans. The APs without this clause either refer to another document aiming at the gradual integration of a regional market into the EU's Internal Energy Market[39] or simply do not touch on this issue at all, such as the PS AP. It is important to note that this issue of legal approximation is none of the APs' "priorities for action" but rather a medium-term goal as so defined, for example in the JO AP.

A second central concern of APs is regional cooperation in the field of energy.[40] Its scope varies of course from country to country, but in general, three regions can be distinguished, namely Maghreb, Mashreq and the Southern Caucasus. The first comprises the understanding of the EU's external energy policies Algeria, Morocco and Tunisia; the second covers the territories of Egypt, Lebanon, Jordan, Syria and Turkey; and the third encompasses Armenia, Azerbaijan and Georgia. Most of the EU's initiatives in the field of energy follow these regional borders. It goes without saying that cooperation with the EU as a region is also an important element in each of the APs.

Conclusion

Analyzing the APs, energy has undoubtedly become one of the priorities of the EU's relations with the ENP countries; something not to be expected with the provisions on energy in the EMAs in mind. Cooperation between the EU and most ENP countries goes far beyond what has been agreed in the relevant bilateral agreements.

Having another closer look at the APs, slight differences between the APs of the Southern Caucasus countries and the other ENP partners can be discerned. According to the first paragraph of the 'priorities for action' section of the three APs with the Southern Caucasus countries, "[t]he Action Plan sets out a comprehensive set of priorities in areas within *and beyond* the scope of the Partnership and Cooperation Agreement" (emphasis added).[41] None of the other APs include this phrase and declare that their areas of cooperation remain within the scope of the EMAs. It is highly doubtful that the – if only political – obligation to continue approximation of laws and policies is covered by the EMAs, which still are the only legally binding documents regulating the bilateral relationships between the ENP countries and the EU. The initial idea of the APs was to set concrete targets for each of the countries to make the achievement of the bilateral agreements' objectives easier. In the field of energy, however, one could surmise that the objectives of the EMAs have already been achieved and the APs seem to introduce the beginning of a new era of cooperation, which is also marked by the objective to "advance the approximation of [. . .] legislation, norms and standards to those of the European Union"[42] in specific areas such as energy.

This unexpected autonomy of the APs does not give rise to any additional legally binding obligations to be fulfilled by the ENP countries. If the APs' priorities are observed by the ENP countries, this will probably lead to complaisant country reports and more ambitious priorities in subsequent APs. Sanctions for non-compliance are only stipulated in the EMAs, PCAs and the ENPI regulation.[43] However, these sanctions can only be applied in cases of violations of essential elements or fundamental principles defined in the respective documents; and energy is – quite naturally and as already discussed – not one of them. Thus, on one hand, the priorities of the APs are a specification of the bilateral agreements' content, but on the other, they also include offers to go further and to increase cooperation in areas not covered by the underlying bilateral agreements.[44] An important aspect to remember is that the APs are adopted by the respective association or cooperation councils which are made up of representatives of both parties to an agreement.

As an interim conclusion, it is essential to realize that political conditions are of great importance in all forms of conditionality wielded by the EU in its external relations. However, close cooperation in the field of energy does not presuppose compliance with the values outlined in Article 1 ENPI Regulation or in any of the bilateral agreements. Thus, analyzing the issue of energy in the ENP context suggests that the EU does not follow its own rules on conditionality in cases in which vital interests of the EU are involved. Another explanation could be that the APs are not taken too seriously by the ENP countries regarding the priority of those priorities; this result, however, is highly doubtful because the APs are the only ENP documents bearing the features of joint ownership. As just outlined, the stake in the EU's Internal Energy Market within the ENP framework does not seem to be conditional on the fulfilment of any conditions but the ability to comply with EU energy legislation; the same is true for the EnC. Membership in the EnC is not connected to any form of political conditionality, at least not concerning issues other than energy. Recalling the enlargement of the EnC to Moldova and Ukraine underlines this argument: the only condition that needed to be fulfilled by both candidates was the enactment of gas laws, which were in compliance with Directive 2003/55/EC. The actual implementation of these laws is, of course, a remarkable achievement for both countries, but at the same time, it is a truism that the enactment of a law does not necessarily effectuate the underlying principles if not implemented properly and by capable authorities.

The Energy Community as a model for the ENP?

What Prodi had in mind when talking about "the possibility of developing new structures [. . . in the form of] innovative concepts such as institutions co-owned by the partners"[45] was probably not the EnC, but it should be stressed that the notion of the ENP also included "a stake in the EU's Internal Market and further integration and liberalization".[46] The EnC combines these two ideas, and from what has been established, the current legal relations with the ENP countries in question do not conflict with membership in the EnC. In this vein, the European Commission announced a feasibility study on the "potential benefits of putting in place a common legal framework for the neighbourhood in the field of energy"[47] already in 2006. It also stressed the need for a coherent external energy policy of the EU which also included the notion of a pan-European energy community "building on the Energy Community Treaty with partners in South-East Europe, as well as the development of the EU-Maghreb electricity market and the EU-Mashrek gas market [. . .]".[48]

This approach was supported by the council[49] as well as by the European Parliament.[50] Two years later, the Euro-Mediterranean Parliamentary Assembly, one of the institutions established within the Euro-Mediterranean Partnership (EMP) framework, supported the idea of the establishment of a Euro-Mediterranean energy community modelled on the EnC concept.[51] This idea was also taken over by the European Parliament in its resolution on the commission's Second Strategic Energy Review.[52] The commission explicitly expected that the countries of the Black Sea and the Caspian Basin would apply for membership in the EnC, adding that this would "boost new supplies from central Asia to the EU".[53] Thus, the EnC has already been used several times as an example or as an instrument for future developments in the EU's external energy policy in recent years. These declarations serve as a starting point for the following ideas.

It is important to recall that no legal obstacles, neither within the ENP nor within the EnC framework, would make the enlargement of the EnC to any of the ENP countries impossible. The only limitation according to the TEnC is that only neighbouring third countries may apply for observer status (Article 96 (1) TEnC). It is necessary to apply a broad interpretation to the term *neighbouring third country* in this context including at least Azerbaijan and Armenia. Furthermore, there is no indication whatsoever in the TEnC that every observer cannot become a contracting party at some point, provided it complies with the accession criteria, which are not determined in the TEnC but seem to be only related to energy. Thus, the ENP's principle of equal opportunities is in fact not contradicted by EnC law because every ENP country can apply for EnC membership if it so desires. An affirmation of this finding is even found in the TEnC itself: recital 5 refers to the EMP and the ENP.

Having reaffirmed that concerning the current framework there are no legal objections to an extension of the EnC to the ENP countries, more political considerations shall follow. The Western Balkans' participation in organizations of enhanced (sectoral) multilateralism constitutes a preparation for EU membership and at the same time a foretaste of what membership is going to be like in terms of benefits and obligations. In the case of the ENP, however, the reasons for the participation in a form of enhanced (sectoral) multilateralism such as the EnC will most probably be of a different nature. The distinction between the Western Balkans' contracting parties and the ENP countries is the clear membership perspective of the first and the rather blurry or non-existent possibility to join the EU for the latter. Especially for the second but also for the first group, EnC membership could constitute an alternative to EU membership. Instead of taking over the entire body of the EU's *acquis*, they could confine themselves to integration with the EU in selected sectors only, for example such as energy.

However, none of the ENP countries without any EU membership perspective has expressed any particular keenness to participate in the EnC. Accession to the EnC for most of them would stand for considerable effort in terms of energy market reform. At the same time, the benefits that would come with EnC membership are, compared to those for the Western Balkans, rather limited. Their ability to take over the relevant pieces of the *acquis* is not going to have a positive impact on their EU accession negotiations and foreign investments in the energy sector will not necessarily increase. Those countries which are of particular significance for the EU's Internal Energy Market are going to benefit from substantial investments in their energy sectors either way, no matter what kind of rules are applied in those countries. Furthermore, compared to the bi- and multilateral initiatives, the EnC is characterized by a relatively high degree of transparency; considering the very limited information available on the energy issues on some of those countries, it is very unlikely that they would highly value increased transparency.

Prange-Gstöhl identified three motives for countries with no or only a vague membership perspective to agree to deeper energy integration with the EU. He calls them "identification motive", "independence motive", and "economic motive". The first is the driver for countries which are eager to demonstrate their capability to assume the obligations of EU membership even if without a (current) membership perspective. The second motive applies to countries seeking greater independence from a regional supreme power in their neighbourhood; Russia's predominance in the former Soviet Republics may not be underestimated.[54] The third, and, for the EnC's case, most important motive focuses on increased economic exchange with the EU if common norms, rules and standards are introduced, which is expected to lead to significant economic gains in a second step.[55] The pursuit of any of those aims by seeking membership in the EnC is, of course, even more appealing because of and facilitated by the lack of political conditionality within the EnC framework.

However, these theoretical remarks need to be subjected to a reality check. It need not be forgotten that despite the ENP drawing a clear dividing line between the (potential) EU candidates and the countries to the east of the EU, the latter have the same "right" to apply for membership in the EU pursuant to Article 49 TEU-L as the countries of the Western Balkans. The integration of the energy markets of the "Eastern ENP countries" with the Internal Energy Market would indeed close the gap between the Western Balkans and the EU's Eastern neighbours at least for one sector. Despite bearing the risk for them of remaining in this sectoral "near membership" forever, the accessions of Moldova and Ukraine showed that deeper integration in the field of energy is – for whatever motive – wanted and countries are willing to abide by the rules defined even if they remain rule takers. They countries seem to consider EnC membership – just as in the case with the SAP countries – as preparation for EU membership. What has to be kept in mind, however, is that the SAP countries from the very beginning of their rapprochement to the EU were considered as (potential) candidate countries and this is clearly not the case for Ukraine and Moldova. For good reasons they were put into the group of ENP countries, a characterization which could make them remain members of organizations of enhanced sectoral multilateralism for good. An option, which – in the light of what was said earlier – could also lead to the achievement of the described aims and definitely a viable alternative for the EU's enlargement policy. Furthermore, the establishment of forms of enhanced sectoral multilateralism such as the EnC could give the EU the opportunity to tidy up and consolidate the manifold legally non-binding bi- and multilateral (energy) initiatives with neighbouring countries. Additionally, including more and more countries into a legally binding framework also fosters legal security and certainty. It would also make it easier for the ENP to achieve the objectives set and would thus add weight to the entire policy.

Still, the emergence of second-class members is intrinsic to the notion of enhanced sectoral multilateralism; in the EnC is a group of countries making the rules and another group which has no decision-making power but the same obligation to implement and abide by these rules. This might not be considered as a problem on its face; nevertheless, no country appreciates being worse off than another group of countries. Thus, the "rule takers" might well at some point discontinue implementing new legislation. This issue already leads to the next problem – homogeneity or rather lack thereof.

The fragmentation of the EU's Internal Energy Market constitutes one of the most important issues to deliberate on before considering further enlargements of the EnC and applying the EnC model in other sectors or regions. The situation on the Western Balkans' energy markets differs considerably from the ones in the EU member states. It is undisputed that the intended level of liberalization has not yet been achieved at EU level either, although most member states had several decades to comply with the EU's idea of an Internal Energy Market. The energy markets, the national legal orders, and the overall situation in most of the ENP countries, however, are again manifestly dissimilar to those in the contracting parties. Including yet another couple of national "interpretations" of the EU's Internal Energy Market legislation into the EnC's framework, which does not provide for any mechanism to ensure homogeneity regarding the application of EU legislation, appears to be "risky business". The commission, however, seems to aim at exactly that. It does not seek to export all standards and regulatory frameworks of the EU's energy legislation to the third countries, but aspires to "have substantively similar levels of market access and adoption of equivalent standards".[56] If that is indeed the primary goal, enlargement of the EnC is not the right instrument for its achievement; for this venture, it seems to be much more apt to use bilateral initiatives or to incorporate this goal into the APs. A well-designed form of enhanced sectoral multilateralism and all the necessary efforts that come with it are – as shown by the EEA – only worthwhile if the underlying intention is the establishment of a true single regulatory space.

The issues mentioned here can very briefly be only food for thought for further discussions on how the EnC's concept of enhanced sectoral multilateralism can influence the EU's external energy relations. In summary, it can, however, be concluded that there are no legal obstacles to combine the ENP with forms of enhanced sectoral multilateralism similar to the EnC. However, both the EU and the prospective members should consider such a commitment carefully.

Notes

1. The findings, interpretations and conclusions expressed in this chapter are those of the author and do not reflect the positions of the institution.
2. Adam Łazowski, 'Enhanced multilateralism and enhanced bilateralism: Integration without membership in the European Union' (2008) 45 *CMLRev* 1433, 1436.
3. Directive 2003/54/EC concerning common rules for the internal market in electricity and Directive 2003/55/EC concerning common rules for the internal market in natural gas [2003] OJ L 176/57.
4. Article 1 Decision 2008/03/MC-EnC of 1 December 2008.
5. See www.energy-community.org/portal/page/portal/ENC_HOME/ENERGY_COMMUNITY/Legal/EU_Legislation.
6. European Commission, Proposal for a Council Decision on the signing by the European Community of the Energy Community Treaty and Proposal for a Council Decision on the conclusion by the European Community of the Energy Community Treaty, COM(2005) 435, 3.
7. Energy Community Secretariat, *Annual Report on the Implementation of the Acquis under the Treaty Establishing the Energy Community* (2012).
8. Energy Community Secretariat, *loc cit*, fn 6, 1.
9. Michael Pollitt, "Evaluating the evidence of electricity reform: Lessons for the South East Europe (SEE) market" (2009) 17 *Utilities Policy* 13, 14.
10. Andris Piebalgs, "External projection of the EU internal energy market", SPEECH/06/712, 20 November 2006.
11. Christophe Hillion, 'The EU's Neighbourhood Policy towards Eastern Europe' in Alan Dashwood and Marc Maresceau (eds), *Law and Practice of EU External Relations* (CUP 2008) 309, 309.
12. European Commission, *Wider Europe – Neighbourhood: A New Framework for Relations with our Eastern and Southern Neighbours*, COM (2003) 104, 4.
13. European Commission, *Report from the Commission on the stabilisation and association process for South East Europe Third Annual Report*, COM (2004) 202, 5.
14. Christophe Hillion and Alan Mayhew, 'The Eastern Partnership – something new or window-dressing' (2009) Sussex European Institute Working Paper 109, 20.
15. European Commission, *Wider Europe* COM (2003) 104, 8.
16. "The Eastern Partnership (2009) Sussex European Institute Working Paper 109, 19.
17. European Commission, *A Strong European Neighbourhood Policy*, COM (2007) 774, 777.
18. European Commission, *On Strengthening the European Neighbourhood Policy*, COM (2006) 726, 728.
19. PCA between the European Communities and their Member States, of the one part, and the Republic of Armenia, of the other part [1999] OJ L 239/3 (AM PCA).
20. PCA between the European Communities and their Member States, of the one part, and the Republic of Azerbaijan, of the other part [1999] OJ L 246/3 (AZ PCA).
21. PCA between the European Communities and their Member States, of the one part, and Georgia, of the other part [1999] OJ L 205/3 (GE PCA).
22. EMA establishing an association between the European Communities and their Member States, of the one part, and the State of Israel, of the other part [2000] OJ L 147/3 (IL EMA).

23. Euro-Mediterranean Interim Association Agreement on trade and cooperation between the European Community, of the one part, and the Palestine Liberation Organization (PLO) for the benefit of the Palestinian Authority of the West Bank and the Gaza Strip, of the other part [1997] OJ L 187/3 (PS EMA).
24. EMA establishing an Association between the European Community and its Member States, of the one part, and the People's Democratic Republic of Algeria, of the other part [2005] OJ L 265/2 (DZ EMA).
25. EMA establishing an Association between the European Communities and their Member States, of the one part, and the Arab Republic of Egypt, of the other part [2004] OJ L 304/39 (EG EMA).
26. EMA establishing an Association between the European Communities and their Member States, of the one part, and the Hashemite Kingdom of Jordan, of the other part [2002] OJ L 129/3 (JO EMA).
27. EMA establishing an Association between the European Community and its Member States, of the one part, and the Republic of Lebanon, of the other part [2006] OJ L 143/2 (LB EMA).
28. EMA establishing an association between the European Communities and their Member States, of the one part, and the Kingdom of Morocco, of the other part [2000] OJ L 70/2 (MA EMA).
29. Euro-Mediterranean Agreement establishing an association between the European Community and its Member States, on the one part, and the Syrian Arab Republic, on the other part (SY EMA). For the text, see, for example, European Commission, Proposal for a Council Decision on the conclusion of a Euro-Mediterranean Association Agreement between the European Community and its Member States of the one part, and the Syrian Arab Republic, of the other part COM(2004) 808, 10.
30. EMA establishing an association between the European Communities and their Member States, of the one part, and the Republic of Tunisia, of the other part [1998] OJ L 97/2 (TN EMA).
31. This core provision is found in different versions depending on the time of the conclusion of the EMAs in Article 53 EG EMA, Article 74 JO EMA, Article 54 LB EMA, Article 57 MA EMA, and Article 57 TN EMA. To a certain extent, Article 48 PS EMA also shares this common core.
32. Eurostat, *Energy, transport and environment indicators – 2012 edition* (2013), 42.
33. The EU decided to suspend all preparations in relation to new bilateral cooperation programs and the on-going bilateral programs under ENPI and MEDA instruments (Council Conclusions on Syria, 3091st Foreign Affairs Council meeting, Brussels, 23 May 2011).
34. GAER Council of 16 June 2003 and of 14 June 2004.
35. Christophe Hillion, 'The EU's Neighbourhood Policy towards Eastern Europe' in Alan Dashwood and Marc Maresceau (eds), *Law and Practice of EU External Relations* (CUP 2008) 309, 319–321.
36. European Commission, Communication from the Commission to the Council on the Commission Proposals for Action Plans under the European Neighbourhood Policy, COM(2004) 795, 3.
37. Instead of listing the title of each AP, suffice it to provide their source at eeas.europa.eu/enp/documents/action-plans/index_en.htm. Abbreviations in the text will follow the following form of, for example AM AP for the Action Plan of Armenia.
38. Item 4.6.2 AM AP, item 4.6.2 AZ AP, item 2.3. b) EG AP, item 4.6.2 GE AP, item 2.5.6. IL AP, item 2.5. (53) JO AP, item 2.5. (b) LB AP, item 2.5. (67) MA AP and item 2.6 (63) MD AP.

39. Item 2.5. (67) MA AP and item 2.5. (61) TN AP refers to the memorandum of understanding on the gradual integration of the electricity markets of the Maghreb countries into the EU's internal electricity market.
40. Items 4.1.3 and 4.6.2 and priority area 8 AM AP, priority area 8 and item 4.6.2 AZ AP, item 2.3. b) EG AP, priority areas 5 and 8 GE AP, item 2.5.9. IL AP, item 2.3. (29) JO AP, item 2.5. (b) LB AP, items 2.1 (12) MA AP, item 2.4 (20) PS AP and items 2.1. (10) and 2.5. (61) TN AP.
41. The AP with Israel includes a similarly worded clause, but the EU's relations with Israel are already much more advanced compared with the other littoral countries of the southern Mediterranean basin.
42. Para 8 Introduction EG AP, para 6 Introduction JO AP, para 8 Introduction LB AP, para 5 Introduction MD AP, para 7 Introduction MA AP, para 8 Introduction PA AP, para 7 Introduction TN AP.
43. Regulation (EC) No. 1638/2006 of the European Parliament and of the Council of 24 October 2006 laying down general provisions establishing a European Neighbourhood and Partnership Instrument [2006] OJ L 310/1.
44. Also see, for example, Recital (5) Recommendation on the implementation of the EU-Georgia Action Plan attached to Council Decision on the position to be adopted by the European Communities and their Member States within the Cooperation Council established by the Partnership and Cooperation Agreement between the European Communities and their Member States, of the one part, and Georgia, of the other part, with regard to the adoption of a Recommendation on the implementation of the EU-Georgia Action Plan, 14406/06, 8 November 2006.
45. Romano Prodi, 'The wider Europe – a proximity policy as the key to stability', SPEECH/02/619, 6 December 2006.
46. European Commission, *Wider Europe*, COM (2003) 104, 4.
47. European Commission, *A Strong European Neighbourhood Policy*, COM (2007) 774, 7.
48. European Commission, *Green Paper: A European Strategy for Sustainable, Competitive and Secure Energy*, COM(2006) 105, 16.
49. Council of the European Union, 'Transport, Telecommunications and Energy (Press Release)', 7009/06, 14 March 2006.
50. See items 60 and 74 European Parliament resolution on a European strategy for sustainable, competitive and secure energy – Green paper, P6_TA(2006)0603.
51. Euro-Mediterranean Parliamentary Assembly, 'Final Declaration of the Presidency' (Athens, 28 March 2008), 7.
52. Item 54 European Parliament, 'European Parliament resolution of 3 February 2009 on the Second Strategic Energy Review', P6_TA(2009)0038, 3 February 2009.
53. European Commission, *Communication from the Commission to the European Council: External energy relations – from principles to action*, COM(2006) 590, 5.
54. Gerhard Mangott and Kirsten Westphal, 'The Relevance of the Wider Black Sea Region to EU and Russian Energy Issues' in Daniel Hamilton and Gerhard Mangott (eds.), *The Wider Black Sea Region in the 21st Century: Strategic, Economic and Energy Perspectives* (Center for Transatlantic Relations, 2008) 147, 157.
55. Heiko Prange-Gstöhl, 'Enlarging the EU's internal energy market: Why would third countries accept EU rule export?' (2009) 39 *Energy Policy* 5296, 5297.
56. European Commission, *Communication from the Commission to the Council and the European Parliament on the Development of Energy Policy for the Enlarged European Union, its Neighbours and Partner Countries*, COM (2003) 262, 5.

References

Council of the European Union, Transport, Telecommunications and Energy (Press Release), 7009/06, 14 March 2006.

Energy Community Secretariat (2012), *Annual Report on the Implementation of the Acquis under the Treaty Establishing the Energy Community.*

Euro-Mediterranean Agreement establishing an association between the European Communities and their Member States, of the one part, and the State of Israel, of the other part [2000] OJ L 147/3.

Euro-Mediterranean Agreement establishing an Association between the European Community and its Member States, of the one part, and the People's Democratic Republic of Algeria, of the other part [2005] OJ L 265/2.

Euro-Mediterranean Agreement establishing an Association between the European Communities and their Member States, of the one part, and the Arab Republic of Egypt, of the other part [2004] OJ L 304/39.

Euro-Mediterranean Agreement establishing an Association between the European Communities and their Member States, of the one part, and the Hashemite Kingdom of Jordan, of the other part [2002] OJ L 129/3.

Euro-Mediterranean Agreement establishing an Association between the European Community and its Member States, of the one part, and the Republic of Lebanon, of the other part [2006] OJ L 143/2.

Euro-Mediterranean Agreement establishing an association between the European Communities and their Member States, of the one part, and the Kingdom of Morocco, of the other part [2000] OJ L 70/2.

Euro-Mediterranean Agreement establishing an association between the European Community and its Member States, on the one part, and the Syrian Arab Republic, on the other part [not yet signed].

Euro-Mediterranean Agreement establishing an association between the European Communities and their Member States, of the one part, and the Republic of Tunisia, of the other part [1998] OJ L 97/2.

Euro-Mediterranean Interim Association Agreement on trade and cooperation between the European Community, of the one part, and the Palestine Liberation Organization (PLO) for the benefit of the Palestinian Authority of the West Bank and the Gaza Strip, of the other part [1997] OJ L 187/3.

Euro-Mediterranean Parliamentary Assembly, Final Declaration of the Presidency, Athens, 28 March 2008).

European Commission (n.d.), Syrian Arab Republic Country Strategy Paper 2007–2013 and National Indicative Programme 2007–2010.

European Commission, *A Strong European Neighbourhood Policy*, COM(2007) 774.

European Commission, *Communication from the Commission to the Council and the European Parliament on the Development of Energy Policy for the Enlarged European Union, its Neighbours and Partner Countries*, COM(2003) 262.

European Commission, *Communication from the Commission to the Council on the Commission Proposals for Action Plans under the European Neighbourhood Policy*, COM(2004) 795.

European Commission, *Communication from the Commission to the European Council: External energy relations – from principles to action*, COM(2006) 590.

European Commission, *Green Paper: A European Strategy for Sustainable, Competitive and Secure Energy*, COM(2006) 105.
European Commission, *On Strengthening the European Neighbourhood Policy*, COM(2006) 726.
European Commission, *Proposal for a Council Decision on the signing by the European Community of the Energy Community Treaty and Proposal for a Council Decision on the conclusion by the European Community of the Energy Community Treaty*, COM(2005) 435.
European Commission, *Report from the Commission The Stabilisation and Association process for South East Europe Third Annual Report*, COM(2004) 202.
European Commission, *Wider Europe – Neighbourhood: A New Framework for Relations with our Eastern and Southern Neighbours*, COM(2003) 104.
European Parliament, Resolution of 3 February 2009 on the Second Strategic Energy Review, P6_TA(2009)0038.
European Parliament, Resolution on a European strategy for sustainable, competitive and secure energy - Green paper, P6_TA(2006)0603.
Eurostat (2013), *Energy, transport and environment indicators – 2012 edition.*
Hillion, C. (2008) "The EU's Neighbourhood Policy towards Eastern Europe", in A. Dashwood and M. Maresceau (eds.), *Law and Practice of EU External Relations.* Cambridge: CUP, pp. 309–333.
Hillion, C. and Mayhew, A. (2009) *The Eastern Partnership – something new or window-dressing.* Sussex European Institute Working Paper 109, Sussex, UK.
Łazowski, A. (2008) "Enhanced multilateralism and enhanced bilateralism: Integration without membership in the European Union", *CMLRev* 45: 1433–1458.
Mangott, G. and Westphal, K. (2008) "The Relevance of the Wider Black Sea Region to EU and Russian Energy Issues", in D. Hamilton and G. Mangott (eds.), *The Wider Black Sea Region in the 21st Century: Strategic, Economic and Energy Perspectives.* Washington: Center for Transatlantic Relations, pp. 147–176.
Partnership and Cooperation Agreement between the European Communities and their Member States, of the one part, and the Republic of Armenia, of the other part [1999] OJ L 239/3.
Partnership and Cooperation Agreement between the European Communities and their Member States, of the one part, and the Republic of Azerbaijan, of the other part [1999] OJ L 246/3.
Partnership and Cooperation Agreement between the European Communities and their Member States, of the one part, and Georgia, of the other part [1999] OJ L 205/3.
Piebalgs, A. (2006) 'External projection of the EU internal energy market' SPEECH/06/712.
Pollitt, M. (2009) 'Evaluating the evidence of electricity reform: Lessons for the South East Europe (SEE) market', *Utilities Policy* 17: 13–23.
Prange-Gstöhl, H. (2009) 'Enlarging the EU's internal energy market: Why would third countries accept EU rule export?', *Energy Policy* 37(12): 5296–5303.
Prodi, R. (2006) 'The Wider Europe – A Proximity Policy as the Key to Stability' SPEECH/02/619.

Chapter 10

Beyond regional cooperation

The Energy Community as an instrument of the EU external energy policy?

Natasha Hroneska[*]

In September 2011, the European Commission published a communication on security of energy supply and international cooperation – "The EU Energy Policy: Engaging with Partners beyond Our Borders".[1] The communication was welcomed as the first consolidated attempt to enhance the European Union (EU) external energy policy. In it, the European Commission highlighted the necessity for integration of the neighbouring countries in the EU's internal energy market, referring primarily, to the Energy Community of South-East Europe (hereafter, Energy Community), as the best model for the countries of the EU's immediate neighbourhood willing to implement the relevant EU legislation and to become part of the European energy system.[2]

The Energy Community was established in 2005, when the Energy Community Treaty (ECT), was signed between the EU and nine countries Contracting Parties of the Western Balkans (WB).[3] The ECT has obliged the contracting parties to transpose the relevant EU *acquis communautaire* to their energy sectors, in order to create a stable regulatory and market framework, a single regulatory space for trade in energy, and to enhance the security of supply. In one word, the Energy Community was modelled as an extension of the EU energy policy to the countries of the Western Balkans.

In 2011, six years after its establishment, the Energy Community was assessed as a success story for the EU and a positive model for regional cooperation on energy matters.[4] Hence, the European Commission has called for deepening and extending the validity of the Energy Community beyond 2016, expanding it thematically and geographically to other countries of the EU's neighbourhood, committed to accept the principles of liberal energy markets and to implement the relevant EU energy legislation.[5] This process has already started with the first enlargement of the Energy Community to Moldova and Ukraine in 2010 and 2011, respectively, which was the first step towards "de-Balkanization" of the Energy Community and export of the EU energy norms to other EU neighbours.[6]

Taking the previously mentioned as a starting point, this chapter analyzes to what extent the Energy Community is a viable model for regulating the energy relations of the EU with the countries of its Eastern and Southern neighbourhood. The working hypothesis is that the Energy Community has grown beyond the context of regional cooperation, into a model of the EU's energy relations with neighbouring countries and an instrument of the EU external energy policy. The theoretical background of the analysis is the concept of "normative power Europe". Through the Energy Community, the EU's energy norms have been promoted to third countries, thus reinforcing the EU's role of "normative power" in energy.

In the instalment, the perspectives and the limitations to the replication of the Energy Community in the EU's neighbourhood are discussed. The focus of the analysis is on two important aspects, the deepening of the integration of the Energy Community, and the diverse nature of the relations between the EU and the countries of its Eastern and Southern neighbourhood. The findings demonstrate that the level of cooperation and integration of the EU neighbouring countries with the EU is one of the main determinants for the export of the EU energy norms through the extension of the Energy Community.

The Energy Community: Model of EU "normative power"

Energy policy and "normative power Europe"

The attempts to develop the European external energy policy have opened a room for various theoretical considerations. Traditionally, the intergovernmentalists and the neo-functionalist have had different explanation of the process of development of the EU external energy policy. The former argue that the EU cannot have a common external energy policy because the EU's external energy relations are only a sum of the foreign policies of its Member States in the energy area.[7] On the other hand, neo-functionalists believe in supranationalism in the energy policy. According to this approach, the European Commission plays a crucial role, being a driver for development of external energy policy.[8] In this chapter, an alternative approach, developed in recent years, is adopted – the "normative power Europe" concept. Contrary to the intergovernmentalists and the neo-functionalists, this theoretical approach explains the EU's "actorness" by conferring the main importance to the ideas and norms.[9] It explains how, through the export of energy norms, regulations and institutions to third countries, the EU is developing its external energy policy.

Applying the concept of "normative power Europe" to the energy area is a challenging task. For a long time, the energy policy was an example of the limited character of the EU in the exertion of its normative power and the political will to do so.[10] Energy was considered as a particular policy area where the role of the EU as a global actor is hindered by the "capability–expectation" gap.[11] However, the energy security issues and the EU's quest for stronger 'actorness' have accelerated the process of consolidation of the European energy policy, as well as the process of developing its external dimension.

Deriving from the original idea of Ian Manners of "Europe as a normative power", the "normative power" concept argues that the global role of the EU extends well beyond the traditional conceptions of the Union as a civilian-military power dichotomy.[12] Manners argues that the EU's role in the international relations can be best conceived through the external promotion of internal norms and values, embedded in the main EU documents and treaties.[13] With the norm reinforcement and norm diffusion, the EU can better present and legitimate itself in the global system. The promotion of the norms in international relations, according to Manners, is taking place through six norm diffusion channels: contagion, informal diffusion, procedural diffusion, transference, overt diffusion and cultural filter.[14]

Throughout the last two decades, norms regarding security of energy supply, energy market liberalization and sustainable development have been established within the EU Treaties, as well as in different strategic documents – directives, regulations, decisions and communications.[15] Moreover, the three consecutive energy packages for liberalization of the energy market have also paved the way towards "Europeanization" of the energy policy.[16] The Lisbon Treaty, for the first time, has set a legal basis for development of the EU energy policy, duly reconfirming the importance of a functional EU internal energy market, and opening the way towards a proactive external energy policy.[17]

Today, the EU "normative power" in the energy policy is exported through different policy instruments regulating the EU's energy relations with third countries. The most visible and most successful instrument is the Energy Community.

Energy Community and "norm diffusion"

The Energy Community was created to facilitate the transposition of EU's energy norms and values to the countries of the Western Balkans. The original goal of the Energy Community was to reform the energy sectors of the WB countries, in order to gradually prepare them for EU membership, as well as to re-integrate the WB region in the EU internal energy market.[18]

Identified as a process of procedural norm diffusion, the Energy Community has established a highly institutionalized relationship and structure between the EU and the countries of the WB.[19] Hence, developing a holistic approach, the Energy Community has built an innovative institutional setting, which includes a ministerial council, a permanent high level group, an advisory for an Energy Community Regulatory Board and a secretariat, with the role of supporting and of monitoring the ECT implementation, as well as the dispute-settlement mechanism for treaty's enforcement.[20]

The legal structure of the Energy Community is an overarching and progressively increasing one. Through the stable regulatory framework, the EU norms of security of supply, energy efficiency, sustainable development, market competition and transparency are diffused to the contracting parties.[21]

Binding the countries to transpose the *aquis communautaire* in these areas, the EU has been "externalizing" its energy norms outside its borders. The contracting parties, on the other side, have voluntarily accepted to implement the EU energy *aquis*, by a pre-defined deadline, following country specific assessment road maps.

Regarding electricity, the WB countries have agreed to transpose the EU directives on the common rules of the internal electricity market, the security of electricity supply and the cross-border exchange of electricity.[22] The main goal is to create a single regional electricity market, through further legal harmonization and application of technical standards for interconnectivity and cross-border trade. This would allow for better integration of the contracting parties among each other and with the EU internal energy market.

Similar conditions apply to the natural gas market, where the contracting parties agreed to transpose the provisions regulating the internal European gas market, the conditions for access to natural gas transmission networks, and provisions for security of gas supply.[23] Regarding renewables, by transposing and implementing the *aquis* on renewable energy, the contracting parties have agreed to increase the share of renewable energy sources for electricity production, according to the national indicative targets, and to promote biofuels in transport.[24] In addition, the directives and regulations for energy efficiency oblige the contracting parties to transpose and implement the EU rules in the area of end-use energy efficiency and energy services, energy performance of buildings and labelling.[25]

Hence, the Energy Community has played a momentous role for the EU and for the WB countries. Among the other, it has helped to (1) reform the legal energy framework of each of the contracting parties, which had committed to draft new energy laws and secondary legislation, transposing the EU internal energy norms; (2) establish market rules in the region – independent regulatory bodies and market access codes; and (3) generate investment opportunities and political commitment for integration of the WB in the EU internal energy market.

The Energy Community beyond 2016: what is at stake?

In 2011, six years after its establishment, the Energy Community was assessed as a "mature organization, which provides a solid institutional framework for cooperation, mutual support and exchange of experiences and therefore serves as a model for regional cooperation on energy matters".[26] Making a positive evaluation of the Energy Community, the European Commission has opted for application of the model of the Energy Community to the EU's energy relations with countries of its wider neighbourhood.[27] Last year, recognizing the positive evaluation of the work and achievements of the Energy Community, as well as the attractiveness of the model, the Ministerial Council decided to extend the validity of the ECT for another 10 years, until 2026. [28]

Exploring the likelihood that the Energy Community becomes a potential model for the EU's energy relations with its Eastern and Southern neighbours, two parameters are analyzed: the prospect of further deepening of the Energy Community and the nature of the EU's relations with the countries of its wider neighbourhood.

Deepening of the Energy Community

Deepening of the integration of the Energy Community is *sine qua non* for a stronger convergence of the Western Balkans energy markets with the EU internal energy market. At the current stage of integration the Second Energy Package has been successfully transposed in almost all contracting parties of the Energy Community and the implementation of the Third Energy Package is underway. However, many of the countries are lagging behind with the implementation of crucial energy provisions, such as timely opening of the electricity and gas markets, or full liberalization of energy prices.[29] The implementation of the *aquis* is not equally achieved in all contracting parties, and some of them still need to make fundamental changes to their primary and secondary energy legislation.[30] Moreover, the market opening is the "elephant in the room" for more integration of the Energy Community. The creation of regional energy markets has still not become a reality – electricity cannot be traded regionally, and gas continues to flow east-west. The opening of the regional electricity wholesale market still remains burdensome, whereas independent regulators and cost-reflective energy prices are still directly or indirectly interfered by the state.[31]

The deepening of the integration of the Energy Community is on track, making slow but steady progress, yet with no easy deadline to set. In the absence of appropriate sanctions deterring a contracting party from breaching the *acquis*, the full application of the EU rules is the major impediment for a more integrated Energy Community.[32] Further deepening of the integration of the Energy Community depends on the effective implementation of the EU energy *aquis* – the Third Energy Internal Market Package, effective law enforcement, the finalization of the regional energy market and the creation of a favourable and predictable investment climate.[33]

Nature of the EU's relations with its neighbours

Regarding the second parameter, two factors influence the process of extension of the EU's role as a "normative power" beyond the Western Balkans: (1) the level of integration with the EU and (2) the geopolitics of energy supply.

Level of integration with the EU

Looking at the model of the Energy Community, the buzzword is the "EU membership perspective". Linking the ECT to the EU enlargement policy, the Energy Community has been perceived as an added value of the EU enlargement to the WB.[34] Therefore, the role of the EU as a "normative power" in the countries of the WB has been facilitated by the enlargement perspective. Given that the (non-)compliance with the accepted obligations of the ECT is echoed in the regular progress reports by the European Commission, the EU membership presents an additional incentive for the WB countries to adopt and implement the EU energy norms. Offering a sound membership perspective to the WB countries, the EU has established high level of political, legal and institutional integration with the WB region:

> For Contracting Parties that have reached the status of EU candidates or potential candidates, the level of national implementation and enforcement of the relevant Energy Community *acquis* will be considered decisive in the negotiations for accession to the European Union.[35]

On the other side, the EU's energy relations with the countries of the European Neighbourhood Policy (ENP) are not that comprehensive.[36] The ENP is not an enlargement policy, but a bilateral policy of the EU for privileged partnership with neighbouring countries ready to embrace common values.[37] Hence, a certain level of norm diffusion through the EU contractual framework is present, but there are no binding obligations for transposition of the EU energy *aquis*. The EU incentivizes the ENP countries, in return for a greater progress on the relevant reforms, through enhanced market access, more integration into the European programmes and networks and increased EU financial and technical assistance.[38] The EU's "silver carrot" – everything but EU membership – offered to the ENP countries, does not attract strong political commitment for accepting and adopting EU's strict regulatory and legal energy provisions, such as in the case of the Energy Community.

In the East, the energy relations between the EU and South Caucasus countries (Armenia, Georgia and Azerbaijan) are regulated with a comprehensive cooperation framework, consisting of political and legal instruments, which, to a certain degree, allows for transference norm diffusion in the energy policy. In the ENP Action Plans with these South Caucasus countries, gradual harmonization of their energy legislation with the EU energy *aquis* is recognized and aims towards further convergence with the EU internal energy market.[39]

In the south, the EU's energy relations with the neighbouring countries have been part of a wider strategy, which aims to bring about political and economic reform in the southern Mediterranean. The main focus of the cooperation is on trade liberalization and integration of the Mediterranean region in the EU single market, while the energy norms are incorporated in the bilateral and legally binding association agreements.[40] However, as the region remains one of constant political turmoil and instability, the slow implementation of the association agreements has had negative reverberation in the energy integration of the Mediterranean region.

Geopolitics of energy supply

No less important factor is the geopolitics of energy supply and the national energy interests of the EU neighbours – energy security providers. This factor has a crucial influence in the energy relations with the EU's closer and wider neighbours in Central Asia, North Africa and the Middle East, rich with energy resources. The position of these countries regarding energy issues and energy cooperation, is usually driven by power politics (such as in Central Asia) or by the risks of instability for the political regimes (such as in North Africa and the Middle East).

Central Asia is one of the priority regions for the diversification of the EU gas supply. However, the bilateral partnerships and the Strategy for Central Asia offer a loose framework for cooperation and provide limited level of harmonization of the legal, regulatory and institutional framework.[41] In the energy sector, by providing access to its large and relatively stable energy market, as well as offering financial and technical assistance, the EU is permanently engaging itself in energy dialogue with the Central Asian countries. At the same time, the EU has been assisting the Central Asian countries to gradually open their energy markets to foreign investments, and to adopt market-based rules and regulations.[42] However, stronger EU leverage in this region is limited. In conditions of availability of easily accessible export markets in Asia, national economic interests of Central Asian countries prevail over following strict regulations in their energy sector, imposed by the EU. Furthermore, the geopolitics of energy supply in Asia is influencing to a great extent the development of the energy sectors of these countries. Rising energy markets of large consuming countries such as China and India are constantly changing the geopolitics of supply (both oil and gas) in this region.[43] New downstream projects to other countries in East and South-East Asia do not include conditionality clauses, but offer direct economic benefit. Hence, the former are of a greater interest for the Central Asian governments than is having a restrictive and legally binding energy cooperation framework with the EU.

Regarding the instability of political regimes, Syria, Egypt and Libya are the most recent cases where EU norms have lost "power". Political turmoil and armed conflicts in these countries have significantly held back or even halted the cooperation framework with the EU, including the energy cooperation.[44] Moreover, lack of democratic reforms in certain countries, such as in Belarus, has been an obstacle for more substantial EU's involvement in energy cooperation, while the bilateral relations are kept at the level of "policy of critical engagement".[45]

Prospect for extension of the Energy Community

The factors analyzed earlier explain some of the main limitations to the extension of the EU energy norms beyond the "borders" of the Energy Community. However, several particular cases relativize these factors and demonstrate the potential of the EU "normative power" in the energy relations with its neighbours.

The best example is the latest enlargement of the Energy Community to Ukraine and Moldova, countries of the ENP. When the two countries became members of the Energy Community, they had legally committed themselves to transpose the EU energy norms, without having a clear EU membership perspective.[46] One of the explanations lies in the core of the ENP – sharing common values between the EU and the ENP countries and the preparedness of the ENP countries to extend these values. The ENP is not an enlargement policy, but at the same time, by not prejudging how the relationship of the EU and its neighbours may develop in the future, it leaves an open area for cooperation.[47] Another explanation can be tracked in the Commission's Communication, which states that "the widening of the Energy Community could be considered for countries that [. . .] *demonstrate both willingness and ability to implement relevant EU legislation*" (emphasis added).[48] Moldova and Ukraine have voluntarily expressed their willingness to become members of the Energy Community and have shown considerable political commitment in transposing strict EU rules in their energy sectors, without having an EU membership perspective.[49]

Another positive example of the prospect to extend the Energy Community to other EU neighbours is the case of the two South Caucasus countries, Georgia and Armenia. Both countries have recently upgraded their status in the Energy Community. In February 2014, Georgia started negotiations with the European Commission for accession to the Energy Community, whereas Armenia has been an observer country to the Energy Community since 2011.

Turkey is a special case. The country is an observer to the Energy Community since its establishment, and ever since Turkey has not expressed willingness to join the Energy Community. On one side, the integration level between Turkey and the EU is noteworthy. Turkey signed a Customs Union with the EU in 1996 and opened membership negotiations with the EU in 2005. However, the EU membership perspective of Turkey remains uncertain, influenced by the national politics of the EU member states and by their positions towards the future enlargement of the EU. On the other side, the geopolitics of energy supply is influencing Turkey's energy policy choices. Being an important transit country and an energy hub between Asia and Europe, Turkey is championing the geographical position of a key player on the Euro-Asian "energy chessboard". This gives the Turkish government a strong political leverage in the energy relations with the EU, and less incentives to confine to the EU's normativeness in the energy policy.

Taking into consideration the previous analysis, we can conclude that the energy norm diffusion to the Eastern and Southern EU's neighbourhood has already started, with the enlargement of the Energy Community to Ukraine and Moldova. However, this is a slow and burdensome process, influenced by the geopolitical setting and the nature of the political relations between the EU and its neighbouring countries. The main challenge for the

EU is to incentivize its neighbours to transpose the relevant energy norms and to implement EU regulations, without enforcing well-developed institutional framework for energy cooperation and without offering them a fully fledged EU membership.

Conclusion

This chapter explored to what extent the Energy Community, the most successful model of diffusion of EU norms in the energy policy, is a potential model for the EU's energy relations with its southern and eastern neighbours. Taking into consideration the process of deepening of the Energy Community and the different nature of the relations between the EU and its neighbours, the analysis identified two general limitations to the replication of the Energy Community in the EU's wider neighbourhood – first, the level of integration between the EU and its neighbours and, second, the geopolitics of energy supply.

Unlike the Western Balkan countries, which are offered clear EU membership perspective and thus have developed high political, legal and institutional integration with the EU, the energy relations with the ENP countries – offered all but EU membership – remain incomprehensive. The norm diffusion in the EU's neighbourhood has been taking place through different legal and political instruments, which do not reflect the same level of integration with the EU, as in the case of the Energy Community. Moreover, (lack of) willingness and political commitment to import the EU energy norms among the ENP countries is a factor that could either speed up or postpone the burdensome process of extension of the EU energy norms to the ENP countries.

The similar applies regarding the geopolitics of energy supply, considered as an issue of highest national interest in some of the countries of the Caspian, North Africa and Middle East region. These countries, rich with energy resources, are using their energy policy as an instrument of "power politics" in their relations with third parties, as well as with the EU. For that reason, the EU's efforts for extension of its energy norms and values cannot be proportionally reverberated in the EU's energy relations with these countries.

Besides many limiting factors, the Energy Community is still the most attractive model of energy cooperation with the EU. It is a "reference point for countries willing and able to join the EU regulatory framework, and act in accordance with these pro-competitive principles",[50] and has a potential to be extended to the EU's neighbourhood. The best examples are the enlargement of the Energy Community to Ukraine and Moldova, as the first non-Balkan members of the Energy Community, as well as the forthcoming accession of Georgia, which is not physically interconnected with any of the Energy Community contracting parties – for neither electricity nor gas.

The Energy Community is a potential model of norm diffusion in the EU's neighbourhood. Yet, it should gradually deepen its integration and should strengthen the implementation process of the EU *aquis*. Moreover, for a successful extension of the Energy Community, as an instrument of the EU external energy policy, the European Union has to re-evaluate the existing policy, and to establish new clear goals and perspectives for the character and the level of the overall integration process with its neighbouring countries. The regional approach is one of the areas worth considering. The Energy Community aims to create a single regional energy market in the Western Balkans. Therefore, new initiatives for cooperation, such as the Association of the Mediterranean Energy Regulators and the Mediterranean transmission system operators (which envisage creation of a free-trade energy area) should be grasped as potential instruments for the future energy relations between the EU and its neighbours.

Notes

* The findings, interpretations and conclusions expressed in this chapter are those of the author and do not reflect the positions of the institution.

1. European Commission, *Communication from the Commission to the European Parliament, the Council, the European Economic and Social Committee and the Committee of the Regions. On Security of Energy Supply and International Cooperation – "The EU Energy Policy: Engaging with Partners beyond Our Borders"*, COM (2011) 539 final, Brussels, 07.09.2011.
2. Ibid, pp. 6–7.
3. The ECT was signed after two consecutive rounds of negotiations and their outcomes in forms of so-called Athens Memoranda on Electricity and Gas in 2003 and 2004, and went into force on 1 July 2006. The original contracting parties of the Energy Community were Albania, Bosnia and Herzegovina, Bulgaria, Croatia, Montenegro, Republic of Macedonia, Romania, Serbia and UNMIK on behalf of Kosovo. Since the accession of Croatia to the EU on 1 July 2013, its status changed from contracting party to participant.
4. European Commission, *Report from the Commission to the European Parliament and the Council under Article 7 of Decision 2006/500/EC (Energy Community Treaty)*, COM (2011) 105 final, Brussels, 10 March 2011, p. 3.
5. European Commission, *Communication from the Commission*, p. 8.
6. Energy Community website, retrieved 17.09.2013, from www.energy-community.org/portal/page/portal/ENC_HOME/ENERGY_COMMUNITY.
7. A. V. Belyi, "EU External Energy Policies: A Paradox of Integration" in Jan Orbie (ed.), *Europe's Global Role: External Policies of the European Union*, Ashgate, Farnham, 2008, p. 205.
8. See the theoretical overview in S. Renner, "The Energy Community of Southeast Europe: A Neofunctionalist Project of Regional Integration", *European Integration Online Papers*, 13, 2009, pp. 1–21.
9. I. Manners, "Normative Power Europe: A Contradiction in Terms?", *Journal of Common Market Studies*, 40(2), 2002, p. 238.
10. S. Wood, "The European Union: A Normative or Normal Power? ", *European Foreign Affairs Review*, 14(1), 2009, pp. 113–128.
11. More about the "capability–expectations" gap in the EU, see C. Hill, "The Capability-Expectations Gap, or Conceptualizing Europe's International Role", *Journal of Common Market Studies*, 31(3), 1993, pp. 305–328.
12. F. Dûchene, "Europe's Role in World Peace" in Richard Mayne (ed.), *Europe Tomorrow: Sixteen Europeans Look Ahead*, London, Fontana, 1972, pp. 32–47.

K. Smith, "Still 'Civilian Power EU'?", European Foreign Policy Unit Working Paper 2005/1, London School of Economics. H. Bull, "Civilian Power Europe: A Contradiction in Terms?" *Journal of Common Market Studies,* 21 (1), 1982, pp. 149–164.

13. I. Manners, "Normative Power Europe," p. 236.
14. Contagion is unintentional diffusion of ideas from the EU to other political actors, exporting its model in regional integration, by the virtue of its existence. Informal diffusion is a range of strategic EU communications. Procedural diffusion involves the institutionalization of a relationship between the EU and a third party. Transference diffusion takes place during trade and aid between the EU and third countries. Overt diffusion occurs as a result of the EU's physical presence in other countries, whereas the cultural filter is the interaction between the construction of knowledge and the creation of political and social identity by the subjects of norm diffusion. Ibid., pp. 244–245.
15. In H. Prange-Gstöhl, "Enlarging the EU's Internal Energy Market: Why Would Third Countries Accept EU Rule Export?", *Energy Policy*, 37(12), 2009, p. 5296.
16. A. V. Belyi, "EU External Energy Policies", pp. 203–216.
17. Title XXI, Article 194. European Union, "Consolidated Versions of the Treaty on European Union and the Treaty on the Functioning of the European Union" [The Lisbon Treaty], *Official Journal of the European Union,* C 83, 30 March 2010.
18. The Energy Community has three levels of ambition: national, regional and pan-European. In the short term, it aims at creating open and transparent national energy markets based on stable regulatory and market frameworks capable of attracting investments in power generation and networks; in the medium term, an integrated energy market should be in place across the region which allows for cross-border trade in energy, guarantees energy supply and takes into consideration climate and social aspects; in the long term, the regional market should be fully integrated in the European Union's internal energy market. European Commission, *Report from the Commission*, p. 2.
19. Prange-Gstöhl, "Enlarging the EU's Internal Energy Market", p. 5299.
20. Energy Community website, retrieved 10.09.2013, from www.energy-community.org/portal/page/portal/ENC_HOME/ENERGY_COMMUNITY/Institutions.
21. See Title II, Chapters II, III and V of the Energy Community Secretariat, *Treaty establishing the Energy Community*, 2005, pp. 6–9.
22. Energy Community website, retrieved 20.09.2013, from www.energy-community.org/portal/page/portal/ENC_HOME/ENERGY_COMMUNITY/Legal/EU_Legislation.
23. Ibid.
24. Ibid.
25. Ibid.
26. European Commission, *Report from the Commission*, p. 3.
27. Ibid, p. 11.
28. Energy Community Secretariat, *Meeting Conclusions 11th Energy Community Ministerial Council*, Belgrade, 24 October 2013, p.3. Retrieved 04.04.2014 from www.energy-community.org/pls/portal/docs/2388178.PDF.
29. European Commission, *Report from the Commission*, pp. 7–12.
30. Most pressing is that this applies to Albania and Ukraine in electricity and to Bosnia and Herzegovina in gas. Energy Community Secretariat, *Energy Community Annual Implementation Report,* September 2012, p. 10, retrieved 20.09.2012, from www.energy-community.org/pls/portal/docs/1770178.PDF.

31. Energy Community Secretariat, *Energy Community Annual Implementation Report*, September 2013, pp. 10–12. Retrieved 04.04.2014, from www.energy-community.org/pls/portal/docs/2304177.PDF.
32. Ibid.
33. Jerzy Buzek, at the Conference on the Future of the Energy Community in Brussels, 20 March 2014. *Information on the work of the High Level Reflection Group of the Energy Community*, Energy Community website, 20 March 2014. Retrieved 05.04.2014, from www.energy-community.org/pls/portal/docs/2962024.PDF.
34. The establishment of the Energy Community in 2006 was a continuation of the wider integration process in the Balkans known as the Athens Process. The aim of this process was to facilitate the reconciliation of the Western Balkans countries, after a decade of wars and bloodshed, through regional cooperation.
35. European Commission, *Report from the Commission*, p. 8.
36. The ENP framework is proposed to the 16 of EU's closest neighbours – Algeria, Armenia, Azerbaijan, Belarus, Egypt, Georgia, Israel, Jordan, Lebanon, Libya, Moldova, Morocco, Palestine, Syria, Tunisia and Ukraine. See the European External Action Service (EEAS) website, retrieved 18.09.2013, from http://eeas.europa.eu/enp/about-us/index_en.htm.
37. Ibid.
38. The ENP addresses issues regarding the political dialogue and reforms, economic and social cooperation and development, trade-related issues and market and regulatory reform, and cooperation in justice and home affairs. The sectoral approach covers sectors such as transport, energy, information society, environment, research and development
39. Negotiations are underway to conclude an Association Agreement with Armenia, including deep and comprehensive free trade area (DCFTA). The EU and Azerbaijan are currently negotiating an association agreement to succeed the partnership and cooperation agreement, which will significantly deepen Azerbaijan's political association and economic integration with the EU. From the EEAS website.
40. The available cooperation instruments within the framework of the ENP are: regular energy dialogue, development of common energy networks and regulatory cooperation. R. Youngs, *Energy Security. Europe's New Foreign Policy Challenge*, Routledge, Oxon, 2009, pp. 57–63.
41. European Council, *The European Union and Central Asia: The New Partnership in Action,* June 2009, retrieved 11.02.2012, from http://eeas.europa.eu/central_asia/docs/2010_strategy_eu_centralasia_en.pdf.
42. Ibid, p. 24.
43. Natural gas use in non–Organisation for Economic Co-operation and Development (OECD) Asia increases by an average of 3.3 per cent annually, from 13.9 trillion cubic feet (tcf) in 2010 to almost triple that amount – 36.3 tcf in 2040. Further expansion of production in Central Asia will depend on securing markets and transit routes to reach markets in China. Turkmenistan is just beginning to develop its recently reassessed giant South Yolotan-Osman field, with much of the production likely to be exported by pipeline to China. In 2012, China received imports from Uzbekistan for the first time, in addition to flows on the same pipeline from Turkmenistan. EIA, *International Energy Outlook 2013,* pp. 47–60, retrieved 24.09.2013, from www.eia.gov/forecasts/ieo/nat_gas.cfm.
44. For example, the EU's response to the repression of the protests in Syria; Consilium, *Factsheet, The European Union and Syria*, Brussels, 4 June 2013, retrieved 22.09.2013 from www.consilium.europa.eu/uedocs/cms_data/docs/pressdata/EN/foraff/128379.pdf.

45. EEAS website, *Countries – Belarus,* retrieved 21.09.2013 from http://eeas.europa.eu/belarus/index_en.htm.
46. H. Prange-Gstöhl, "Enlarging the EU's Internal Energy Market", p. 5299.
47. EEAS website, *What Is the European Neighbourhood Policy?*, retrieved 18.09.2013, from http://eeas.europa.eu/enp/about-us/index_en.htm.
48. European Commission, *Communication from the Commission*, p. 7.
49. One of the motivations behind this choice is the security and economic benefits for Moldova and Ukraine from the participation in the EU internal energy market, which provides security of supply, reforms and new investments in their energy sectors.
50. Jaroslav Neverovič, the Lithuanian energy minister at the informal meeting of EU energy ministers in Vilnius, 19–20.09.2013. *The Common EU Energy Market: Not Only a Competitive European Economy but also Consumer-Friendly Prices*, Lithuanian EU Presidency official website, 20 September 2013, retrieved 23.09.2013, from www.eu2013.lt/en/news/pressreleases/the-common-eu-energy-market-not-only-a-competitive-european-economy-but-also-consumer-friendly-prices.

References

Belyi, A. (2008) "EU External Energy Policies: A Paradox of Integration", in Jan Orbie (ed.), *Europe's Global Role: External Policies of the European Union.* Farnham: Ashgate, pp. 203–216.

Bull, H. (1982) "Civilian Power Europe: A Contradiction in Terms?", *Journal of Common Market Studies,* 21(1): 149–64.

Consilium, *Factsheet, The European Union and Syria,* Brussels, 4 June 2013. Retrieved September 22, 2013, from www.consilium.europa.eu/uedocs/cms_data/docs/pressdata/EN/foraff/128379.pdf.

Duchêne, F. (1972) "Europe's Role in World Peace", in Richard Mayne (ed.), *Europe Tomorrow: Sixteen Europeans Look Ahead.* London: Fontana, pp. 32–47.

EIA (2013) *International Energy Outlook 2013.* Retrieved September 24, 2013, from www.eia.gov/forecasts/aeo/pdf/0383%282013%29.pdf.

Energy Community Secretariat (2005) *Treaty establishing the Energy Community.* Retrieved October 31, 2010, from www.energy-community.org/pls/portal/docs/530177.PDF.

Energy Community Secretariat (2012) *Energy Community Annual Implementation Report.* Retrieved 20.09.2012, from www.energy-community.org/pls/portal/docs/1770178.PDF.

Energy Community Secretariat (2013) *Energy Community Annual Implementation Report.* Retrieved 04.04.2014, from www.energy-community.org/pls/portal/docs/2304177.PDF.

Energy Community Secretariat (2013) *Meeting Conclusions 11th Energy Community Ministerial Council.* Retrieved 04.04.2014, from www.energy-community.org/pls/portal/docs/2388178.PDF.

Energy Community Website. Retrieved 17.09.2013, from www.energy-community.org/portal/page/portal/ENC_HOME/ENERGY_COMMUNITY.

Energy Community Website. Retrieved 20.09.2013, from www.energy-community.org/portal/page/portal/ENC_HOME/ENERGY_COMMUNITY/Legal/EU_Legislation.

Energy Community Website. *Information on the work of the High Level Reflection Group of the Energy Community*, 20 March 2014. Retrieved 05.04.2014, from www.energy-community.org/pls/portal/docs/2962024.PDF.

European Commission (2011a) *Communication from the Commission to the European Parliament, the Council, the European Economic and Social Committee and the Committee of the Regions. On security of energy supply and international cooperation – "The EU Energy Policy: Engaging with Partners beyond Our Borders"*, COM(2011) 539 final, Brussels.

European Commission (2011b) *Report from the Commission to the European Parliament and the Council under Article 7 of Decision 2006/500/EC (Energy Community Treaty)*, COM(2011b) 105 final, Brussels.

European Council (2009) *The European Union and Central Asia: The New Partnership in action.* Retrieved 11.02.2012, from http://eeas.europa.eu/central_asia/docs/2010_strategy_eu_centralasia_en.pdf.

European External Action Service Website. www.eeas.europa.eu

European External Action Service Website, *Countries – Belarus.* Retrieved 21.09.2013, from http://eeas.europa.eu/belarus/index_en.htm.

European Union (2010) "Consolidated Versions of the Treaty on European Union and the Treaty on the Functioning of the European Union", *Official Journal of the European Union*, C 83.

Hill, C. (1993) "The Capability-Expectations Gap, or Conceptualizing Europe's International Role", *Journal of Common Market Studies*, 31(3): 305–328.

Lithuanian EU Presidency Website, *The common EU energy market: not only a competitive European economy but also consumer-friendly prices*, 20 September 2013. Retrieved 23.09.2013, from www.eu2013.lt/en/news/pressreleases/the-common-eu-energy-market-not-only-a-competitive-european-economy-but-also-consumer-friendly-prices.

Manners, I. (2002) "Normative Power Europe: A Contradiction in Terms?", *Journal of Common Market Studies*, 40(2): 235–258.

Prange-Gstöhl, H. (2009) "Enlarging the EU's internal energy market: why would third countries accept EU rule export?", *Energy Policy*, 37(12): 5296–5303.

Renner, S. (2009) "The Energy Community of Southeast Europe: A neofunctionalist project of regional integration", *European Integration Online Papers*, 13. http://eiop.or.at/eiop/index.php/eiop/article/viewFile/2009_001a/88

Smith, K. (2005) *Still "civilian power EU"?*, London: London School of Economics. European Foreign Policy Unit Working Paper, 2005/1.

Wood, S. (2009) "The European Union: a normative or normal power?", *European Foreign Affairs Review*, 14(1): 113–128.

Youngs, R. (2009) *Energy Security. Europe's New Foreign Policy Challenge.* Oxon: Routledge.

Part III

Case studies: Implications from the export of the EU model

Chapter 11

European legal space and application of the European Union's "energy acquis" in Ukraine and Moldova

*Roman Petrov**

The entry into force of the Treaty of Lisbon (TEU) on 1 December 2009 considerably enhanced the role of the EU as a global actor which aspires to promote its own democratic, economic and legal values beyond its borders. The objective of this process is twofold. On one hand, it is to enhance the position of the EU as a global player worldwide. On the other hand, it is to ensure the functioning of friendly and secure neighbourhood around the EU's borders.

These processes have received vivid interest and attention among experts and academics. Contemporary academic research on the subject focuses on the political and legal implications of the EU's enhanced international capacity which leads to promotion of European democratic and market economy values to third countries, Europeanization of third countries institutional and legal structures and creation of European Legal Space.[1] These processes are being exercised via sectoral multilateralism and differentiated integration within the EU. Sectoral multilateralism presumes closer legal cooperation between the EU and third countries, beyond traditional approximation of laws, for instance, by voluntary application of the EU sectoral acquis by a third country. Usually, sectoral multilateralism is achieved with help of sectoral bilateral agreements between the EU and third countries, such as the European Economic Area Agreement (EEA) and EU bilateral agreements with Switzerland. There is no unified opinion on what motives push third countries to pursue voluntary multilateral integration with the EU.[2] However, one may argue about the evident positive impact of the multilateralism phenomena on the process of democratization in the countries of Central and Eastern Europe.[3] Nevertheless, these trends cannot be viewed as exclusive but as complementary; therefore, it could be said that the creation of the European Legal Space is not possible without successful Europeanization of third countries.[4] One of the means of successful Europeanization of third countries is their gradual access to the EU Internal Market and, as a consequence, adoption of the EU "sectoral acquis".[5] One of the best Europeanization examples is accession of third countries to the Energy Community (EnC). After acquiring membership in the EnC, third countries undertake commitments to implement and effectively apply the EU "sectoral acquis" without being a full member of the EU.[6]

This chapter does not deal with the multidimensional and complex phenomenon of Europeanization in its whole scope, but endeavours to clarify only one aspect, the Europeanization of the eastern neighbouring countries through their participation in the EnC and application of its rules and standards. The simple acknowledgment of the EU's role as a global player and a promoter of the Europeanization beyond its borders tells us little about its use and interrelationship with other more practical issues. One of these issues is acceptance of the EU's sectoral acquis by third countries – addressees of the EU's global influence. In other words, it is important to look at the practical application of EU law beyond its borders. The conceptual focus of this contribution is on the linkage of an EU's external policies and actual application of the EU "energy acquis" by third countries which do not have immediate perspective of joining the EU in the future – countries of the Eastern Partnership.

The external dimension of the EU's energy policy

The Treaty of Lisbon emphasized the prominence of the energy sector among other EU policies by introducing a new legal base and outlining security of supply, competitiveness, and sustainability as major objectives of the EU energy policy.[7] These objectives were further elaborated in the Europe 2020 Strategy[8] which sets out ambitious plans for the EU to become a smart, sustainable and inclusive economy by 2020. One of the most prominent objectives of the Europe 2020 Strategy is to strengthen the external dimension of the EU energy market (new energy supply sources and routes for the EU by 2020, integrating energy markets and regulatory frameworks with the EU's neighbours).

The external dimension of the EU energy policy found its further articulation in the European Commission's Communication "The EU Energy Policy: Engaging with Partners beyond Our Borders" (hereafter, the Communication)[9] and in the Council conclusions on strengthening the external dimension of the EU energy policy on 24 November 2011 (hereafter, the Council Conclusions). The former (the Communication) calls on the EU and its member states to build an integrated energy market with all neighbouring countries based on regulatory convergence which is to be done taking as a reference the EnC framework. The Communication states that the EnC's "regulatory scope should be progressively extended and combined with more effective implementation and enforcement, as well as concrete assistance to reform [the neighbouring countries] markets".[10] Setting up free trade areas between the EU and third countries is linked to membership of the latter in the EnC and their ability to implement relevant EU energy acquis. The latter (the Council Conclusions) emphasizes that "regulatory cooperation and convergence with our neighbours has a key role to play in order to build a wide energy market while ensuring a level playing field. Such regulatory cooperation should take into account the diversity of EU's neighbours and their own energy policy objectives." One of the top priorities of the Council Conclusions is "encouraging full and timely implementation and enforcement of the acquis, as well as the removal of technical barriers, aiming at the creation of an Energy Community-wide energy market".[11]

Thus, the preceding documents recognize the EnC as a key pillar of the external dimension of the EU energy policy. From a legal point of view, the EnC represents a unique product which sets up a quasi supranational legal system modeled on the EEA with some elements of legal homogeneity between the contracting parties. The EnC framework represents a unique chance to implement and apply dynamic sectoral EU acquis within third countries' legal systems. The Treaty establishing the Energy Community (TEnC) was signed in Athens, Greece, on 25 October 2005, and entered into force on 1 July 2006. From the outset the EnC's objective was to establish a common regulatory framework between the EU and the participating third countries and to promote the EU's "energy acquis" (electricity, natural gas and petroleum products) beyond the EU borders. The relevant acquis to be adopted by the contracting parties also cover some elements of the EU acquis in areas of environment and competition. The EnC embraces the EU (according to the TEnC any EU member state can be represented in the EnC institutions [Ministerial Council, the Permanent High Level Group and the Regulatory Board]; hitherto, only 17 EU member states used this privilege)[12] and 9 third countries (Albania, Bosnia and Herzegovina, Croatia, Macedonia, Montenegro, Serbia, Moldova, Ukraine and the United Nations Interim Administration Mission in Kosovo). Most of these countries are either EU candidate countries (Croatia, Macedonia, Montenegro and Serbia) or EU potential candidates (Albania and Bosnia and Herzegovina). Ukraine and Moldova are participants of the European Neighbourhood Policy (ENP) and Eastern Partnership (EaP).[13] These countries strive to set up closer political and economic relations with the EU through concluding association agreements and setting up free-trade areas. There is also a group of observer countries (Armenia, Georgia, Norway and Turkey) with the right to attend the meetings of the EnC institutions without taking part in discussions and voting. Whereas Turkey is a long-standing EU candidate country, Armenia and Georgia are participants of the ENP and EaP. Thus, in the meantime, the EnC is the only regional organization that covers the countries which are full members of the EU, candidate countries, associate countries and the neighbouring countries.

The EnC institutional framework is composed of a Ministerial Council, the Permanent High Level Group, the Regulatory Board and the secretariat. It resembles the institutional structure of the EEA but, unfortunately, on a much lesser scale. The major decision-making and judicial body of the EnC is the Ministerial Council which is composed of representatives of the contracting parties. The Ministerial Council takes measures, provides general policy guidelines, and adopts procedural acts. The Permanent High Level Group (PHLG) assists the Ministerial Council with its work (prepares the work, reports, take measures if empowered by the Ministerial Council, follows dynamic acquis and recommends adoption of new acquis). The Regulatory Board advises the Ministerial Council and PHLG on details of statutory, technical and regulatory rules and makes recommendations in the case of cross-border disputes between regulators. The secretariat provides administrative support to the Ministerial Council and PHLG, is responsible for reviewing implementation of the EU energy acquis and submits regular progress reports to the Ministerial Council.

A dispute settlement mechanism in the EnC[14] bears certain resemblance to the EU's infringement procedure. However, it does not envisage a permanently functioning court, as is the case in the EEA. Under the TEnC, any "concerned" party, the Regulatory Board and the Secretariat (on complaint of a contracting/concerned party or on its own motion) may bring a case of non-compliance by an EnC member state with the EU energy acquis to the attention of the Ministerial Council. In case of breach, the Ministerial Council may issue a binding decision and, at the last resort (in cases of serious and persistent breaches), may suspend certain rights of a party to the TEnC.

The Commission's Communication "The EU Energy Policy: Engaging with Partners beyond Our Borders" unequivocally states that "[t]he Energy Community Treaty is the reference point for the majority of the EU's neighbours willing to be a part of the European energy system" and a prerequisite for concluding a Free Trade Agreement with the EU. Therefore, it can be suggested that any further rapprochement of third countries with the EU and closer integration with the EU Internal Market is conditional on successful regulatory convergence within the EU sectoral acquis. For this purpose, the EU has either already engaged most of its neighbouring countries to join the EnC or encouraged others to participate in the EnC's activities as observers.[15] Joining the EnC implies gradual integration of the neighbouring countries' energy sector to the EU energy market, fulfillment of the membership commitments in the EnC, and, consequently, implementation and application of the EU energy acquis within the neighbouring countries' legal systems.

Scope of the EU energy acquis to be implemented via the EnC framework

The eastern neighbouring countries are some of the key energy partners of the EU. For instance, transit networks of Ukraine provide about 20% of the EU's gas supply and, therefore, play a prominent role in the overall EU's energy security. Azerbaijan plays the key role of an alternative energy supplier for the EU from the Caspian area. Armenia and Georgia are important transit countries for oil and gas from the Caspian basins to the EU. Oil and gas pipelines connecting the Caspian basin with Turkey, thus bypassing Russia and Iran, will increasingly become a strategic alternative energy corridor for the EU. Currently Armenia and Georgia experience political and economical vulnerability vis-à-vis their big neighbour because of their increasing dependence on imports of energy from Russia. Thus, enhancement of energy security through further progress in energy sector reform and regulatory development, as well as progress in the development of energy efficiency and renewable energy should help these countries in acquiring energy independence.

As it follows from the Communication the major focus of the energy cooperation between the EU and its eastern neighbouring countries is the participation of the latter in the EnC.[16] The first eastern neighbouring country to join the EnC was Moldova. It completed its accession negotiations on 29 April 2009 and acceded to the EnC on 17 March 2010.[17] Lengthy negotiations on accession of Ukraine to the EnC were completed on 7 October 2009. The protocol of Ukraine's accession to the Treaty on the Energy Community (hereafter, Protocol on accession) was signed on 24 September 2010, and, consequently, ratified by the Parliament (Verkhovna Rada) of Ukraine on 15 December 2010.[18] Thus, as of 2011, Moldova and Ukraine had obtained full membership in the EnC which implies that these eastern neighbouring countries are under a legal obligation to implement the EU energy acquis in the areas of gas, electricity, nuclear energy, oil and renewable energy. The eastern neighbouring countries with observer status (Georgia and Armenia) are under soft commitments to ensure regulatory convergence within the EU energy acquis.

Scope of the EU energy acquis to be implemented and applied by the eastern neighbouring countries

The Protocols on accession impose legally binding obligations on Moldova and Ukraine to implement specific scope of the EU energy acquis by four waves for Moldova and five waves of deadlines for Ukraine. Furthermore, Ukraine undertook legal commitments: (1) to abide to principles of the EU "competition acquis",[19] (2) to follow generally applicable standards of the European Community on operating energy network systems,[20] and (3) to adopt "security of supply statements describing diversity of supply, technological security, and geographic origin of imported fuels".[21]

However, commitments of the eastern neighbourhood countries under the TEnC are not limited solely to the timely implementation of the static EU energy acquis into their legal systems that is defined at the moment of signing the Protocol of Accession. The TEnC offers unique and unprecedented opportunities and challenges for the eastern neighbouring countries' legal system to follow and apply the EU "fundamental acquis" and the EU "dynamic acquis". In the following, I briefly look at various types of the EU acquis to be adopted by Moldova and Ukraine.

Application of fundamental principles of EU Law by the eastern neighbouring countries

The TEnC imposes legal obligation on the contracting parties to follow the principles of sincere cooperation and non-discrimination. Article 6 TEnC provides that "[t]he Parties shall take all appropriate measures, whether general or particular, to ensure fulfillment of the obligations arising out of this Treaty. The Parties shall facilitate the achievement of the Energy Community's tasks. The Parties shall abstain from any measure which could jeopardize the attainment of the objectives of this Treaty." This provision resembles an almost identical provision of Article 4(3) TEU on the principle of sincere cooperation which is considered by the European Court of Justice (ECJ) as a fundamental principle of EU law.[22] According to EU law the principle of sincere cooperation means that EU member states shall take all appropriate measures to ensure the fulfilment of the obligations arising from the EU founding treaties and resulting from action taken by the institutions of the EU.

The principle of non-discrimination is enshrined in Article 7 TEnC and prohibits "any discrimination" in relations between the parties. This provision resembles Articles 3(3) TEU and Article 10 TFEU which prohibit all discrimination on grounds of nationality, sex, racial or ethnic origin, religion or belief, disability, age or sexual orientation in the EU.[23] It does not imply that the scope of the principle of sincere cooperation and non-discrimination should be understood as identical to the scope of these principles in the EU founding treaties. The exact clarification of the scope of these principles is yet to be done by the EnC's institutions and dispute settlement authorities. However, one may presume that the interpretation and application of the fundamental principles of the EnC's legal order will take place in line with reception of the EU fundamental principles under the legal framework of the European Economic Area and the EU–Swiss sectoral cooperation.[24]

Obligation of the eastern neighbouring countries to follow the EU dynamic energy acquis

The EU acquis reflects the dynamic, or, sui generis, nature of the EU legal order. In this respect, the dynamism of the EU legal order entails its never-ending evolution, under the pressure of various internal and external factors, such as the need for closer economic development inside the EU, and the enhancement of security and political stability along EU borders. The dynamism of the EU legal order is based on acquired common rules, practices and values, which are embraced by the complex notion "*acquis communautaire*". The same approach could be applied to the EU dynamic nature of the EU sectoral acquis.

The TEnC envisages the binding and soft commitments of the EnC and its contracting parties to ensure the implementation and application of the EU energy acquis. The TEnC deals with two types of the EU energy acquis. The first type is the "pre-signature" EU energy acquis specified in the text of the TEnC and its annexes at the time of signature of the TEnC. The second type is the EU "dynamic energy acquis" which appeared after the TEnC entered into force. With regard to the former case, Article 5 TEnC provides that the EnC "shall follow the acquis communautaire (described in Title II [sectoral acquis]) adapted to both the institutional framework of this Treaty and the specific situation of each of the Contracting Parties, with a view to ensuring high levels of investment security and optimal investments". In our opinion it implies that not only the EnC common institutions but also the contracting parties of the EnC are under obligation to follow the "pre-signature" EU energy acquis in their legal orders (provided that the EnC common institutions issued binding decision in accordance with Article 89 TEnC). However, this obligation is balanced by a considerable discretion which allows the implementation of the "pre-signature" EU energy acquis to be done "taking into account both the institutional framework of this Treaty and the specific situation of each of the Contracting Parties".[25] It means that the binding obligation to implement and apply the "pre-signature" EU energy acquis could be hindered (revised, delayed or postponed) due to some inter-institutional arrangements within the EnC or problems of either political or economic or legal nature in the EnC's contracting parties.

With regard to the latter case, Chapter VII TEnC "The Adaptation and Evolution of the Acquis" sets up a right but not an obligation of the EnC to follow the EU dynamic energy acquis: "The Energy Community *may* [RP] take Measures to implement amendments to the acquis communautaire described in this Title, in line with the evolution of European Community law".[26] To fulfill this objective the Permanent High Level Group shall "discuss the development of the acquis communautaire described in Title II on the basis of a report that the European Commission shall submit on a regular basis".[27] Furthermore, the EnC's common institutions "*shall* [RP] interpret any term or other concept used in this Treaty that is derived from EC Law in conformity with the case law of the ECJ and General Court. When no interpretation from those Courts is available, the Ministerial Council *shall* [RP] give guidance in interpreting this Treaty. It may delegate that task to the PHLG. Such guidance shall not prejudge any interpretation of the acquis communautaire by the ECJ or General Court at a later stage".[28]

In my opinion the soft character of the obligation of the EnC to follow the EU dynamic energy acquis does not downgrade its importance for the EnC legal order. Indeed, the provisions of the TEnC on adoption of the

EU dynamic energy acquis do not have the same binding effect as relevant provisions on homogeneity in the EEA.[29] However, the provisions of the TEnC on adoption of the EU dynamic energy acquis are much stronger than, for instance, corresponding provisions in the bilateral agreements between the EU and Switzerland. Notwithstanding the lack of any legal requirements and binding obligations in the EU–Swiss sectoral agreements, Switzerland achieved impressive results in adopting EU dynamic acquis in areas of bilateral sectoral cooperation.[30] Therefore, I argue that far-reaching general objectives of the TEnC (for example, objectives "to establish among the Parties and integrated market in natural gas and electricity. . . . to create a stable regulatory and market framework . . . and single regulatory space") and the fact that majority of the EnC's contracting parties have already followed the EU dynamic energy acquis implies a paramount importance for the EnC to ensure timely and efficient implementation and application of the EU dynamic energy acquis within its legal order and legal orders of its contracting parties.

In accordance with the principle of sincere cooperation enshrined in Article 6 TEnC the EnC's contracting parties are bound to follow the EU dynamic energy acquis through decisions and guidance of the EnC common institutions as well as interpretations of the EnC dispute settlement decisions.[31] In other words, if the EnC's common institutions issue binding decisions aimed at implementation of the EU dynamic energy acquis, the EnC's contracting parties bear a legal obligation to do so or to face the possibility of sanctions provided by the TEnC.[32] One of the recent examples of the need to implement the EU dynamic energy acquis is the so-called "Third Package" of the EU energy acquis.[33] Adoption of the EU new legislative package in the field of energy in 2009 required reactions by the EnC institutions. On 24 September 2009, the Ministerial Council adopted Recommendation 2010/02/MC-EnC, suggesting the voluntary implementation of the "Third Package" by the EnC's contracting parties. A legally binding decision on the implementation of the "Third Package" EU energy acquis was adopted on 6 October 2011,[34] and, thereby, the EnC's contracting parties, including Moldova and Ukraine, will be under a legal obligation to implement and apply the Third Package EU energy acquis in their national legal orders.[35]

Obligation of the eastern neighbouring countries to follow decisions of the EnC's institutions and dispute settlement

The EnC institutional framework is empowered with many tools to ensure the extension of the EU energy acquis to the legal system of the contracting parties. In particular, Moldova and Ukraine are bound by legally binding decisions and non-binding recommendations of the EnC's institutions and must do their best to ensure the recommendations' effective application.[36] Legally binding decisions of the EnC's institutions must be implemented within the specified time; failure to do so may lead to determination by the Ministerial Council of a serious and persistent breach of a party's obligations and to suspend certain rights of this party, including the suspension of voting rights and exclusion from meetings or mechanisms in the TEnC.[37]

The procedure for dispute settlement in the EnC resembles the dispute settlement in the EU. According to Articles 90–93 TEnC the secretariat or the Regulatory Board may bring a case of non-compliance before the Ministerial Council. Even private bodies may approach the Secretariat with complaints about non-compliance by the contracting parties. The Ministerial Council considers the case and may suspend voting rights, exclude from meetings and mechanisms a guilty party. The Ministerial Council creates an advisory committee composed of three lawyers who bear functions similar to advocates general at the ECJ (deliver opinions on cases of non-compliance). In accordance with the EnC Secretariat "dispute settlement proceedings constitute the means to clear away key obstacles to full implementation of the acquis and to actively support the emerging of competitive markets based on the application of harmonized rules" and "the existence of a workable dispute settlement procedure constitutes and important element in providing the legal certainty needed to attract investment to the Energy Community Contracting Parties".[38]

The EnC's dispute settlement procedure, especially, the right of private bodies to complain to the EnC's institutions about non-compliance of the contracting parties, brings forward considerable challenges for the Moldovan and Ukrainian legal systems. First, the EnC secretariat may initiate non-compliance procedures against Moldova and Ukraine for failure of their governments to fulfill their obligations under the TEnC. Second, the Moldovan and the Ukrainian judiciary will be faced with the necessity of not only applying the relevant EU dynamic energy acquis but also ensuring the compliance of Ukrainian and Moldova laws within its own judgments.

Implementation of the EU energy acquis into the legal system of eastern neighbouring countries

The question for consideration is whether the EU energy acquis is fully applicable in legal orders of the neighbouring countries. What restraints of a legal and political nature may prevent the neighbouring countries from applying the EU energy acquis? I argue that the correct implementation and effective application of the EU sectoral acquis depends on two aspects, the constitutional law of the eastern neighbouring countries and positive reception of general principles of EU law and the EU energy acquis by executives and judiciaries in the eastern neighbouring countries.

Constitutional foundations of application of the EU energy acquis in Ukraine and Moldova

Implementation and application of the EU energy acquis within the legal systems of Ukraine and Moldova is governed by their national constitutional laws. The TEnC is an international treaty which was duly ratified by the parliaments of Ukraine (Verkhovna Rada; on 15 December 2010) and Moldova (on 31 March 2010) and, therefore, became part of national legislation of these countries.[39] According to the Protocols on accession, Ukraine and Moldova are "entitled to all rights granted to the Contracting Parties and is subject to all obligations imposed on the Contracting Parties by the Treaty and by all Decisions and Procedural Acts adopted in application of the Treaty since its entry into force".[40] The national constitutions of Ukraine and Moldova provide that in case of conflict of the TEnC provisions with their national legislation (excluding national Constitutions), the former prevails.

However, commentators emphasize certain difficulties in applying international law within the legal systems of eastern neighbouring countries. The major obstacle is the reluctance of the judiciary in the eastern neighbouring countries to apply and effectively implement international law sources in own judgments.[41] For example, the Ukrainian courts refer mainly to international agreements which are duly signed and ratified by the Verkhovna Rada and which are self-executing within the Ukrainian legal system. Even in these cases, the correct application of international agreements is not guaranteed. It happens because one of the most important impediments for the application of international law by the Ukrainian judiciary is the correct understanding of these international conventions by national judges.

The TEnC and relevant elements of the EU energy acquis are relatively new to the Ukrainian and Moldovan legal systems. Undoubtedly, the TEnC is a duly ratified international agreement which is part of the Ukrainian and Moldovan legal systems and prevails over conflicting national legislation but not the Constitutions of Ukraine and Moldova. However, either additional national law or ruling of the respective constitutional courts are required to ensure duly application of legally binding decisions of the EnC's institutions in order to avoid possible conflicts with the constitutions of Ukraine and Moldova. Furthermore, the Ukrainian and Moldovan judges must be well aware about the scope and legal nature of the TEnC and the EU energy acquis and their legal implications for the national legal systems in order to apply it correctly in own judgments. At the time of writing, practice on application of the TEnC and relevant EU energy acquis by judiciaries in Ukraine and Moldova was quite scarce. One of the first references to the EU energy acquis can be found in judgment of the Ukrainian Regional Appeal Administrative Court in *Shidenergo v. State Tax Authority*[42] case. Therein the Regional Appeal Administrative Court recognized the TEnC as part of the legal system of Ukraine and acknowledged the commitment of

Ukraine to implement Directive 2001/80 as a mitigating circumstance before the state tax authority.

Scarce practice on application and references to the TEnC and the EU energy acquis could be explained by some factors of socio-legal nature. Among them are limited familiarity of the judiciaries in Ukraine and Moldova with the TEnC and the EnC's legal order as well as the absence of claims on behalf of Ukrainian and Moldovan nationals based on provisions of the TEnC and the EU energy acquis. However, one may deduce general trends of possible future application of the TEnC and the EU energy acquis from existing practice of applying EU acquis in Ukraine and Moldova. In the following, I look at recent and future implementation and application of the EU energy acquis by judiciaries, governments and administrative agencies in Ukraine and Moldova.

Application and implementation of the EU energy acquis in forthcoming association agreements between the EU and Ukraine and the EU and Moldova

Chances for efficient implementation and application of the EU energy acquis by the eastern neighbouring countries will be significantly enhanced in case of the establishment of associations between these countries and the EU. Association agreements between the EU and third countries have become one of the most recognisable brands of the EU external policy. In particular, this relates to the countries of the EU's eastern neighbourhood (Ukraine, Moldova, Belarus and the Caucasus countries) which have either already started negotiations on association agreements with the EU, or are about to do so in the nearest future.

The new generation of the EU association agreements with the EU's eastern neighbours will substitute outdated partnership and association agreements which were concluded between 1994 and 1998. The first eastern neighbourhood country to start negotiations on an association agreement was Ukraine. The negotiations were launched in September 2008 and successfully completed in December 2011. The EU–Ukraine Association Agreement (AA),[43] was signed (not the whole argument but only its part on political cooperation and concluding provisions) on 21 March 2014, is the first of a new generation of AAs to be concluded between the EU and the Eastern Partnership countries (Ukraine, Moldova, Belarus, Armenia, Azerbaijan and Georgia). According to Herman Van Rompuy, president of the European Council, it is "the most advanced agreement of its kind ever negotiated by the European Union". The EU–Ukraine AA essentially aims to deepen the political and economic relations between Ukraine and the EU through the establishment of an enhanced institutional framework and innovative provisions on regulatory and legislative approximation.

The EU–Ukraine AA is the most voluminous and ambitious among all EU association agreements with third countries. It comprises 7 titles, 28 chapters, 486 articles, 43 annexes on about 1,000 pages. This is a comprehensive mixed agreement based on Article 217 TFEU. There are many novelties introduced to this agreement. Most prominent of them are strong emphasis on comprehensive regulatory convergence between the EU and Ukraine and possibility for application of the EU sectoral acquis within the Ukrainian legal order.

Of particular significance of the EU–Ukraine AA is the ambition to set up a deep and comprehensive free trade area (DCFTA), leading to Ukraine's gradual and partial integration in the EU internal market. Accordingly, the AA belongs to the selected group of "integration-oriented agreements", that is agreements including principles, concepts and provisions which are to be interpreted and applied as if the third State is part of the EU. It is argued that the EU–Ukraine AA is unique in many respects and, therefore, provides a new model of integration without membership. Two of the Eastern Partnership countries (Moldova and Georgia) have completed the process of negotiation of the AA with the EU and are likely to initialize the texts of these agreements at the Eastern Partnership Summit in Vilnius in November 2013.

In my opinion the signature of an association agreement with the EU should trigger serious internal reforms in the eastern neighbouring countries. First, future association agreements will serve as a template for further political and economic reforms in these countries. The obligation to share the EU's common democratic values will imply regular monitoring by the EU institutions. Thereby this should prevent the eastern neighbouring countries from undemocratic practices. Establishing a deep and comprehensive free-trade area will open new markets for the eastern neighbouring countries' products on the EU's internal market and will enhance internal competition within the EU neighbourhood. The new joint institutions set up under the framework of association agreements will help to pursue the programme of approximating the laws with the help of its binding decisions. However, it is important that the eastern neighbouring countries take full advantage of the reciprocity under the association agreements with the EU, and ensure their nationals' access to the freedoms of the EU internal market, and the visa-free regime with the EU. Eventually, the principle of reciprocity should lead to the neighbouring countries' informal participation in the EU's decision-making process.

The process of effectively implementing the association agreements will constitute the greatest challenge for Ukraine and Moldova. It has to prove their adherence to the EU's common democratic and economic values, and ensure the proper functioning of their deep and comprehensive free trade areas. The latter objective may be achieved only under the condition of establishing truly competitive market economies and the adoption of international and EU legal standards. Similar to the TEnC, Ukraine and Moldova will be bound by decisions of the dispute settlement body established by an association agreement. Following the widely used practice in the EU's external agreements, any future association agreements with the countries of the Eastern Partnership may contain so-called evolutionary and conditionality clauses. These are provisions in the EU's external agreements with specific objectives (for instance, granting a visa-free regime, the status of a candidate country to join the EU), the attainment of which is conditional either on certain actions on behalf of a party to an agreement (such as the elimination of trade barriers and uncompetitive practices) or the effective functioning of democratic and market-economy standards (such as free and fair elections and fighting corruption).

The future association agreements between the EU and the eastern neighbouring countries will offer many incentives for efficient implementation and application of the EU energy acquis by the eastern neighbouring countries. For instance, future "evolutionary" and "conditionality" clauses in the association agreements will envisage an obligation by the eastern neighbouring countries to fulfill commitments under the TEnC (to apply the fundamental principles of EU law and the EU energy acquis, to follow dynamic EU energy acquis and others).[44] Besides, entering into force of the association agreements and their subsequent impact on legal systems of the eastern neighbouring countries will trigger better reception and application of the EU energy acquis by judiciaries and executives in Ukraine and Moldova. In other words, Ukrainian and Moldovan judges will be more inclined to apply not only "static" EU energy acquis but also "dynamic" EU energy acquis as well as case law of the ECJ because of political and economic integration oriented objectives of the future association agreements.

Conclusion

To conclude, I set out a number of considerations which lead us to believe that the EU energy acquis should and will find frequent application within the legal system of the eastern neighbourhood countries.

The first consideration is that the growing prominence of the EU energy policy has considerably influenced the EU relations with its eastern neighbours. The external dimension of the EU energy policy accelerates and encourages the eastern neighbouring countries to adopt the EU energy acquis and to actively engage into the institutional and legal framework of the EnC. Ukraine and Moldova were among the first eastern neighbouring countries to join the EnC and thereby to bind themselves to the EU energy acquis contained in the TEnC. The TEnC offers many legal challenges for the Ukrainian and Moldovan legal systems as for application of the EU energy acquis. Apart from the legal commitments of implementing the EU energy acquis the TEnC makes Ukraine and Moldova abide by fundamental principles of EU law, decisions of the EnC common institutions and the EnC's dispute settlement.

The second consideration is that the accession of Ukraine and Moldova to the EnC implies far-reaching challenges for legal systems of these countries. Ukrainian and Moldovan judiciaries and executives will face a necessity of implementing and applying not only the EU energy acquis defined at the moment of signing of Protocols on accession to the EnC but also fundamental principles of EU law, dynamic EU energy acquis, and decisions of the EnC's institutions. However, it is argued that the legal systems in Ukraine and Moldova are not yet ready to absorb all elements of the dynamic EU acquis contained in the TEnC. The reserved attitude of judiciaries in Ukraine and Moldova towards application of international law may inhibit efficient implementation of the EU energy acquis by these countries. It is hoped that the judiciaries in Ukraine and Moldova will show a great degree of legal activism through voluntary application of the EU acquis in their decisions in order to ensure the success of this process. Association agreements with the EU impose binding commitments on Ukraine and Moldova to implement a vast scope of the EU sectoral acquis. Therefore, the establishment of association relations between the EU and Ukraine and Moldova may considerably enhance the effectiveness of implementation and application of the EU static and dynamic energy acquis in these countries.

The third consideration is that the promotion of the EU energy acquis to the eastern neighbouring countries via the EnC offers several lessons for future EU external policies towards the East. The first lesson – the Europeanization of the EU's eastern neighbouring countries depends on many local circumstances. Success of the process of Europeanization does not apply only to adoption of legal norms and rules by third countries but depends on true sharing of common democratic values, application of fundamental principles of EU law and dynamic EU acquis. The EU must employ its best internal and external tools and incentives to promote friendly perception and application of European legal standards and common democratic values by judiciaries, governments and administrative agencies in third countries. Nationals in third countries must be well aware of rights offered to them under EU external agreements like association agreements and the TEnC. The second lesson – promotion of the EU sectoral acquis via the EnC is a timely and useful exercise towards promotion of the whole scope of the EU sectoral acquis beyond the EU's borders. Indeed, experience gained by the EnC may be successfully adopted by other EU external projects such as new communities in fields of air and land transport, environment, education and food safety. Issues such as direct effect of provisions of founding treaties of such communities, dispute settlement procedure, application of the EU dynamic acquis and interaction with the EU's framework agreements (association, partnership) could be learned from the EnC's case.

I conclude with the suggestion that the newly emerging EnC's legal order offers many challenges of a legal and political nature for the EU and third countries. With no doubt, the EnC will play a role of a laboratory working on better and deeper engagement of third countries into expanding European Legal Space.

Notes

* The findings, interpretations and conclusions expressed in this chapter are those of the author and do not reflect the positions of the institution.

1. M. Cremona, "The Union as a Global Actor: Roles, Models and Identity", 41 *Common Market Law Review* (2004) 553–573. A. Bendiek and H. Kramer, "The EU as a 'Strategic' International Actor: Substantial and Analytical Ambiguities", 15 *European Foreign Affairs Review* (2010) 453–474. P. Leino and R. Petrov, "Between 'Common Values' and Competing Universals; The Promotion of the EU's Common Values through the European Neighbourhood Policy", 15 *European Law Journal* (2009) 654–671.
2. For example, see H. Prange-Gstöhl, "Enlarging the EU's Internal Energy Market: Why Would Third Countries Accept EU Rule Export?", 39 *Energy Policy* (2009) 5296, 5297. The author identifies three motives for countries with no or only a vague membership perspective to agree to deeper energy integration with the EU ("identification motive", "independence motive" and "economic motive").
3. L. Morlino and W. Sadurski (eds.), *Democratization and the European Union,* (London: Routledge, 2010).
4. For more on Europeanization, see J. Olsen, "The Many Faces of Europeanisation", 40(5) *Journal of Common Market Studies* (2002) 921–952. C. Radaelli, "Europeanisation: Solution or Problem?", 8(16), EIoP (2004), available at www.eiop.or.at/eiop/pdf/2004–016.pdf, last visited 25 August 2013.
5. Henceforth, I apply the notion of "EU acquis" instead of "EU law" to emphasize the comprehensive and complex nature of the EU legal heritage. More on the scope of the EU acquis see R. Petrov, "Exporting the Acquis Communautaire into the Legal Systems of Third Countries", 13 *European Foreign Affairs Review* (2008) 33–52.
6. S. Padgett, "Energy Co-operation in the Wider Europe: Institutionalizing Interdependence" 49 *Journal of Common Market Studies* (2011) 1065–1087. R. Karova, "Rationale behind the Establishment of the Energy Community" (2010) 2010/14 EUI LAW Working Papers, available at http://papers.ssrn.com/sol3/papers.cfm?abstract_id = 1699146, last visited 25 August 2013.
7. Article 194 TFEU.
8. Adopted by the European Council on 17 June 2010.
9. Communication of the European Commission, "On Security of Energy Supply and international cooperation – 'The EU Energy Policy: Engaging with Partners beyond Our Borders'", COM (2011) 539 final, Brussels. 07.09.2011.
10. Communication of the European Commission, "On Security of Energy Supply and international cooperation – 'The EU Energy Policy: Engaging with Partners beyond Our Borders'", COM (2011) 539 final, Brussels, 07.09.2011.
11. Council conclusions on strengthening the external dimension of the EU energy policy on 24 November 2011, at p. 2.
12. Austria, Bulgaria, Cyprus, Czech Republic, France, Germany, Greece, Hungary, Italy, the Netherlands, Poland, Romania, Slovakia, Slovenia and the UK.
13. The ENP Communication from the Commission to the Council and the European Parliament, "Wider Europe – Neighbourhood: A New Framework for Relations with our Eastern and Southern Neighbours", COM (2003) 104 final, Brussels. European Commission, *Communication European Neighbourhood*

Strategy Paper, COM (2004) 373 final, Brussels, 12 May 2004, offers the neighbouring countries a privileged relationship with the EU that is based on a mutual commitment to European common values (democracy and human rights, rule of law, good governance, market economy principles and sustainable development). Ultimate objectives of the ENP are political association and deeper economic integration, increased mobility and more people-to-people contacts and access to the EU Internal Market. However, the achievement of the above objectives depends on the extent to which the European common values are shared by the neighbouring countries. Today, the ENP framework embraces 16 of the EU's closest geographical and "political" neighbours – Algeria, Armenia, Azerbaijan, Belarus, Egypt, Georgia, Israel, Jordan, Lebanon, Libya, Moldova, Morocco, Occupied Palestinian Territory, Syria, Tunisia and Ukraine. The EaP is the EU's policy towards the countries of Eastern Europe and Southern Caucasus. The EaP is a specific Eastern dimension of the ENP. Launched in May 2009 at the Prague Summit, the EaP fosters the necessary conditions to accelerate political association and further economic integration between the EU and Armenia, Azerbaijan, Belarus, Georgia, Moldova and Ukraine; Communication from the Commission and the European Parliament to the Council, "Eastern Partnership" COM (2008) 823 final, Brussels, 3 December 2008.

14. Articles 90–93 TEnC.
15. Hitherto, only two neighbouring countries have obtained the full membership in the EnC (Moldova, from 2010, and Ukraine, from 2011); Armenia (from 2011) and Georgia (from 2007) have the observer's status.
16. The Energy Community is an international organization established between the EU and third countries for the purpose of extending the EU Energy Market beyond its borders. The treaty establishing the Energy Community was signed in Athens, Greece on 25 October, and entered into force on 1 July 2006. More information is available at www.energy-community.org, last visited 25 August 2013.
17. The Parliament of Moldova ratified the Protocol on Accession to the EnC on 31 March 2010.
18. Law of Ukraine No. 2787-VI of 15 December 2010.
19. Article 19 TEnC, Articles 101, 102, 106, 107 TFEU.
20. Article 22 TEnC.
21. Article 29 TEnC.
22. Joint Cases C-36–37/97 *Hilmar Kellinghusen v. Amt für Land-und Wasserwirtschaft Kiel and Ernst-Detlef Ketelsen v. Amt für Land-und Wasserwirtschaft Husum* [1998] ECR I-6337. Case C-213/89, *R v. Secretary of State for Transport, ex parte Factortame* [1990] ECR I-2433.
23. Case 5/67 *Beus v. Hauptzollamt München* [1968] ECR 83.
24. Petrov, "Exporting the Acquis Communautaire", 33–52; and R. Petrov, "The Dynamic Nature of the Acquis Aommunautaire in EU External Relations", 18(2) *European Review of Public Law* (2006) 741–771.
25. Article 24 TEnC.
26. Article 25 TEnC.
27. Article 53(f) TEnC.
28. Article 94 TEnC.
29. S. Breitenmoser, "Sectoral Agreements between the EC and Switzerland: Contents and Context", 40 *Common Market Law Review* (2003) 1137–1186.
30. F. Maiani, "Legal Europeanization as Legal Transformation: Some Insights from Swiss 'Outer Europe'", EUI Working Papers MWP 2008/32; R. Schwok, *Switzerland-European Union. An Impossible Membership?* (Brussels: Peter Lang, 2009).

31. Article 89 TEnC.
32. Articles 89–93 TEnC.
33. This is a legislative package adopted by the EU in 2009. It envisages further liberalization of the internal gas and electricity market in the EU and which substitutes Directive 2003/54/EC and Regulation 1228/2003 with new Directive 2009/73/EC, Regulations 715/2009 and 713/2009.
34. Decision of the Ministerial Council of the Energy Community (Mc/06/10/2011).
35. Article 89 TEnC.
36. Article 76 TEnC.
37. Article 89 TEnC.
38. See Report by the Energy Community Secretariat on Dispute Settlement, www.energy-community.org/pls/portal/docs/724180.PDF, last visited 23 February 2014.
39. Article 9 of the Ukrainian Constitution of 1996 provides that "[i]nternational treaties that are in force, agreed to be binding by the Verkhovna Rada [Ukrainian Parliament] of Ukraine, are part of the national legislation of Ukraine. The conclusion of international treaties that contravene the Constitution of Ukraine is possible only after introducing relevant amendments to the Constitution of Ukraine". Full text in English is available at http://gska2.rada.gov.ua/site/const_eng/constitution_eng.htm, last visited 25 August 2013. Article 8 of the Moldovan Constitution of 1994 provides that "[t]he Republic of Moldova pledges to respect the Charter of the United Nations and the treaties to which she is a party, to observe in her relations with other states the unanimously recognized principles and norms of international law. The coming into force of an international treaty containing provisions contrary to the Constitution shall be preceded by a revision of the latter". Full text in English is available at http://confinder.richmond.edu/admin/docs/moldova3.pdf, last visited 25 August 2013.
40. Article 1(2) of the TEnC.
41. R. Petrov and P. Kalinichenko, "The Europeanization of Third Country Judiciaries through the Application of the EU Acquis: The Cases of Russia and Ukraine", 60 *International & Comparative Law Quarterly* (2011) 325–353. This happens mainly because of (1) the belief that international case law is not relevant to civil law systems, (2) the translation of case law and jurisprudence and (3) lack of translation of case law into Ukrainian to help judges adapt their decisions to best European standards. Furthermore, the Verkhovna Rada of Ukraine is not always expedient in solving conflicts between ratified international agreements and national legislation.
42. Judgment of the Appeal Administrative Court of the city of Donetsk on 01 November 2011, No. 2a/0570.
43. Proposal for a Council Decision on the conclusion of the Association Agreement between the European Union and its Member States, of the one part, and Ukraine, of the other part, of 17 May 2013, COM(2013)0290.
44. Article 278 of the EU–Ukraine AA gives unquestionable priority to the approximation and application of the EU energy acquis in the EnC over conflicting provisions of the EU acquis.

Chapter 12

Legal issues relating to cross-border gas pipeline projects

A case study

Bagdagul Kaya Caner[1]

Natural gas is a commodity dependent on the infrastructure. Therefore, not only the producer of natural gas but also whoever owns or hosts the natural gas pipelines will have a strong bargaining power. In the event that the source country and the consuming country are not neighbors, transport of natural gas from the source to the consumer necessitates transit pipelines which pass through the territory of a third sovereign state. New transit pipelines are necessary in the near future since world energy demand is expected to rise.[2] This is also the case for EU where domestic natural gas production is declining.

Approximately ninety percent of the natural gas reaching the EU is transported through pipelines passing through the territories of other sovereign states. At the moment the Russian gas is transported to the EU via Ukraine and Belarus. This means that the natural gas flow from Russia depends on these transit countries' legal regimes. For this reason, unless there is a unified transit regime, every different county where the gas pipeline passes will constitute another source of legal risk concerning transit passage.

Natural gas pipelines link producers and consumers in a long-term legal relationship, require long-term investments, give late paybacks, are very vulnerable to security threats and easily are affected by political relations.[3] More pipelines will be needed in the future from Russia, the Middle East and the Caspian Region which is landlocked – in other words does not possess access to open seas. Therefore, according to Stevens, "exports of oil or gas in any volume will require transit pipelines".[4]

In terms of EU's security of supply, the importance of Turkey's geostrategic position in energy transit has been acknowledged more clearly in recent years. This significance became more apparent in the supply disruptions due to Russian–Ukrainian transit disputes regarding natural gas transportation in 2006 and 2009.[5] Turkey plays an important role as a transit country for most of the projects which were designed to decrease the EU's import dependency to countries such as Russia by hosting transit pipelines for the supplies from Caspian and Middle Eastern countries.

Several pipeline projects involve Turkey as a transit country. Among these projects, the Nabucco Pipeline Project (Nabucco) has attracted attention.

There are several reasons for that; first, the magnitude of the project is remarkable both in terms of the natural gas it will transport and of the countries it will pass though. Second, although Nabucco intends to bring natural gas from the Middle East and Caspian regions, it is designed as a pure pipeline project; that is to say, it is not directly linked to a supply contract. All of these reasons make Nabucco more interesting than conventional cross-border or transit pipeline projects.

In the first section, this chapter discusses the legal issues related to cross-border or transit pipeline projects. In the second section, Turkey's legal regime as a reliable country for energy investments is discussed. In the third section, Nabucco is analyzed both as a pipeline project and as a legal regime.

Legal issues related to transit projects

National sovereignty versus freedom of transit via pipelines

Shaw states that "The states lie upon the foundation of sovereignty. Sovereignty itself is founded upon territory".[6] Jurisdiction can be deemed as the crucial feature of sovereignty and entails judicial, legislative and administrative competence.[7]

As per transit issues, save where otherwise provided in international treaties,[8] states exercise their full sovereign powers in giving permission to any transit pipeline and defining the legal regime regarding transit passage.

According to Stevens, "A transit pipeline is defined as an oil or gas pipeline which crosses another 'sovereign' territory to get its throughput to market. Thus for any transit pipeline to be built requires an agreement made with the government of the sovereign entity".[9] This agreement ought to be interpreted as a form of consent or permission given by the sovereign state from which a transit pipeline will pass through. Even under the United Nations Convention on Law of the Sea (UNCLOS III), the transit passage of the landlocked states via gas pipelines necessitates a further agreement between the related states.[10]

The consent of the state on transit passage may be given under a treaty between two or more states. Otherwise, there may already be national laws and regulations laying down the rules of transit with which the consent of a state on permitting transit passage is already manifested or given in general terms.

Freedom of transit via pipelines as treaty obligations

Treaties can be defined as

> written agreements whereby the states participating bind themselves legally to act in a particular way or to set up particular relations between themselves. It is possible to divide the treaties into two; first one can be

called as law making treaties which are intended to have general or universal relevance, the second type can be called the treaty-contracts which apply between two or a small number of member states.[11]

Several law-making treaties, among other things, also address the issues relating to transit of oil and gas via pipelines. On the other hand, treaty contracts are most commonly used in creating transit pipeline regimes on project basis. They are commonly used both as source of consent from the states regarding a certain transit passage, and as the source for detailing the transit regime.

Transit via pipelines under GATT

The General Agreement on Tariffs and Trade (GATT) of 1947 can be defined as a law-making treaty regarding removing barriers of international trade of goods. Selivanova explains that "originally there was a common perception that GATT rules did not apply to trade in energy".[12]

Some, state Stevens, argue that "during the negotiations setting up the World Trade Organization's (WTO) predecessor, the General Agreement on Tariffs and Trade (GATT), in 1947 there was a 'gentlemen's agreement' to exclude oil and gas from the rules. For example, it can be argued that Saudi Arabia would never have joined WTO, nor would Russia be seeking membership, if the rules did apply to oil and gas".[13] However, there is growing consensus that WTO rules apply to goods in general including energy products.

Either way, supporters of both opinions agree that WTO rules are not well designed to cater for the needs of the energy sector. According to Selivanova, "Traditionally, WTO rules have been devised in a manner to address import barriers to a larger extent than export barriers. In the energy sector, however, the trade restrictions are more pertinent to export barriers. The concerns over security of energy supply lead to relatively low import duties. The barriers in trade in energy are thus very different from those in other sectors – they tend to be on the export side, rather than the import side."[14]

This unsuitability of WTO rules to the energy sector provokes arguments on how to interpret the rules of WTO regarding freedom of transit of energy products which are transported via fixed infrastructures such as pipelines.

Some, according to Goldthau and Witte, argue that GATT Article V on freedom of transit "establishes a general rule of freedom of transit including energy transit based on non-discrimination, it does not address issues that are crucial for energy flows, such as non-interruption of transit flow and non-impediment of building new infrastructure if available capacity is not sufficient. For the latter an effective investment framework is needed".[15]

Energy transit requires construction of pipelines, which necessitates a great amount of investment. These pipelines are designed for a specific volume which, most of the time, is related with the initial natural gas supply agreement made with the producer state. When the transit gas demand arises, then the freedom of energy transit becomes linked not just to the non-discriminatory use of existing infrastructure but also creates additional transit capacity.[16]

Although there have been situations in which energy transit has been abused by the transit state to charge excessive transit fees or to disrupt the transit flow, until now GATT Article V has never been invoked to address disputes on energy transit via fixed pipelines. As stated by Selivanova, "The problem is especially acute for landlocked states as they do not have an option of sea transportation and are absolutely dependent on transit. The lack of WTO disputes with respect to transit may reflect the fact that most of such problems have arisen outside the WTO".[17]

Otherwise, it might also be due to the fact that, the states are reluctant to acknowledge or they reject to adhere that GATT Article V on freedom of transit is also applicable to transit of energy products which are transported via fixed infrastructures.

Transit under the Energy Charter Treaty

In the beginning of 1990s there had been fragmentation of the legal jurisdictions after the collapse of the Soviet Union and dissolution of former Yugoslavia. These legal jurisdictions included the new states which were the potential transit countries for the pipeline projects which supply natural gas to the EU.[18] Under these circumstances, in order to create an investment-friendly environment for energy transit and investment Energy Charter was initiated as a political declaration. This process resulted in an international agreement known as the Energy Charter Treaty (ECT).

The ECT foresees a transit regime targeting the security of supply and a guarantee of supply of transit investments of the contracting states. Provisions of the ECT regarding transit foresee transit freedom, non-discrimination and national treatment obligations regarding the energy transit and aim to reduce the risks of transit. In addition, it prohibits actions to prevent the construction of the transit lines. Also, participating states are obliged to undertake necessary regulations in order to facilitate the principle of freedom of transit and to facilitate transportation of energy available to guarantee the flow of energy. On the other hand, even in the case of a dispute between contracting countries, they are obliged to guarantee the flow of energy. The relevant provisions of the ECT also foresee instruments regarding the settlement of disputes.

European communities[19] are party to the ECT. The ECT has been ratified by almost all European countries except Norway, Russia and Belarus, which are very important players in cross-border natural gas trade. On the other hand, most North African countries are also outside of the Energy

Charter process. Although Russia was provisionally applying the treaty, Russia declared that it will stop the provisional application and it no longer desires the ratification of the Treaty.[20]

Because the freedom of transit was not addressed properly in the GATT and the WTO, Article 7 of ECT was an attempt to provide more concrete and detailed application of these provisions in energy.[21] Article 7 of ECT on transit is as follows: "Each Contracting Party shall take the necessary measures to facilitate the Transit of Energy Materials and Products consistent with the principle of freedom of transit and without distinction as to the origin, destination or ownership of such Energy Materials and Products or discrimination as to pricing on the basis of such distinctions, and without imposing any unreasonable delays, restrictions or charges."

It is argued that ECT Article 7 favors the principle of state sovereignty over freedom of transit in line with the main character of the ECT which is based on the principle of state sovereignty. As a result, the ECT is favoring territorial sovereignty over the freedom of transit. Stated Waelde, "As a consequence, the ECT only requires that the 'Contracting Parties shall take all necessary measures to facilitate transit of Energy Materials and Products . . .' This definition is rather loose, as it is not clear what is covered by the wording 'measures to facilitate'".[22]

Article V of the GATT foresees that "there shall be freedom of transit", whereas Article 7 states that "contracting parties shall take the necessary measures to facilitate transit . . . consistent with the principle of freedom of transit". According to Waelde, "Although it is assumed that this provision is a direct reference to Article V of the GATT, it has to be concluded that the wording used in the ECT is not as strong as in Article V of the GATT. Moreover, as members of GATT virtually have no experience with the way in which Article V GATT operates, it cannot be expected that Contracting Parties can draw any conclusions from established practice".[23]

In this framework, the Draft Transit Protocol to the Treaty of the Energy Charter was initiated. This protocol aims to strengthen the implementation of the provisions regarding the transit and to reduce operational risks about the transportation of energy. Although negotiations on the Transit Protocol began in 2000, the final text on the talks has not been finalized. The scope of the Draft Protocol covers cost-based tariffs in a non-discriminatory and transparent framework of transit regimes of, and provisions on, the management of constraint networks.

Notwithstanding the related provisions of treaties which the related state is a party to, every state is free to regulate its legal regime in terms of energy investments and transit. This legal regime may also contain provisions regarding the transit fee/royalty which may be demanded by a state for merely giving permission transit passage. This transit fee is demanded in addition to the transport charge of the transit. As Stevens clearly states, "the key difference with a transit pipeline is, as already outlined, that because of the 'sovereign' nature of governments there is no overarching jurisdiction that is effectively enforceable".[24] On the contrary, in other practices, and in particular in the EU, the regulations for transit can be similar to the ones which are applied to national transmission pipelines.

The transit fee demanded by the country in this regard varies from country to country and often depends on the country's bargaining power. The transit of natural gas transmission between the member countries of the EU no longer has this structure. Namely after the Directive No. 2003/55/EC's[25] entry into force, the difference between the EU transit transmission of natural gas and the national transmission activity, are under the same conditions. In this context, agreements according to the 91/296/EEC directive dating 31 May 1999, "Natural Gas Transit through Transition Networks", are still valid. However, in agreements signed after the 2003/55/EC directive, the possibility of discrepancy between the national transmission and transit transmission has been eliminated. That is to say, it is no longer possible for an EU member state to demand a transit fee. The only amount that will be paid is to the pipeline owner or operator as the cost of transmission. The converging of transit and transmission activities of the EU is thought to be a natural situation. This is due to the terms of the common EU market where the natural gas transmission activities are carried out without borders.

The transit of natural gas can be performed either via an independent/exclusive transit pipeline or via using or connecting with the national transmission network of the host state. The issue of the connection of transit natural gas pipelines and the country's own national network brings some other legal problems resulting in disruption to the flow of transit of natural gas. In this case, if there is more withdrawal of gas from the national network transit, flow may be affected. As stated by Stevens, "Domestic use of oil or gas offtake may inhibit the transit country from aggressive behaviour for fear of losing supplies – although this assumes its pressure would be such as to cause throughput to cease. However, this can play both ways since the transit country can refuse to pay, but continue to lift. Supplies cannot be halted without cutting off consumers such as Russia Ukraine Case."[26]

Legal predictability

There are two major issues about the regulations concerning the legal regime of transit. The first one is certainty in the legal regime that the transit pipeline investment will be subjected to. The second one is stability in this legal regime as long as the project continues.

The legal requirements and the permissions need to be known clearly before investment decision of the transit pipeline will be made. Transit pipeline projects are subject to various different regulations in every jurisdiction they pass though. These regulations vary from environmental regulations to provisions regarding land use or from insurance to tax issues.

Legal predictability is also relevant for other national regulations, such as the environmental obligations or the process of vesting of land use rights. These processes necessitate lengthy legal procedures; therefore legal predictability is a necessary prerequisite. For example, environmental impact assessments of the whole planned pipeline route need to be completed. Also depending on national laws, land usage rights need to be granted for the pipelines. These national regulations might affect the overall timing of the project or the obligations of the transit investors to third parties.

The certainty and stability in the legal regime is important not only with regard to viability of the project but also with regard to financing of the project, because most of the major gas pipeline investments need debt financing in order for the project to be realized. These financial institutions include the legal risks in their risk assessments.

The change in predetermined or open public policies or politics may result in non-recovery of sunk cost or inability in retaining forecasted profits. The mentioned policy or politics can be either at a national level or at an EU or international level. The changes may be a result of international or national conjecture instead of economic reasons. These reasons will increase the risk premiums related with the debt financing offered by banks, credit agencies and multinational creditors.[27]

Protection of transit investments via effective dispute resolution

Without effective dispute resolution mechanisms, both the energy investments in general and transit cannot be secured. The disputes regarding cross-border natural gas transport may arise in several circumstances, such as disputes over transit fees, decrease of the gas flow or breach of the specific provisions of the transit agreements.

Transit disputes can have more severe effects than other energy investment disputes because they can result in disruption of gas flow, and for this reason, disputes rising from transit issues ought to be sorted out in a very short time span.

The type of dispute resolution may vary according to the legal framework the under which the cross-border/transit project is foreseen. That is to say, the dispute resolution provisions may be foreseen as a part of the treaty contract which is enacted between two states for a specific transit project; in the absence of such a treaty these provisions might be foreseen under private law contracts which were concluded between host state/host state legal entity and the entity demanding transit passage. Alternatively, multilateral treaty provisions on dispute resolution can be utilized.

The ECT has different dispute settlement mechanisms regarding disputes over the investment, trade, transit or competition and environmental issues.

However, since the ECT lays down different approaches for different disputes, the effectiveness of these different dispute settlement mechanisms varies. This effectiveness can be measured by how many times they are used in related disputes. ECT dispute resolution mechanism regarding host state-investor disputes has been resorted to in more than 25 disputes; however, its dispute resolution mechanism on transit has not been popular until now.[28]

According to ECT Article 7(7), the special dispute settlement mechanism for transit only applies following the exhaustion of all relevant contractual or other dispute resolution remedies previously agreed to between the contracting parties to the dispute or between entities subject to contracting parties' control or jurisdiction:

> The conciliation process is a "fast-track" process. Within 120 days from receipt by the Secretary-General of notification of a transit dispute, the parties to the dispute have to come to an agreement to resolve the dispute. If not, the conciliator recommends a resolution and decides the interim tariffs and other terms and conditions to be observed until the dispute is resolved (ECT)Article 7(7)(c)). Contracting Parties undertake to observe and ensure that the entities under their control or jurisdiction observe the conciliator's decision for 12 months, unless the dispute is resolved sooner.[29]

Azaria contends that the conciliation procedure on transit disputes "has not yet been invoked although there have been several transit disputes that might have triggered its application",[30] including the Russia–Ukraine disputes in 2006 and 2009, although there were other reasons involved in non-application of these provisions. The first dispute only lasted four days and at the time of second dispute Russia declared that it would stop the provisional application of the ECT which Russia signed, however, didn't ratify according to its national law.[31]

Turkey as a reliable legal regime for transit energy investments

It can be argued that possessing natural resources or being in a potential transit geo-strategic position is essential, however not enough to trigger energy trade or attract investments. It is necessary to maintain both political and legal stability in order to create a reliable investment climate.[32] Turkey is a party to ECT and a member to WTO.

On the other hand, Turkey's accession negotiations to EU started in October 2005. Even before that Turkey had initiated extensive reforms in its energy sector by way of liberalizing its energy markets in line with the EU legislation on internal energy market.

In this framework, in 2001, the Natural Gas Market Law[33] entered into force with the objective of developing a fair, stable and competitive energy markets functioning in a transparent and non-discriminatory manner and ensuring independent regulation and supervision in these markets. Accordingly, legal entities are required to obtain licenses in order to be able to engage in any market activities. In line with the EU *aquis* on energy, third party access is to transmission, distribution networks, Liquefied Natural Gas (LNG) terminals and storage facilities is foreseen under tariff regulation of Energy Market Regulatory Authority (EMRA). The liberalization process in the Turkish energy market has been a confidence-building step regarding energy investments.

Although Turkish energy market legislation is in line with EU *aquis*, unlike EU law, there is still a difference between the transit of natural gas and the transmission of natural gas under Turkish legislation. Transit pipeline projects are mainly triggered by project-based intergovernmental agreements.

Turkey has always had strategic importance in transporting oil and gas to European markets. The Baku Tbilisi Ceyhan (BTC) pipeline, which started operation in 2005, constitutes a good example of this. The BTC pipeline, around 1,750 kilometres in length, transports crude oil from offshore oil fields in the Caspian Sea to the Turkish coast of the Mediterranean from where the crude is further shipped via tankers to European markets.[34] The BTC pipeline project also had inter-governmental agreements and host governmental agreements signed in November 1999.[35]

Likewise, the idea of the South European Gas Corridor attracted great attention. The South European Gas Corridor would allow for the resources of the Caspian basin, Russia, the Middle East, Southern Mediterranean countries and other international sources to flow through Turkey to European markets. The importance of Turkey in this regard has been apparent once again in the communication from the European Commission on 'Long term Infrastructure Vision for Europe and Beyond', dated 14 October 2013.[36]

As the initial step of creating the South European Gas Corridor, the Interconnector Turkey–Greece ("ITG") pipeline was envisaged. Following the completion of the pipeline, according to the agreement between Turkish BOTAS and Greek DEPA, the first gas delivery to Greece started in November 2007. Additionally, an agreement between the Italian EDISON and the Greek DEPA was signed on 31 December 2006 in order to expand the Turkey–Greece Natural Gas Pipeline Project to Italy via a pipeline passing through the Adriatic Sea. DEPA and EDISON established Poseidon Co., as a joint venture on 13 June 2008.[37] The trilateral Intergovernmental Agreement (IGA) was signed for the Turkey–Greece–Italy Interconnector (ITGI) on 26 July 2007. The ITGI project is forecasted to begin in 2018.[38]

Again within the Southern European Gas Corridor, Nabucco, a pipeline planned to pass through Bulgaria, Romania, Hungary and Austria, demonstrates another transport route from Turkey to Europe. The Nabucco project is dealt with in detail later. The Intergovernmental Agreement of Nabucco Project was signed on 13 July 2009 in Ankara, Turkey. Additionally, the project support agreements were signed between the Nabucco companies and the responsible ministries of the five transit countries on 8 June 2011 in Kayseri, Turkey.[39]

Moreover, an intergovernmental agreement relating to the sale of Azeri gas to Turkey and the transit passage of natural gas originating from Azerbaijan across Turkey through the development of a pipeline was signed on 25 October 2011 between Turkey and Azerbaijan. Subsequently, on 26 June 2012, an intergovernmental agreement of the Trans-Anatolia Pipeline (TANAP) project was signed between Turkey and Azerbaijan and the host government agreement of the TANAP project was signed between Turkey and the TANAP project entity.[40]

Furthermore, on June 2013, the consortium developing Azerbaijan's Shah Deniz gas field has chosen the Trans Adriatic Pipeline (TAP) over Nabucco West which was proposed as the shorter version of original Nabucco project starting from the Turkish–Bulgarian border. The TAP project, first envisioned more than a decade ago by a Swiss company EGL, is expected to start flowing in 2019.[41] TAP's shareholding is composed of BP (20%), SOCAR (20%), Statoil (20%), Fluxys (16%), Total (10%), E.ON (9%) and Axpo (5%). TAP plans to transport natural gas from the giant Shah Deniz II field in Azerbaijan, via Greece and Albania, and across the Adriatic Sea to southern Italy, and farther to Western Europe.[42]

Nabucco pipeline project

Nabucco was regarded as one of the priority projects of EU in the Trans-European Networks.[43] Additionally, the Nabucco project was identified in the European Council of 24 March 2007 as one of the priority projects that will benefit from an EU coordinator.[44] This EU action clearly manifested the EU support regarding the project.

Nabucco as a different pipeline project

The pipeline was designed to link the sources which are available at the eastern borders of Turkey, to Baumgarten in Austria via Bulgaria, Romania and Hungary. The pipeline has been estimated to be between 3,300 and 4,400 kilometres long. The pipeline's annual capacity has been expected to be 31 billion cubic metres (bcm). The route was intended to start at the eastern borders of Turkey, then cross Turkey with an up to 2,730-kilometre-long pipeline; the pipelines crossing Bulgaria, Romania, Hungary and Austria will be up to 412, 468, 384 and 47 kilometres long, respectively. According to the Nabucco International Company; initially the construction would have started in 2012 and the first gas would have flowed by 2015 and the gas would have been supplied from the Caspian region, the Middle East and Egypt. The total investment required for the realization of the project was approximately 7.9 billion euros.[45] Łoskot-Strachota states the following:

> Gas pipeline projects are usually attached to big, long-term supply contracts. Pipelines are built by the companies that supply the gas or the ones that buy it, or both, as is the case with Nord Stream (a joint venture between Gazprom, on the one hand, and German, Dutch and French gas distribution companies, on the other). Nabucco is different. The Nabucco consortium is made up of companies from the countries that lie along the pipeline's prospective route.[46]

When it comes to transit, Stevens explains that "many of the projects are in effect a series of joint ventures including a mix of private companies, government and state-owned enterprises. This joint-venture approach in part reflects a desire to spread the risks inherent in such projects".[47] Similarly, in order to realize the project the initial talks took place in December 2002 between OMV Gas & Power GmbH (Austria) and BOTAŞ (Turkey), which is the state-owned vertically integrated natural gas company in Turkey. The discussion with MOL (Hungary), TRANSGAZ (Romania) and BULGARGAZ (Bulgaria) followed. Subsequently, in 2004 the Nabucco Gas Pipeline International Gmbh was founded as an Austrian-based company between the previously mentioned five companies of the five states the pipelines were intended to pass through. On the other hand, Nabucco International shall establish one Nabucco National Company in each of the host states. These national companies shall not deal with selling of the capacities in the pipelines they shall only deal with technical operation of the pipelines in their respective jurisdictions. The capacity of the pipeline would be sold in Austria by Nabucco International Company solely, under the one-stop-shop principle.

In late 2007, a search for a new shareholder was initiated. A new shareholder would not only contribute to the political support to the project but would also mean a new financial source for the project, because Nabucco's financing strategy is said to be based on 30% equity and 70% debt financing. RWE has been chosen after rather keen competition (with the French Gaz de France among the rivals), which proves that the project is attractive to European companies.[48] In February 2008 the German company RWE became the sixth partner of Nabucco Gas Pipeline International Gmbh. At the moment, Bulgarian Energy Holding (Bulgaria), BOTAŞ (Turkey), MOL (Hungary), OMV (Austria), RWE (Germany) and TRANSGAZ (Romania) are equal shareholders in the Nabucco Gas Pipeline International Gmbh.

As mentioned previously, Nabucco is quite different from the conventional cross-border project in which, first, a supply agreement of the natural gas is enacted between the source state and the consumer state. It can be said that Nabucco can be defined as a pure pipeline project in which the Nabucco

Gas Pipeline International Gmbh will only sell the pipeline capacity, not the gas in it. In this framework, it expected from the pipeline users to find the gas that will be transported by Nabucco. Nabucco Gas Pipeline International Gmbh will sell the capacity according to one-stop-shop principle. Under this idea, 50% of the total pipeline capacity will be auctioned amongst shareholders of Nabucco Gas Pipeline International Gmbh. The remaining 50% of capacity will be reserved for third-party access. To enhance market liquidity 10% of capacity will be assigned to short-term contracts. Regarding the capacity that has already been bought, in order to avoid capacity hoarding, effective congestion management including the "use-it-or-lose-it" principle will be also be applied.

Nabucco as a different legal regime

Although, the project was initiated in 2002, it was not possible to enact an intergovernmental agreement until 2009. Additionally, the final investment decision of the project has not been made yet. At the moment, the construction of the pipeline is supported by the Nabucco intergovernmental agreement[49] signed in Ankara in July 2009, which harmonizes the legal framework and grants stable and equal transport conditions for all partners and customers.[50] According to Barysch, "By March 2010, all of them (potential host states) had ratified the IGA and some had started the environmental impact assessments required for the project to go ahead."[51]

The Nabucco IGA stipulates the enactment of further agreements, namely the Nabucco Project Support Agreements between Nabucco International Company and each host state separately. According to Article 2 of the Nabucco IGA:

> Nabucco project support agreement shall mean a contract governed by private law and as far as the specific circumstances of a specific party require also by public law that is concluded in support of the agreement between a state party and Nabucco International Company and the relevant Nabucco National Company. On issues including but not limited to regulatory conditions investment protection and other issues as defined in the project support agreement.

There are two main reasons why Nabucco necessitated an IGA despite four of the states involved in the project are already EU member states. First, the portion of the pipeline which will pass through Turkey is more than half the total length, which means most of the legal requirements regarding this pipeline will be out of EU member states' jurisdiction. On the other hand, some of the legal requirements relating to the project are even outside the EU's jurisdiction, which means that the member states have exclusive powers on imposing them. These legal requirements are on issues such as land acquisition and tax structures.

Although it is no longer possible for an EU country to demand transit royalty fees, theoretically that is not the case for Turkey. This is why the IGA was important for Turkey – in order to clarify the transit fee structure. However, in the end, no transit fee was foreseen for any host states including Turkey. Nevertheless, a different kind of income was generated to all host states over the tax structure. This tax structure is based on the pipeline length in the related host country. Had this not been agreed to, Austria would have been in a better of situation because Nabucco International Company is an Austria-based company, and all the gas transported via Nabucco would be sold there under the one-stop-shop principle.

Because Nabucco is a pure pipeline project, it needed to be attractive to the potential clients who will transport their gas via Nabucco. Additionally, in order for the project to be economically feasible in attaining finance from creditors it needed to secure its income. To this end, Nabucco's marketing and feasibility strategy is based on long-term capacity reservations in the pipeline under long-term transmission tariffs.

However, as a crucial part of the EU's liberalization process with regard to the natural gas market, ensuring third-party access to transmission pipelines under cost-reflective, approved tariffs is obligatory within all EU member states. Nevertheless, Article 22 of Directive 2003/55/EC concerning common rules for the internal market in natural gas provides for exemptions from third-party-access obligations. Member states or the national regulators can grant those exemptions, if specific criteria are fulfilled. The European Commission has to be notified of an exemption and, within two months, can request that the decision is amended or withdrawn. Criteria for the exemptions are

- the investment must enhance competition in gas supply and enhance security of supply,
- the level of risk attached to the investment is such that the investment would not take place unless an exemption was granted,
- the infrastructure must be owned by a natural or legal person which is separate at least in terms of its legal form from the system operators in whose systems that infrastructure will be built,
- charges are levied on users of that infrastructure and
- the exemption is not detrimental to competition or the effective functioning of the internal gas market or the efficient functioning of the regulated system to which the infrastructure is connected.

In this framework, Nabucco attained exemptions for 25 years regarding TPA and long-term tariff from all the host EU states, and the exemption decisions were approved by European Commission. However, the exemptions could not work without the application of the same legal regime in Turkey as well, because the project had to be governed by the same legal regime as a whole. That is to say, the total capacity of the pipeline starting in Turkey and ending in Austria needed to be sold in Austria on a long-term basis and with long-term tariffs. Thus, the notion of exemption from TPA and long-term tariff structure was inserted in the Nabucco IGA for the Turkish part as well.

Although the term *transit* is absent in the wording, Nabucco IGA stipulates that each state party shall refrain from actions which might impede, curtail or delay the project. Additionally, states agreed on facilitating the administrative procedures on issues such as land use, grants and other concessions. However, regarding environmental standards only cooperation and coordination between the states was foreseen.

Finally, regarding disputes arising from the application of the Nabucco IGA, constitution of an *ad hoc* tribunal is foreseen. The *ad hoc* tribunal shall be resorted to if an amicable agreement between states cannot be reached within 90 days and this tribunal shall be constituted and conduct proceedings according to ECT Article 27 (7). However, it is more likely that the disputes might arise not between the host states but between Nabucco Company and the host states. It is expected that there shall be detailed provisions in project support agreements which shall be enacted between each state and Nabucco Company. However, in the absence of it, the ECT investment dispute-resolution mechanisms can be utilized in case of a dispute.

Conclusion

Looking at EU's growing demand for natural gas, it is apparent that cross-border and transit pipeline projects will be essential to EU's energy security of supply.

However, unsuitability of the GATT rules on transit of energy via fixed infrastructures poses an important threat to freedom of transit. On the other hand, the inefficiency of transit rules in the ECT makes the treaty less attractive regarding transit disputes. Under these circumstances, the need for project-based treaty contracts in realizing transit/cross-border projects remains as the most popular way to promote and protect transit/cross-border investments.

On the other hand, Turkey is able to provide unique opportunities as a transit country. This is due not only to Turkey's geo-strategic position, which constitutes a bridge between natural gas source countries and the EU, but also to Turkey's track record as a stable legal environment for transit projects. Additionally, Turkey as an accession country to the EU provides a legal framework which is in line with the EU energy legislation.

The Nabucco Intergovernmental Agreement is also a good example of harmonizing the legal regimes in different jurisdictions of the pipeline. Whatever the outcome of Nabucco as a project, the Nabucco intergovernmental agreement should be deemed as a success in harmonizing five different legal regimes.

Notes

1. The findings, interpretations, and conclusions expressed in this chapter are those of the author and do not reflect the positions of the institution.
2. Paul Stevens, "Transit Troubles Pipelines as a Source of Conflict – A Chatham House Report", Royal Institute of International Affairs Publishing (2009), p. 7.
3. Brenda Shaffer, *Energy Politics*, University of Pennsylvania Press (2010), s. 47.
4. Stevens, "Transit Troubles", p. 3.
5. Danae Azaria, "Energy Transit under the Energy Charter Treaty and the General Agreement on Tariffs and Trade", *Journal of Energy & Natural Resources Law* 27, no. 4 (2009), p. 583.
6. Malcolm Nathan Shaw, *International Law*, Cambridge University Press (2003), p. 409.
7. Ian Brownlie, *Principles of Public International Law*, 7th Edition, Oxford University Press (2008), p. 299.
8. Freedom of transit as treaty obligations is discussed in the next chapter.
9. Stevens, "Transit Troubles", p. 1.
10. Kishor Uprety, *Transit Regime fort the Landlocked States*, World Bank (2006), p. 117.
11. Shaw, *International Law*, p. 409.
12. Yulia Selivanova, *The WTO and Energy: WTO Rules and Agreements of Relevance to the Energy Sector*, ICTSD Trade and Sustainable Energy Series Issue Paper No. 1, International Centre for Trade and Sustainable Development, (2007), p. 14.
13. Stevens, "Transit Troubles", p. 25.
14. Selivanova, *The WTO and Energy*, p. 16.
15. Andreas Goldthau and Jan Martin Witte, *Global Energy Governance: The New Rules of the Game*, Brookings Institution Press (2010), p. 56.
16. Ibid., p. 56.
17. Selivanova, *The WTO and Energy*, p. 30.
18. Stevens, "Transit Troubles", p. 4.
19. After the Lisbon Treaty, the European Union became a legal entity.
20. For further information on application of a treaty, see The Vienna Convention on Law of Treaties, Article 25.
21. Matthew D. Slater, *The Energy Charter Treaty: A Brief Introduction to Its Scope and Initial Arbitral Awards, Alternative Dispute Resolution in the Energy Sector*, Mukla Publishing (2009), p. 23.
22. Thomas W. Waelde, *The Energy Charter Treaty: An East-West Gateway for Investment and Trade*, Kluwer Law International (1996), p. 509.
23. Ibid., p. 509.
24. Stevens, "Transit Troubles", p. 2.
25. Directive 2003/55/EC of the European Parliament and of the Council of 26 June 2003 concerning common rules for the internal market in natural gas and repealing Directive 98/30/EC OJ L 176, 15.7.2003, pp. 57–78.
26. Stevens, "Transit Troubles", p. 12.
27. Erik Codrington, "Financing for Oil and Gas Projects Is Heating Up", *Global Finance*, http://findarticles.com/p/articles/mi_qa3715/is_200312/ai_n9306081/, accessed on 10 January 2011.
28. For further information, see Energy Charter Treaty Secretaiat Case Statistics, www.encharter.org/index.php?id=269&L=1%252, accessed on 3 January 2011.
29. Azaria, "Energy Transit", p. 582.
30. Ibid., p. 582.
31. Ibid., p. 584.

32. Stevens, "Transit Troubles", p. 5.
33. Natural Gas Market Law No:4646, dated 18 April 2001, published in the Official Journal (OJ) of the Turkish Republic on 2 May 2001, no:24390.
34. Baku Tiblisi Ceyhan Pipeline Information, www.bp.com/sectiongenericarticle.do?categoryId=9006669&contentId= 7015093, accessed on 23 October 2013.
35. www.bp.com/sectiongenericarticle.do?categoryId=9032198&contentId= 7058942, accessed on 23 October 2013.
36. European Commission, *Communication from the Commission to the European Parliament, the Council, the European Economic and Social Committee and the Committee of the Region's Long-Term Infrastructure Vision for Europe and Beyond*, Brussels, 14.10.2013 COM (2013) 711 Final.
37. Poseidon Pipeline Information, www.edison.it/en/media/press-releases/2008–01–25-edison-and-depa-poseidon-co.-a-company-established-to-built-the-italy-greece-natural-gas-pipeline-will-be-operational-in-march.shtml, accessed on 23 October 2013.
38. BOTAS International Projects, www.botas.gov.tr/icerik/tur/projeler/yurtdisi.asp, accessed on 25 October 2013.
39. Ibid.
40. Ibid.
41. Ibid.
42. www.trans-adriatic-pipeline.com, accessed on 28 October 2013.
43. Decision No. 1364/2006/EC, see footnote 35.
44. "Commissioner Piebalgs Recalls European Commitment to Nabucco Project in a Conference Held in Budapest", 14 September 2007, http://europa.eu/rapid/pressReleasesAction.do?reference=IP/07/1338, accessed on 25 December 2010.
45. Nabucco Gas Pipeline International Company and Project Information, www.nabucco-pipeline.com/portal/page/portal/en/press/Facts%20_Figures.
46. Agata Łoskot-Strachota, "Nabucco vs. South Stream – Rivalry over Balkan Gas Pipelines", *CES Commentary*, no. 3, ,1 9.03.2008, Centre For Eastern Studies, p. 2.
47. Stevens, "Transit Troubles", s. 5.
48. Łoskot-Strachota, "Nabucco", p. 1.
49. Ratified with the Law no: 5956 dated 4/3/2010, the English version was published in the Turkish Republic Offical Journal no. 27601 dated 04/06/2010.
50. www.nabucco-pipeline.com/portal/page/portal/en/Home/the_project.
51. Katinka Barysch, "Should Nabucco Be Shelved", 2010, p. 2, www.cer.org.uk/pdf/pb_Nabucco_5may10.pdf, accessed on 25 December 2010.

References

Azaria, Danae. "Energy Transit under the Energy Charter Treaty and the General Agreement on Tariffs and Trade", *Journal of Energy & Natural Resources Law* 27, no. 4 (2009), pp. 559–596.

Baku Tiblisi Ceyhan Pipeline Information, www.bp.com/sectiongenericarticle.do?categoryId=9006669&contentId=7015093.

Barysch, Katinka. "Should Nabucco Be Shelved", 2010, www.cer.org.uk/pdf/pb_Nabucco_5may10.pdf.

BOTAS International Projects, www.botas.gov.tr/icerik/tur/projeler/yurtdisi.asp

Brownlie, Ian. *Principles of Public International Law*, 7th Edition, Oxford University Press (2008).

Codrington, Erik. "Financing for Oil and Gas Projects Is Heating Up", *Global Finance*, http://findarticles.com/p/articles/mi_qa3715/is_200312/ai_n9306081/.

"Commissioner Piebalgs Recalls European Commitment to Nabucco Project in a Conference Held in Budapest", 14 September 2007, http://europa.eu/rapid/pressReleasesAction.do?reference=IP/07/1338.

Energy Charter Treaty Secretariat Case Statistics, www.encharter.org/index.php?id=269&L=1%252

European Commission. *Communication from the Commission to the European Parliament, the Council, the European Economic and Social Committee and the Committee of the Region's Long-Term Infrastructure Vision for Europe and Beyond*, Brussels, 14.10.2013 COM (2013) 711 Final.

Goldthau, Andreas, and Witte, Jan Martin. *Global Energy Governance: The New Rules of the Game*, Brookings Institution Press (2010).

Łoskot-Strachota, Agata. "Nabucco vs. South Stream – Rivalry over Balkan Gas Pipelines", *CES Commentary*, no. 3, 19.03.2008, Centre for Eastern Studies.

Nabucco Gas Pipeline International Company and Project Information, www.nabucco-pipeline.com/portal/page/portal/en/press/Facts%20_Figures.

Poseidon Pipeline Information, www.edison.it/en/media/press-releases/2008–01–25-edison-and-depa-poseidon-co.-a-company-established-to-built-the-italy-greece-natural-gas-pipeline-will-be-operational-in-march.shtml.

Selivanova, Yulia. *The WTO and Energy: WTO Rules and Agreements of Relevance to the Energy Sector*, ICTSD Trade and Sustainable Energy Series Issue Paper No. 1, International Centre for Trade and Sustainable Development (2007).

Shaffer, Brenda. *Energy Politics*, University of Pennsylvania Press (2010).

Shaw, Malcolm Nathan. *International Law*, Cambridge University Press (2003).

Slater, Matthew D. *The Energy Charter Treaty: A Brief Introduction to Its Scope and Initial Arbitral Awards, Alternative Dispute Resolution in the Energy Sector*, Mukla Publishing (2009).

Stevens, Paul. "Transit Troubles Pipelines as a Source of Conflict – A Chatham House Report", Royal Institute of International Affairs (2009).

Uprety, Kishor. Transit Regime for Landlocked States, International law and development perspective, World Bank (2006).

Waelde, Thomas W. *The Energy Charter Treaty: An East-West Gateway for Investment and Trade*, Kluwer Law International (1996).

Chapter 13

National independent regulatory authorities as a case for EU rules promotion in the southern Mediterranean region

Carlo Cambini and Donata Franzi[1]

This chapter investigates the extent of the independence and decision-making autonomy of energy regulatory agencies in the southern Mediterranean region. Independent regulatory agencies (IRAs) are relevant in those cases in which a high degree of infrastructure and innovation investments is required to modernize the utility sector. Agencies' independence from political power and sector's stakeholders, as one of the key features characterizing the process of energy liberalization, is widely recognized as a guarantee of country regulatory credibility.

Regarding Middle Eastern and North Africa (MENA) countries, the electricity sector is currently under reform and regulatory agencies have been set up since the beginning of the 2000s in relevant energy producers and transit countries such as Algeria, Egypt, Turkey and Jordan. The reform process follows the increasing relevance of the southern Mediterranean region in terms of renewable sources, solar and wind mainly, at the centre of public and private European initiatives of cooperation such as the Mediterranean Solar Plan and Desertec.

The study mainly refers to findings from two surveys launched during 2012 among Mediterranean regulators. The first one investigates dimensions of IRAs' independence and identifies an independence index. The second one is a perception survey investigating EU rules promotion as viewed by Mediterranean regulators. Data from the first survey are collected from a questionnaire directed to IRAs or other regulatory bodies operating in Egypt, Jordan, Morocco, Tunisia, Turkey and Croatia. The perception survey provides data for Algerian, Egyptian, Jordanian and Turkish IRAs. Turkey and Croatia have been considered as a "control" factor for the other MENA countries. Being both Mediterranean countries involved by EU accession process, concluded in July 2013 for Croatia, these countries have been in the process of harmonizing their regulatory framework on energy sectors with the EU one for long time. These two countries then represent a useful benchmark on how rule harmonization from the EU may affect the implementation of reforms by a non-EU country.

Difficulties related to data collection were mainly related to missing direct contacts with competent authorities, in the Lebanon and Israel case mainly; at the same time, due to the political turmoil in the region at the time of data collection we were not able to address surveys to Libyan and Syrian authorities and energy companies. For these countries, we collected information from publicly available sources. The sample of our analysis is thus representative of the southern Mediterranean region considering their role in terms of energy production and renewables potentials. Moreover, all of them face similar problems relating to infrastructure needs and adaptation to renewables integration. Our study closes the gap in the literature on the independence of regulators in developing countries, due to the missing analysis of the institutional design of regulatory bodies, as well as tools and practices of regulation, in the region.

Rationale behind IRAs' establishment

IRAs' historical evolution

Regulatory agencies have been established for the first time in the US system, with a great emphasis on transport. Defined as commissions, these bodies were designed to be independent, in their decision making from the executive; accountable, in front of the elected bodies (parliaments); and autonomous, in terms of financial resources to be dedicated to their own management and recruiting of experts. As reported in Joskow (2007), the first examples of US commissions were mainly advisory bodies that had to inform the state's incumbents about the railroads sector. With the adoption of the Granger laws in the 1870s in midwestern states, these bodies started to be recognized as having a certain degree of autonomy and decision-making power on prices. At the end of the 1880s, almost 20 states had railroad advisory commissions. The spread to other sectors, however, was slow.

Clemens (1950) reports on the Board of Gas Commissioners established by Massachusetts in 1885; in 1887 its power was extended to electric light companies. Only after the First World War did two-thirds of US states create public commissions for the utilities sector. Stigler and Friedland (1962) and Phillips (1993) analyzed the introduction of regulation in the electricity industry. Since the beginning of 1930s in the US, the gas distribution, electric power, water, telephone and telegraph service prices became managed by regulatory agencies at a federal level: for example, the Federal Communication Commission (FCC) established with the US Communications Act of 1934, the Federal Power Commission (FPC, currently Federal Energy Regulatory Commission) expanded by the Federal Power Act in 1935, and the Public Utility Holding Company Act, 1935, which established the Securities and Exchange Commission (SEC), which has regulatory responsibilities for interstate gas and electric public utility holding companies.

Starting from the US, the institutional model of regulatory agencies spreads worldwide. In Europe, Great Britain was the first country to adopt this model and properly restructure the energy industry (Saal, 2002). At the same time regulatory agencies were developed at the European Union level following the US model (Geradin, 2004). In Latin America, the experience of Chile – where the energy IRA was established in 1978 – has been widely analyzed as crucial both for reforms' sequencing (Newbery, 2001) and for their impact on market evolution. Among developing countries, Latin America and Caribbean cases have been mainly studied (Gutierrez, 2003; Andres, Foster, and Guasch, 2006; Correa et al., 2006; Andres et al. 2007; Andres, Guasch, and Azumendi, 2008). Specifically, Cubbin and Stern (2006) show that, for a panel of electric utilities in developing countries tracked from 1980 to 2001, in those countries where an independent agency has been set up, generation capacity has been improved, confirming the relation between performance of the utility sector and the governance of regulatory institutions. Egert (2009), using industry-level data from 13 Organisation for Economic Co-operation and Development (OECD) countries, shows that incentive regulation implemented jointly with an independent regulator has a sizeable positive impact on investment in network industries; however, when taken separately, the two variables do not display any significant effect.[2] Trillas and Montoya (2011) present an analysis of the evolution of regulatory independent agencies for 23 Latin American and Caribbean countries in the telecommunications industry and show that a higher degree of authorities' independence is associated with higher network penetration. Moving to the European countries, Cambini and Rondi (2011) find that the inception of an IRA has a positive impact on the investment decisions of a large panel of public utilities, including energy, telecoms, transport and water companies.

The spread of regulatory agencies all around the world, together with the features characterizing their independence from the incumbents' discretion, confirm the effectiveness of a regulatory governance model that considers the institutional design of agencies as the element that may offer a first insurance on investment decisions to market operators. Considering the involvement of the Mediterranean region in huge investment projects for electricity generation, from both conventional and renewable sources, the assessment of the institutional design of newly established regulatory agencies remains essential together with the evaluation of regulatory tools and extent of the agencies' powers. The section that follows analyzes the rationale behind the establishment of independent IRAs, developing some hypotheses related to the establishment of agencies in the MENA context.

Rational for IRA establishment and independence in developing countries: The southern Mediterranean case

The rationale behind the establishment of IRAs, in a context where the liberalization of the electricity sector is still missing, is the subject under debate in this section.

The existing literature shows that the diffusion of IRAs in developing countries, coinciding with the liberalization of utilities, is a case of effective regulation. The establishment of IRAs signals the intention to implement sector' reforms, while providing IRA legitimacy with legal backing and increases the regulatory reputation of a country (Cubbin and Stern, 2006). Thus, countries' institutional endowment (Levy and Spiller, 1994) and political cleavages, directly affect the establishment and functioning of IRAs (Jordana and Levi-Faur, 2006). Incumbents' ideological cleavages, mainly, are the determinants of the model of regulation adopted, usually distinguished between market conforming and market controlling (Gallardo and Murillo, 2011).

Considering MENA institutional endowment, a market-controlling model seems to be prevailing. Let's take as an example the problem of time inconsistency in the incumbents' decisions, which is one of the explanatory factors for the establishment of IRAs the literature recognizes. Energy is one of the sectors in which time inconsistency problems arise in association with different and very often contending social interests. If in democratic contexts, the alternation of parties in power reveals such contending interests, in authoritarian regimes the representativeness of these interests is missing. Whereas in democracies, delegating the rules' implementation phase to technical agencies may reduce the instability of the regulatory framework of a country usually associated to the alternation of parties representing different interests, in the majority of MENA countries long-lasting regimes show that such a "risk" of being replaced through democratic alteration of parties in power was almost absent for the past 20 years.

Nonetheless, popular uprisings throughout 2011, as well as the current political situation, reveal vulnerability of authoritarian leaders and, at the same time, consolidated groups of powers.

First, as recent "Arab Spring" events confirm the army is a pivotal power in the region. In the past, MENA leaders "*had strong relations with the army. Often, politicians never ended relationships forged at the end of the colonial period as can be seen in the fact that, in almost every case, the head of the executive power is a former member of the army*" (Franzi, 2013, p. 71, emphasis in original). In the aftermath of recent popular uprisings, the army has a specific role in managing both the political life and the transition phases, as can be seen in Egypt and Tunisia albeit with huge differences.

Second, a bureaucracy widely connected with the ruling elites is almost inevitable in countries without any switching of political parties in power, such as in the majority of MENA countries. Indeed, the lack of sufficient checks and balances between domestic institutions, strong power of incumbents, poor level of rule of law are those institutional factors characterizing the regulatory framework in most of these countries. The close relation between bureaucrats and incumbents makes governments self-binding, limiting the effectiveness of IRA's decision making. The close relationship between bureaucrats and incumbents reinforces the capacity of elites in powers to infiltrate bureaucracy (Gilardi, 2005a). Similarly, the lingering relationship between incumbents and bureaucrats strengthens bureaucratic elites, making bureaucrats among the most relevant players in the region.

Because the MENA economy is mainly publicly driven, bureaucrats are also influent actors in those reforming processes that involve the utility sector. To a certain extent, bureaucrats may infiltrate the elite in power, through the development of knowledge and technical expertise necessary for influencing the implementation of rules and reforming projects; they are the actors who may assure continuity in the sector's management and stability of the regulatory framework also in case of unexpected regime change, such as the ones that occurred during 2011–2012. Both forms of infiltration may favour a marketing-controlling model of regulation.

Measuring southern Mediterranean agencies' independence: Methodology of the study

Our analysis is directed at defining the degree of independence that regulators may have in the MENA context looking mainly at decision-making autonomy, organizational autonomy and accountability measures.

A first survey was launched in order to understand the extent of tools available to regulatory agencies as well as regulators' organizational autonomy and accountability. The survey, directed at both energy regulators and energy companies, enlightens the issues of regulators' independence and the credibility of the regulatory framework in the Mediterranean region. The paragraphs that follow clarify the dimensions constituting the definition of regulators' independence, such as those applied to our case study. The data collection method and the assessment of the independence index method follow this.

Dimensions of IRAs' independence

There are three dimensions identified in the survey that characterize regulators' effective independence:

- the regulatory tools available to IRAs and the decision-making process,
- regulators' organizational autonomy and
- accountability measures.

Considering the tools available to regulators, we refer to the involvement of regulators in aspects that characterize the management and restructuring of the electricity sector: tariff-setting mechanisms, system unbundling and the definition of Third Party Access (TPA) regime. Moreover, regulators' competences regarding the licensing issue, dispute settlement, consumer protection and service quality regulation are considered. In terms of regulators' organizational autonomy, we mainly consider both economic and human resources available to regulators, autonomy in resources management, the personnel policy and internal organization. Focusing on accountability, we analyze the relations between regulators and governments, the formal obligations of regulators towards the executive (the president, the prime minister and ministries) and legislative powers and the eventual involvement of the magistracy in the case of an appeal against the regulators' decisions.

The focus on the three identified dimensions is derived from a thorough analysis of the relevant literature. The first point to assess is the tools available to regulators that are pivotal to characterize IRAs' independence. Limiting the regulators' ability to be actively involved by regulated companies affects the independence of the agency and increases the risk of capture by energy industries. Moreover, in the MENA context, we should be aware of direct relationships that exist between the bureaucratic heads and incumbents, as well as of the state-owned nature of energy industries, which makes political interference more likely. A form of political interference may emerge when the issues at stake are mainly related to tariffs. Consumers' opinions have an impact on incumbent politicians whose main interest is not to lose support and votes while maintaining social order and stability. In this case, companies' managers and politicians share the same interest: "*underpricing to stimulate demand and secure political support*" (Joskow 2005, p. 9, emphasis in original). Thus, regulatory capture by both regulated companies and politicians is what makes regulators subject to particular interests (Parker, 2002). Moreover, a weak regulator may work in close cooperation with the firm that is supposed to be regulated which can lead to discriminatory regimes at the expense of other companies, specifically foreign ones. Almost all MENA countries that have adopted liberalization measures maintain discriminatory regimes to the detriment of foreign investors.

Regarding organizational autonomy, Armstrong and Sappington (2006) state that in order to have optimal and correct regulation, emphasis must be placed on the importance of the resources available to regulators. The available economic resources and, consequently, the possibility to recruit experts as well as properly manage relations with the stakeholders (i.e. energy companies and consumers), are different dimensions of regulators' independence that should be considered when designing an optimal regulatory policy. In contexts in which the resources available are limited, both achieving information and recruiting qualified experts are difficult. A proper budget, which ensures the possibility to pay competitive salaries and to finance an agency's costs, thus, is the first structural condition for obtaining and maintaining independent IRAs.

Finally, a further risk for the independence of IRAs is the misunderstood accountability requirements, which are often viewed as an obligation to ask for the state's permission for any decision that the regulator has to implement, and/or by the mere consultative role the regulator may have when working with specific regulatory tools, such as the definition of tariffs. The problem associated with the accountability of regulators is a very sensitive issue in a world characterized by increasing "agencyfication" (Majone, 1997). Correctly implemented, accountability provisions imply that regulators work within rules codified by elected bodies (the parliaments) in front of whom they must justify their actions without renouncing their operational independence.

Independence index definition

As seen in the studies of Gilardi (2002) and Johannsen, Pedersen and Sørensen (2004), in order to measure agencies' independence, we adopt binary variables (i.e. yes or no) varying between 0, absence or very low degree of independence, and 1, independent agencies. Then, based on the answers received, the index of regulators' independence was defined as a simple average of the scores of the three dimensions of independence that were considered. As in previous studies undertaken by Correa et al. (2006), Brown et al. (2006) and Andres et al. (2007), our analysis defined three sub-indexes for the assessment of both formal and substantial aspects of the regulatory agencies' independence. Specifically, as for Andres et al. (2007), decision-making legitimacy, autonomy and accountability of regulators are the dimensions on which regulatory performance in each country is measured, and they were considered in the definition of sub-indexes. The independence of IRAs is, thus, described in relative terms: each single dimension for which IRAs are independent contributes relationally to the independence of the agency (Hanretty and Koop, 2012). Thus, sub-indexes for each of the three dimensions considered are available, together with an overall independence index. Answers are reported for each relevant dimension.

Data for countries and organization that did not answer the survey (Algeria, Israel and Lebanon) have not been considered in the ultimate index calculation.

The data collection method

Questions in the survey refer mainly to the international standards for electricity sector liberalization mentioned in EU documents and plans for cooperation adopted first in 1996, when the Euro-Mediterranean energy partnership was launched and then are further specified in MENA country reports released by the European Commission in 2007. Questions referring to the regulators' role, the internal organization and budge and the relations with political powers and stakeholders have been defined based on previous studies (Gilardi, 2005a, 2005b; Johannsen, Pedersen and Sørensen, 2004; Larsen et al., 2006; Trillas and Montoya, 2011). All aspects illustrate the dimensions of regulators' independence in the Mediterranean region.

The survey was launched in two phases. The first, undertaken from summer to autumn 2011, circulated among regulators from Egypt and Algeria that took part in a training week arranged by the Association of Mediterranean Regulators for electricity and gas (MedReg) and the Florence School of Regulators at the European University Institute (Florence) in May 2011. Moreover, thanks to contacts from the Energy Regulators Regional Association (ERRA) secretariat, other regulators in the eastern Mediterranean region received the questionnaire. Answers to this survey have been collected for Egypt, Jordan and Turkey. This first phase was conceived as a preliminary test of the survey. The preliminary questionnaire mainly focused on aspects of accountability and organizational autonomy. After considering the emerging interests for the area due mainly to the investments projects for renewables, the survey was revised, paying more attention to the tools available to regulators, as well as their organizational autonomy. Indeed, it was clear that a stable and transparent regulatory framework against administrative and incumbents' discretion matters more for investors than formal measures of independence. Thus, the section on the availability of tools for regulators received more attention than in the preliminary questionnaire. The survey was uploaded on the website SurveyMonkey and launched in January 2012; regulators have been invited by e-mail mainly.

Data summarizing IRAs involved in the study are provided in Table 13.1. All mentioned authorities took part in the survey with the exception of Algerian, Israel and Lebanon authorities, the data of which mainly derives from the existing literature.

Research findings

This section sums up findings available in Cambini and Franzi (2013). Looking at the decision-making autonomy dimension, Jordan IRA emerge from the survey as fully competent for those aspects identified as pivotal to the sector's liberalization: tariff setting, unbundling and TPA. The other well-performing agency is, at this stage of analysis, the Turkish one having full operational autonomy in all investigated issues with the sole exception of tariff setting. On the contrary, for those Mediterranean countries for which an IRA does not exist, the electricity sector's most sensitive aspects are co-managed by public companies and central administration apparatus. With specific attention to the aspects on the base of which credibility of regulatory commitment is measured, tariff setting, we may conclude that IRAs

Table 13.1 IRAs in MENA countries

Relation with the EU	*Country name*	*IRA's name*	*Acronym*	*Other regulatory body*	*Acronym*	*Sector*	*Year set up*	*Number of employees*	*Normative source*
PC	**Egypt**	Egyptian Electric Utility and Consumer Protection Regulatory Agency	EgyptERA			Electricity	2001	70	Presidential Decree
PC	**Jordan**	Electricity Regulatory Commission	ERC			Electricity	2001	85	General Electricity Law No. 64, 2002
CC	**Turkey**	Energy Market Regulatory Authority	EMRA			Electricity Gas Oil LNG	2001	467	Law no: 4628
CC	**Croatia**	Croatian Energy Regulatory Agency	HERA			Electricity Gas Thermal energy	2005	57	Energy Activities Regulation Act
PC	**Morocco**			Office National d'Electricité	ONE	Electricity	1963	8,705	Dahir
PC	**Tunisia**			Ministry of Industry-Directorate general of Energy	MIN	Electricity Gas		20	Function of the ministry
PC	**Algeria**	Electricity and Gas Regulatory Commission	CREG			Electricity Gas	2002	50	Law N. 02-01 February 5, 2002 on electricity and the distribution of gas
PC	**Israel**	Public Utility Authority	PUA			Electricity	2003	30	Electricity law 2003
PC	**Lebanon**			Ministry of Energy and Water	MEW	Electricity Gas			

PC: partner countries involved in the cooperation programmes; CC: candidate countries to EU membership at the time of data collection.

in the Mediterranean area are not truly autonomous or independent in the decision-making process.

Regarding the second dimension of independence assessed in the study – regulators' operational autonomy – IRAs do not emerge as fully autonomous in terms of internal organization and personnel policy. Apart from the Croatian regulator, such competences are mainly shared between the regulator and the legislative power in the Egyptian, Jordanian and Turkish cases. By looking at the organizational structure of the IRAs' board, it can be seen that the dominant model is the Regulatory Council. This aspect has been positively evaluated as collegial boards being more independent than a single chief, usually directly appointed by the head of executive power.

Budget issues are indeed in the hands of executive power for all considered cases. Budget is one of the most important aspects of regulators' organizational autonomy, due to the possibility of using resources independent of political will, including the possibility to appoint experts and qualified human resources. In the cases analyzed, executive power is competent in the definition and approval of the regulator's budget.

Accountability measures depict IRAs as advisory bodies of governments in a region where the electricity sector firms are mainly the state-owned energy companies. Specifically, the survey gives the possibility to choose among advisory, ministry, and independent model. With the exception of Egypt, which chooses the ministry model (IRA is part of the executive body), the other respondents selected the independent model, which do not imply any control by the executive and full autonomy regarding the relation with utilities. But, when checking these answers with those provided under the first dimension of independence, it is possible to see that apart from Jordan, other IRAs involved in the study were not fully autonomous. Thus, it is our opinion that MENA IRAs look more like advisory bodies of the executive than like fully independent bodies; indeed, no compulsory decisions are taken by regulators in key issues, such as tariff setting, but they remain a sort of specialized councillor of governments.

Among the analyzed countries, Lebanon presents the most critical situation due to an unclear legislative framework (sector's reform stopped in 2002 and projects for new laws ruling the sector have never been adopted) and the absence of a regulatory agency competent on licensing and eventually dispute settlement issues. Moreover, Morocco and Tunisia, two of the most relevant countries for investments on renewables, tend to preserve the dominant position of the state companies and limit the potential new entrants. Both of them do not have established an IRA. Eventually, Jordan is the most promising MENA country both in terms of sector's restructuring and agency's autonomy. Indeed, a certain degree of autonomy, more than true independence, is what we can assess in the MENA region.

Such degree of autonomy, registered in our analysis, proves that, consistent with the literature on the IRAs set-up rationale, countries' regulatory

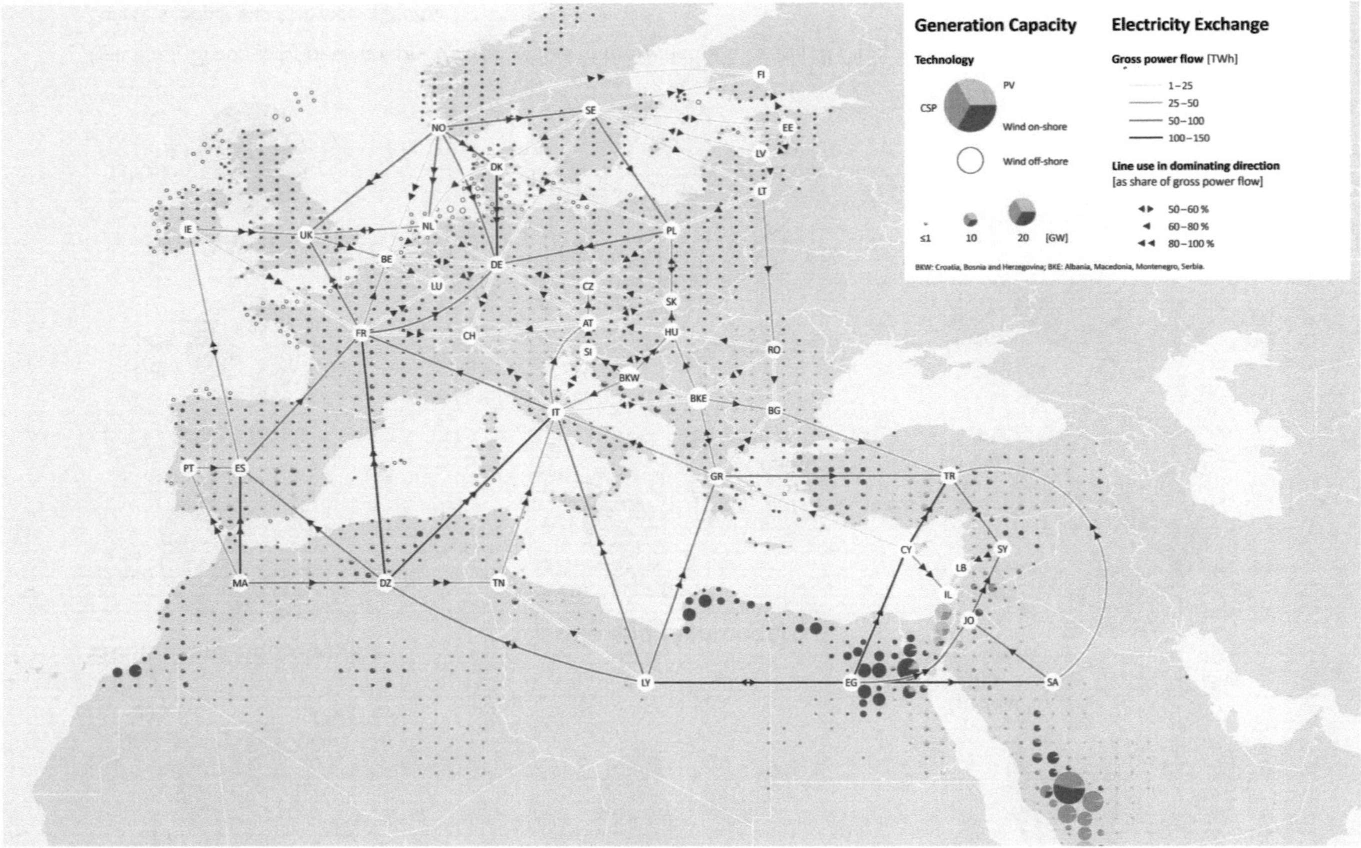

Figure 5.1 Options for renewables and grid infrastructure in MENA countries and around the Mediterranean until 2050

Source: Fraunhofer ISI, Energy Economics Group – TU Wien; Dii GmbH, *Desert Power: Getting Started* (2013).

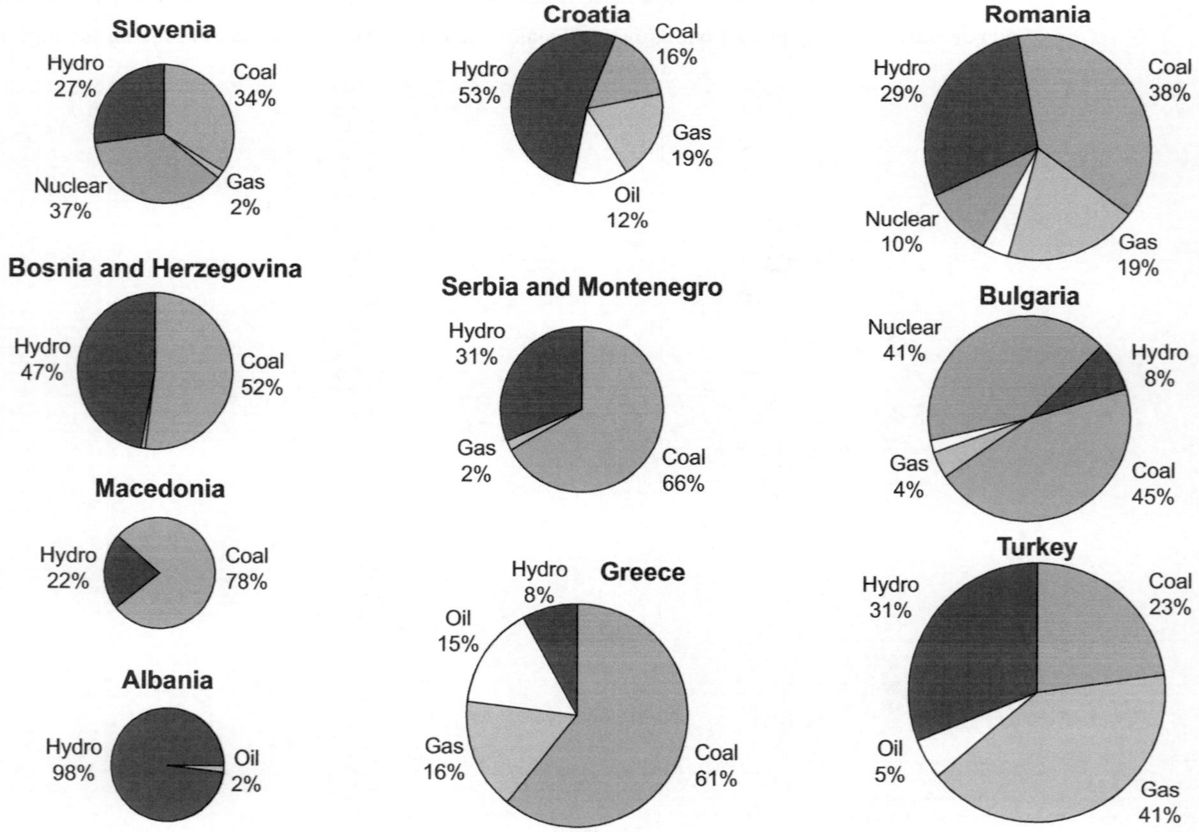

Figure 6.1 Production of electricity by type of fuel in the wider SEE region, 2004

Source: Hooper and Medvedev (2008).

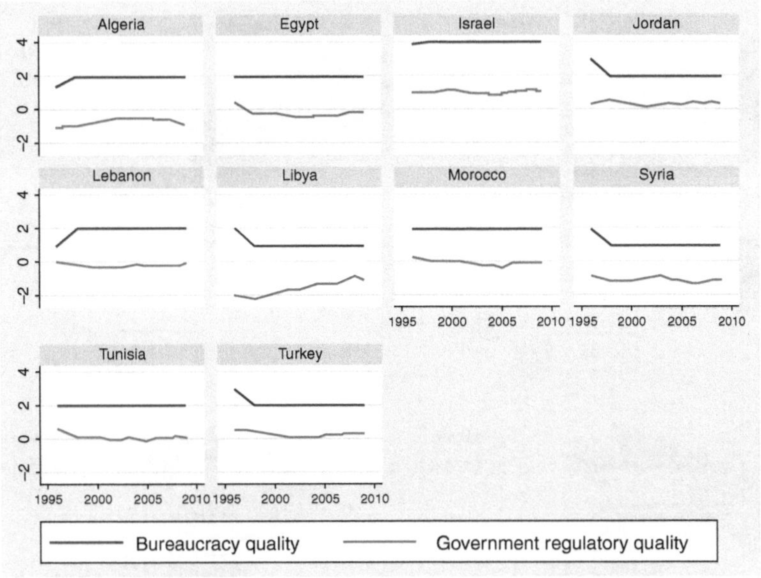

Figure 13.1 Bureaucracy quality and perception of incumbents' regulatory quality

Source: Bureaucracy Quality – International Country Risk Guide; Source: Regulatory Quality – World Bank, Worldwide Governance Indicator.

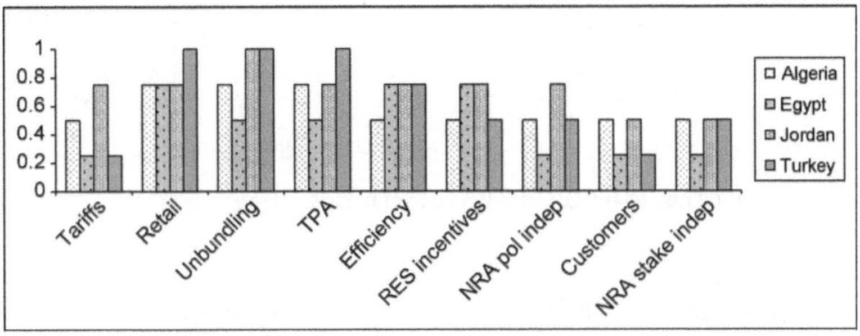

Figure 13.2 EU role in rules adoption

Source: Cambini and Franzi (2013b, p. 15).

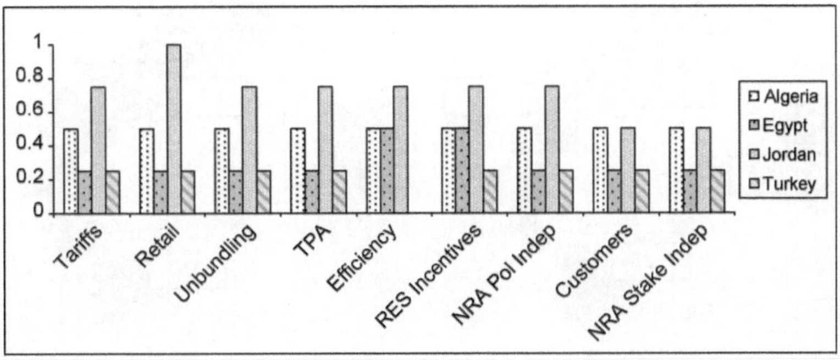

Figure 13.3 MedReg role in rules adoption.

Source: Cambini and Franzi (2013b, p. 17).

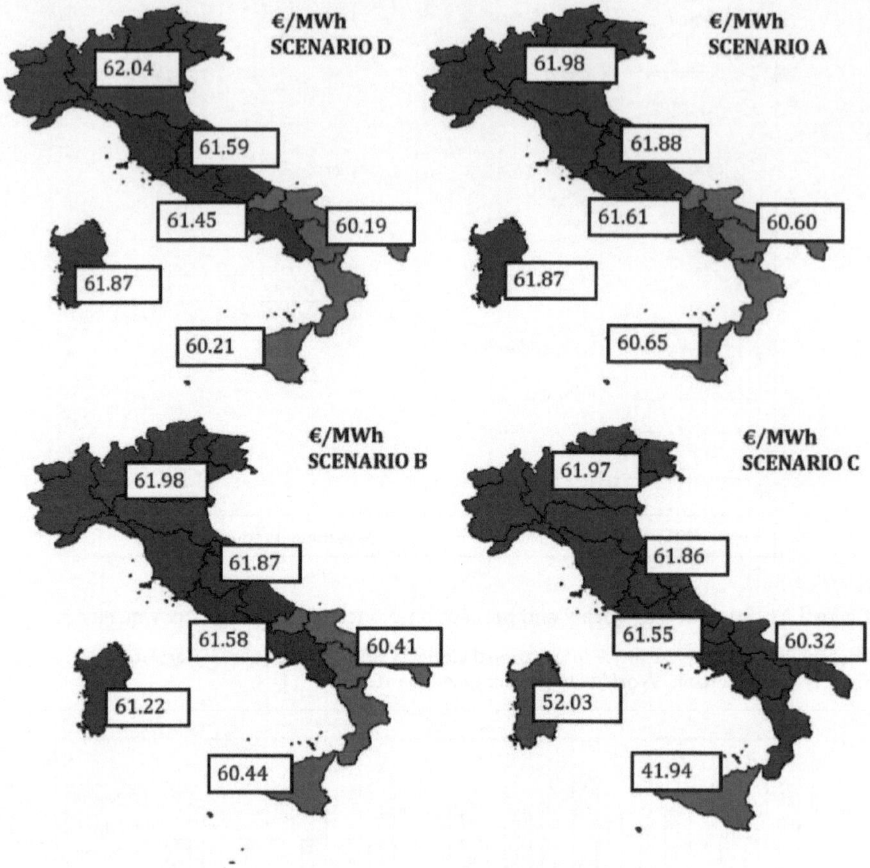

Figure 14.4 Zonal prices (average) in different scenarios

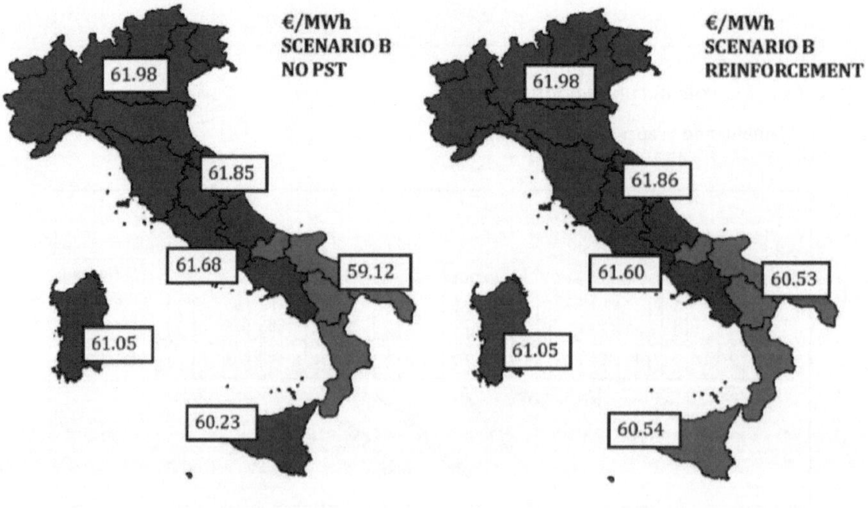

Figure 14.5 Zonal prices (average) in the sensitivity scenario B

commitments are better defined, and the sector better performs, in those countries registering the highest degree of agencies' autonomy and a better performing bureaucracy, than in those having poorly autonomous regulatory bodies. This is also confirmed by data on the perception of incumbents' regulatory quality compared to the quality of bureaucracy (Figure 13.1, color section). The indicators show the relevance of bureaucrats, as we stressed earlier, as actors having a sort of shock-absorber role, mainly in transition phases such as the so-called Arab Spring. The higher the score, the better the function of bureaucracy is. Considering the last 20 years, apart from Israel that obtains the highest value, and Syria and Libya, that are below zero, in all other cases the value is positive and around two points. On the contrary, the perception of incumbents' capability to formulate sound policies in favour of private sector's development (Kaufmann, Kraay and Mastruzzi, 2010) – the index on *Regulatory Quality* – remains in the best cases around zero.

IRAs as model of regulation? Perception of EU rules promotion by MENA regulators

As mentioned in the beginning of the chapter, MENA regulators have been invited to take part in a second survey. The second survey aimed at analyzing the perception of the EU as rules promoter in the energy field. Regulators from four MENA countries – Algeria, Egypt, Turkey and Jordan – participated in the analysis, and two of them – Algeria and Egypt – were also personally interviewed.

Countries involved are relevant due to the different pressure for rules change that the EU may adopt considering the different kind of relations existing at the time of the study (see Table 13.1). Indeed, the study emphasizes the EU's hierarchy pressure for rules adoption as well as the EU's network-based rules promotion (Rhodes, 1996; Lavenex and Schimmelfennig, 2009) or those cases of new modes of governance (Eberlein and Kerwer, 2002; Heritier, 2002) to which MedReg may be compared.

The literature on EU external governance focuses on the external diffusion of the EU acquis, or part of it, because of the pressure applied by the EU in its immediate neighbourhood. Under this approach, replication of the EU system in the neighbouring countries, as well as the achievement of a Euro-Mediterranean Energy Community, is the declared outcome of the promoted process of regional integration. The European Neighbourhood Policy, with the Mediterranean branch, may be conceived as a case of top-down pressure for rules change. Network pressure, on the contrary, is not played by the EU directly but by an organization based on members' commitment; the network being perceived as the place for sharing knowledge and practices of regulation, favouring the progressive and harmonized adoption of common framework of rules. With special attention to networks, we focus on MedReg. MedReg is a membership-based association having the characteristics Rhodes (1996) identified as necessary for networks: voluntary cooperation and membership. It is registered as a non-profit association under Italian law, and its secretariat is based in Milan.

The majority of the questions asked the respondents to express an evaluation of the considered elements on a scale from 0 to 5. To place the countries along a continuum between 0 and 1, all questions using the 0 to 5 scale adopt the following coding system: 5 = 1; 4 = 0.75; 3 = 0.50; 2 and 1 = 0.25; 0 = 0. Main research findings are available in Cambini and Franzi (2014).

Regarding hierarchy pressure, the comparison among countries involved shows that the effectiveness of EU action largely depends on countries' energy interests and relevance. Compared to Egypt, Algeria has different preferences regarding the energy sector regulatory setting, and interviews with regulators show that specific rules for the electricity sector (i.e. unbundling) have been adopted primarily because of the consensus on their effectiveness more than because of the EU recommendation. On the contrary, at the time in which the research was developed, the EU system clearly proves to be relevant for Egypt. The interviews with Egyptian regulators confirm the intention of the policy makers to adopt a system as similar as possible to the EU one. Jordan's survey answers confirm their high appreciation for EU rules and international standards, whereas Turkey only partially confirms the relevance of the membership perspective for the domestic actors' behaviour and the subsequent adoption of EU regulatory framework.

Considering network-based rules diffusion, we asked the respondents to compare indirect rules promotion, primarily through the MedReg network, with the EU's hierarchical pressure. Figures 13.2 and 13.3 (in the color section) synthetically provide for results on such a comparison.

With the exception of Jordan, the MENA countries do not give high scores to MedReg's actions, specifically in the regulatory sector as well as in the promotion of the IRAs' independence. Our findings show that networks are weakened by their perceived lack of legitimacy and by their high structural flexibility.

Concluding remarks

Data collected through the first survey confirm the low degree of independence and decisional autonomy of IRAs in the region. Considering that the majority of MENA countries are characterized by long-lasting regimes, a high degree of bureaucratic infiltration in the regulatory agencies by elites in power was confirmed by data related to the tools available to regulators and the agencies' decision-making power. In fact, MENA incumbents are the dominant actors, together with state-owned companies, in managing specific regulatory tools, such as tariff setting. Jordan is the only country in the MENA region in which the regulatory framework seems more stable than in other MENA countries. When considering the survey data, Jordan appears to be a monarchy highly committed to competitively restructuring the electricity sector.

Findings from the first survey on agencies' independence and organization of the electricity sector need to be combined with those from the second survey aimed at showing the relevance of the EU as rules promoter. The second survey, proving from the perception of EU action from MENA regulators, clearly shows that in terms of regulatory model promotion the EU is perceived as an effective actor mainly by those countries having high energy potentialities but low level of energy (electricity and renewables mainly) deployment, such as in the Egyptian case. Specifically, when comparing the hierarchy and the network pressures, results confirm that the respondents perceive the EU direct rules promotion to be more effective than the network-based one. The same results are observed when MedReg is compared to other organizations, such as ERRA. ERRA pursues the coordination of a different system of rules through training and knowledge exchange. MedReg, indeed, primarily develops studies and policy recommendations. The same objective, thus, is pursued through different methods. How much these differences are perceived by the regulators remains to be investigated.

It is important to stress that in order to be more effective in terms of rules spreading and harmonization, the EU strategy needs to be less centred at the country level and more focused on the energy-corridor dimension. Consistent with previous works on the EU's role as the energy rules promoter in the neighbouring countries (Escribano, 2010; Darbouche, 2011), the focus on energy corridors allows one to take into consideration the policy framework of the EU different countries' energy interests and the relevance of favouring incentives for rule adoption to emerge.

Notes

1. The findings, interpretations, and conclusions expressed in this chapter are those of the authors and do not reflect the positions of the institution.
2. Guasch, Laffont and Straub (2008) studied the impact of regulatory independence in a different context. Using data from 307 transportation and water concession contracts in Argentina, Brazil, Chile, Colombia and Mexico from 1989 to 2000, the authors find that the presence of an IRA lowered the probability of renegotiation by 5% to 7.3%.

References

Andres L., Foster V., and Guasch J.L. (2006) "The Impact of Privatization on the Performance of the Infrastructure Sector: The Case of Electricity Distribution in Latin American Countries". Policy Research Working Paper 3936. Washington, DC: The World Bank.

Andres, L., Guasch, J.L., and Azumendi S.L. (2008) "Regulatory Governance and Sector Performance: Methodology and Evaluation for Electricity Distribution in

Latin America". Policy Research Working Paper 4494. Washington, DC: The World Bank.

Andres L., Guasch, J. L., Diop, M., and Azumendi, S. L. (2007) "Assessing the Governance of Electricity Regulatory Agencies in the Latin American and the Caribbean Region: A Benchmarking Analysis". Policy Research Working Paper 4380. Washington, DC: The World Bank.

Armstrong, M., and Sappington, D.E.M. (2006) "Regulation, Competition and Liberalization", *Journal of Economic Literature*, 44: 325–366.

Brown, A. C., Stern, J., Tenenbaum, B., and Gencer, D. (2006) *Handbook for Evaluating Infrastructure Regulatory Systems*. Washington, DC: The World Bank.

Cambini, C., and Franzi, D., (2013) "Independent Regulatory Agencies and Rules Harmonization for the Electricity Sector and Renewables in the Mediterranean Region", *Energy Policy* 60: 179–191. doi:10.1016/j.enpol.2013.04.078.

Cambini, C., and Franzi, D. (2014) "Assessing the EU Pressure for Rules Change: The Perceptions of Southern Mediterranean Energy Regulators", *Mediterranean Politics,* 19(1): 59–81. doi:10.1080/13629395.2013.809197

Cambini, C., and Rondi, L. (2011) "Regulatory Independence, Investment and Political Interference: Evidence from the European Union". EUI Working paper RSCAS, 42, Florence.

Clemens E.W. (1950) *Economics of Public Utilities*. New York: Appleton-Century-Crofts.

Correa, P., Pereira, C., Mueller, B., and Melo, M. (2006) *Regulatory Governance in Infrastructure Industries: Assessment and Measurement of Brazilian Regulators. Public-Private Infrastructure Advisory Facility*. Washington, DC: The World Bank.

Cubbin, J., and Stern, J. (2006) "The Impact of Regulatory Governance and Privatization on Electricity Industry Generation Capacity in Developing Countries", *The World Bank Economic Review*, 20(1): 115–141.

Darbouche, H. (2011) "Third Time Lucky? Euro-Mediterranean Energy Co-Operation under the Union for the Mediterranean", *Mediterranean Politics*, 16(1): 193–211.

Eberlein, B., and Kerwer, D. (2002) "Theorising the New Modes of European Union Governance", *European Integration Online Papers,* 6: 1–17.

Egert, B. (2009) "Infrastructure Investment in Network Industries: The Role of Incentive Regulation and Regulatory Independence". CESIfo Working Paper 2642, CESIfo Group, Munich (Germany).

Escribano, G. (2010) Convergence towards Differentiation: The Case of Mediterranean Energy Corridors, *Mediterranean Politics*, 15(2): 211–229.

Franzi, D. (2013) *Make Energy from the Sun a Real Growth Perspective*. Saarbrücken (Germany): Lambert Academic Publishing.

Gallardo, C. M., and Murillo M. V. (2011) "Agency Under constraint: Ideological Preferences and the Politics of Electricity Regulation in Latin America", *Regulation & Governance,* 5: 350–367.

Gausch L., Laffont, J. J., and Straub, S. (2008) "Renegotiation of Concession Contracts in Latin America," *International Journal of Industrial Organization*, 26(2): 421–442.

Geradin, D. (2004) "The Development of European Regulatory Agencies: What the EU Should Learn from American Experience", *Columbia Journal of European Law*, 11: 11–52.

Gilardi, F. (2002) "Policy Credibility and Delegation to Independent Regulatory Agencies: A Comparative Empirical Analysis", *Journal of European Public Policy*, 9(6): 873–893.

Gilardi, F. (2005a) "The Formal Independence of Regulators: A Comparison of 17 Countries and 7 Sectors", *Swiss Political Science Review*, 11(4): 139–167.

Gilardi, F. (2005b) "The Institutional Foundations of Regulatory Capitalism: The Diffusion of Independent Regulatory Agencies in Western Europe", *The Annals of the American Academy of Political and Social Science*, 598: 84–101.

Gutierrez, L.H. (2003) "The Effect of Endogenous Regulation on Telecommunications Expansion and Efficiency in Latin America", *Journal of Regulatory Economics*, 23(3): 257–286.

Hanretty, C., and Koop, C. (2012) "Measuring the Formal Independence of Regulatory Agencies." *Journal of European Public Policy*, 19(2): 198–216.

Heritier, A. (2002) *Common Goods: Reinventing European and International Governance*. Lanham, MD: Rowman & Littlefield.

Johannsen, K.S., Pedersen, L.H., and Sørensen, E.M. (2004) "Independent Regulatory Authorities, a Comparative Study of European Energy Regulators", Copenhagen: AKF Forlaget.

Jordana J., and Levi-Faur, D. (2006) "Toward a Latin American Regulatory State? The Diffusion of Autonomous Regulatory Agencies Across Countries and Sectors", *International Journal of Political Administration*, 29: 335–366.

Joskow, P.L. (2005) "Incentive Regulation in Theory and Practice: Electricity Distribution and Transmission Network". CEPR Working Paper. London: Centre for Economic Policy Research.

Joskow, P.L. (2007) "Regulation of Natural Monopolies", in Mitchell Polinsky, A. and Shavell, S. (eds.), *Handbook of Law and Economics*, Vol. II. Amsterdam: Elsevier Science Publishing, pp. 1227–1348.

Kaufmann, D., Kraay, A., and Mastruzzi, M. (2010) "The Worldwide Governance Indicators: Methodology and Analytical Issues", Washington, DC: The World Bank.

Larsen, A., Pedersen, L.H., Sørensen, E.M., and Olsen, O.J. (2006) "Independent Regulatory Authorities in European Electricity Markets", *Energy Policy*, 34: 2858–2870.

Lavenex, S., and Schimmelfennig, F. (2009) "EU Rules beyond EU Borders: Theorizing External Governance in European politics", *Journal of European Public Policy*, 16(6): 791–812.

Levy, B., and Spiller, P. T. (1994) "The Institutional Foundations of Regulatory Commitment: A Comparative Analysis of Telecommunications Regulation", *Journal of Law, Economics & Organization*, 10(2): 201–246.

Majone G. (1997) "From the Positive to the Regulatory State: Causes and Consequences of Changes in the Mode of Governance", *Journal of Public Policy*, 17(2): 139–167.

Newbery, D. M. (2001) "Issues and Options for Restructuring Electricity Supply Industries". DAE Working Paper WP 0210. Cambridge: University of Cambridge, Department of Applied Economics.

Parker, D. (2002) "Economic Regulation: A Review of Issues", *Annals of Public and Cooperative Economics*, 73(4): 439–519.

Phillips, C.F., Jr. (1993) *The Regulation of Public Utilities: Theory and Practice.* Arlington, VA: Public Utilities Report, Inc.

Rhodes, R.A.W. (1996) "The New Governance: Governing without Government", *Political Studies*, 44: 652–667.

Saal, D.S. (2002) *Restructuring, Regulation, and the Liberalization of Privatized Utilities in the UK*. Birmingham: Aston University.

Stigler G.J., and Friedland, C. (1962) "What Can Regulators Regulate: The Case of Electricity", *Journal of Law and Economics*, 5: 1–16.

Trillas, F., and Montoya, M.A. (2011) "Commitment and Regulatory Independence in Practice in Latin American and Caribbean Countries", *Competition and Regulation in Network Industries*, 12(1): 27–56.

Chapter 14

Electricity exchanges with North Africa at 2030

The European and the Italian approaches

Angelo L'Abbate, Gianluigi Migliavacca, Roberto Calisti, Carlo Brancucci Martínez-Anido, Aymen Chaouachi, and Gianluca Fulli[1]

Overview of the Euro-Mediterranean framework

The European Union (EU)'s member states committed themselves to lowering by 20% the greenhouse gas (GHG) emissions (compared to 1990s levels) and are discussing new initiatives and actions targeting a 80% GHG reduction by 2050 [1][2]. It is clear that significant effort and changes have to be considered to achieve the EU's decarbonization objectives while guaranteeing a competitive and secure energy supply. Among others, the EU energy policy road maps and initiatives entail building reinforced international partnerships and physical power interconnectors with neighbouring regions/systems, with a view at diversifying electricity resources portfolio through higher and efficient penetration of renewable energy sources (RES). The uneven geographical distribution of RES potentials, in a scenario of massive renewables integration in the European transmission grid, is expected to lead to increasing north–south power flows mainly triggered by onshore/offshore wind farm installations in the North Seas and south–north power flows mainly brought about by South European and possibly North African solar power plants. In order to accommodate these prospective flows and ensure a better balancing of large-scale RES production across far apart EU and extra-EU regions a reinforced transmission grid – generally named as Supergrid or Electricity Highway [1][12][19][21] – is expected to be developed in the coming decades, also beyond current ENTSO-E (European Network of Transmission System Operators for Electricity) system, thus allowing electricity to flow over long distances and be consumed/stored where more appropriate.

On the northern side of the potential European Supergrid, initiatives such as NSCOGI (North Seas Countries' Offshore Grid Initiative) [22] have been developed in the latest years, investigating several techno-economic, policy and regulatory aspects related to the foreseen grid upgrades in the North Seas region.

On the other hand, on the southern side of the potential European Supergrid, the singularity of the main actors across the Mediterranean basin suggests more extensive investigations that need to cover techno-economic, regulatory and financing issues as well. In fact, by examining the countries around the northern and southern shores of the Mediterranean Sea, some evident differences emerge. In 2010, the GDP (gross domestic product) of the EU countries on the upper coasts of the Mediterranean Sea was 2.5 times higher than the GDP of the remaining Mediterranean countries of the so-called MENA (Middle East and North Africa) area. This ratio becomes 4.7 times if Turkey (an EU candidate country) is excluded. This ratio, yet foreseen to reduce itself to 3.7 times by 2015, is bound to remain significant even in the following years [8].

From the electricity grid perspective, four synchronous areas can be currently recognized around the Mediterranean basin:

- the European continental network, to which north-western Maghreb countries (Morocco, Algeria and Tunisia) are synchronously coupled through undersea cables linking Spain and Morocco,
- the system of the interconnected north-eastern Maghreb (Libya, Egypt) and Mashreq countries (Jordan, Syria and Lebanon),[2]
- the system of Israel and Palestine and
- the network of Turkey.

On completion of ongoing developments, the Turkish system is expected to be fully synchronized with the European continental network by 2014 (a trial parallel operation of the two systems is currently in place) [21].

The Mediterranean region has a very large potential of RES [8]: a massive development of RES generation and related infrastructures across the MENA and the southern European system represents a key target for the EU neighbourhood energy policy, towards the development of MENA countries. In this context, significant initiatives have been set up in the latest years (see e.g. [9][10][11]). However, it has to be remembered that still many, in some cases huge, technical, economic, financial, socio-environmental and political barriers make this process slow and complex. From the technical point of view, the implications for the European networks related to a massive injection of electricity (by RES, such as solar generation, but also by non-RES, such as gas-fuelled generation) from North Africa are not always properly addressed and analyzed. It has to be expected that important bottlenecks arise, prompting significant investments for safely reaching the large consumer centres in Central Europe. New transmission links to strengthen the interconnections in the Mediterranean basin and between central, northern and southern Europe are necessary.

Italy has a particular geographical position, favourable for importing energy both from North Europe and from North Africa and for becoming an important Mediterranean electricity hub. Depending on the prevailing flow, the internal Italian network could be stressed in a completely different manner, leading to different investment needs.

In this framework, the present chapter describes two tightly interrelated approaches, both within a 2030 time horizon for network layout, generation set and system demand evolutions: one from a European perspective and one from a national (Italian) perspective. The former investigates the effect of North African electricity imports on the global European power system (corridor-based approach). The latter, taking account of boundary conditions from the European analysis, performs an in-depth techno-economic investigation of the impact of North African imports on the Italian grid, highlighting system criticalities in Italy (grid-based approach).

The adopted methodology for an integrated approach

The European model: Inputs and scenarios

The EUPowerDispatch model developed by the Smart Electricity Systems and Interoperability group [3], is used to assess the impacts of North African electricity imports on the European system in 2030. The simulation tool provides a minimum cost dispatch model for the European electricity market; the algorithm is based on mixed-integer linear program coded in the General Algebraic Modelling System (GAMS) to solve a minimum cost flow problem (MCFP) that is taking into account both generation and transmission constraints. The objective function is the minimization of the annual variable electricity production costs in the interconnected European power system [8][20].

In the EUPowerDispatch model of the 2030 European system each country is represented by a single node (copper-plate or corridor-based approach) for a total of 32 countries. The model contains 72 interconnections equivalent to cross-border corridors between European countries, and the interconnections between North-Africa and southern Europe (via Spain and Italy) are also included and studied for three different scenarios.

EUPowerDispatch models one year with 52 weekly simulations with one-hour time step. The main inputs of the model for each country (node) include electricity demand (one hour resolution) and installed capacity per energy source type (nuclear, fossil fuels, hydro and renewable energies) characterized by a certain availability factor. In addition, other inputs for the model include European cross-border transmission capacity limits, weather data relevant to the renewable energy generation (including hydro natural inflows), and variable electricity production costs for each type of generation.

The main model outputs include hourly generation levels, cross-border flows, marginal variable electricity production costs and CO_2 emissions [8][20].

In order to simplify the modelling and data collection efforts for representing hydropower systems across Europe, hydropower plants generation has been broken down in two main categories, namely dispatchable and non-dispatchable. The dispatchable hydropower plants are divided in two main sub-categories: pure pumping plants with daily dispatch (75% efficiency) and seasonal storage plants with upper reservoirs fed by natural inflows and managed with seasonal and daily strategies. On the other hand, non-dispatchable hydropower plants typically operate through natural water inflows from rivers (run-of-river). In order to model the flexibility of the hydro plants in the EUPowerDispatch model, ideal conditions, in which start-up, shut-down, ramp-up or ramp-down costs are negligible, have been considered [8][20].

It is clear that one of the most challenging exercises is the modelling of the intermittent renewable sources – wind, photovoltaic (PV) and concentrated solar power (CSP) – on a continental scale. As for the other conventional generation sources, also RES generation is taken into account as aggregated into virtual power plants per each country/node and type; variable RES generation profiles can be obtained through the mapping of meteorological data (offshore/onshore wind speed, solar radiation for one-year time series) applied to the respective installed capacity per country [6][17].

The installed generation capacities for each energy source and the electricity load demand in every European country in this 2030 assessment are based on the ENTSO-E forecast analyses [4][18]. It has to be highlighted that the Italian and German scenarios are characterized by an extremely large deployment of PV installations and the absence of any nuclear generation in 2030 [8][20].

The electricity variable production costs for 2030 are assumed to be equal in every European country considering a CO_2 tax of €35/tonne where the values are mainly based on the TradeWind project [16] with adequate extrapolation to obtain cost scenarios for all generation types, as summarized in Table 14.1 [8][20].

The variable production cost of electricity from North-African is assumed to amount to €41.25/MWh, which is lower than the European fossil-fired sources cost but higher than European nuclear sources. This value takes also into account the fact that the interconnections from North-African, as originally planned, will serve also to dispatch thermal (mainly gas-based) generation; this value is then varied in terms of its position in the merit order with the purpose of performing a sensitivity analysis. In this sense, a scenario with a lower price amounting to €10.52/MWh, due to a higher solar contribution than thermal generation, has been taken into account as well (see Case Studies section).

Table 14.1 Electricity variable production cost per energy source (€/MWh)

Nuclear	*Lignite*	*Hard coal*	*Gas*	*Oil*
11.0	62.9	55.0	61.9	108.7
Mixed fuels	*Hydro*	*Wind*	*Solar*	*Biomass*
114.5	3	2	0	53.1

The projected cross-border transmission system evolution scenarios from the current status up to 2030 in Europe have been built in line with the approaches developed by RSE [5] and first implemented within SUSPLAN and REALISEGRID projects [13][14]. The data used are derived from the figures provided in key references like the ENTSO-E's Ten-Year Network Development Plans (TYNDPs) [12][21]. For the estimation of future cross-border capacity of interconnections within the European power system, due to the lack of information regarding the expected net capacity increase provided by the single expansion project, it was opportunely assumed that only a quota of the theoretically available capacity increase can be effectively considered as net capacity increment due to the existing internal network constraints.

Exceptions to this approach have been made for the Balkan power systems, where the internal congestions severely limit the net available capacity: in those cases, therefore, lower values of new increased capacity of 380 kV lines have been conservatively assumed to be in line with the currently in place capacity limits. On the other hand, for new high-voltage direct current (HVDC) links across the European system, the rated capacity increase has been fully taken into consideration [8][20].

Figure 14.1 shows the assumed level of net cross-border capacities in the ENTSO-E system at 2030.

It is then important to carefully take into account the cross-border transmission losses while considering geographical factors in order to avoid a misrepresentative behaviour of the EUPowerDispatch optimization algorithm resulting in unrealistic loss free flows across the European system. On the other hand, distribution and national transmission losses are included in the nodal hourly loads [8][20].

The Italian model: Inputs and scenarios

The Italian study is tightly and consistently interrelated with the European approach. The assessment of the impact of the electricity imports from North Africa on the Italian system in the 2030 horizon has been based on a transmission planning–oriented approach. The adopted methodology consists in the analysis of grid simulation results in the Italian system for 2030 scenarios related to generation, load demand, fuel costs, CO_2 penalty tax and cross-border interconnections, whereas the internal transmission network (at 220- and 380-kV level, also including equivalents of 132/150 kV portions) is taken into account mainly in its 2020 shape. The purpose is then to investigate how congested the transmission network might be in 2030, given the different boundary conditions, including the new interconnections, also comprising the ones with North Africa, according to the different case studies, while maintaining the internal grid mainly frozen at 2020 [8][20].

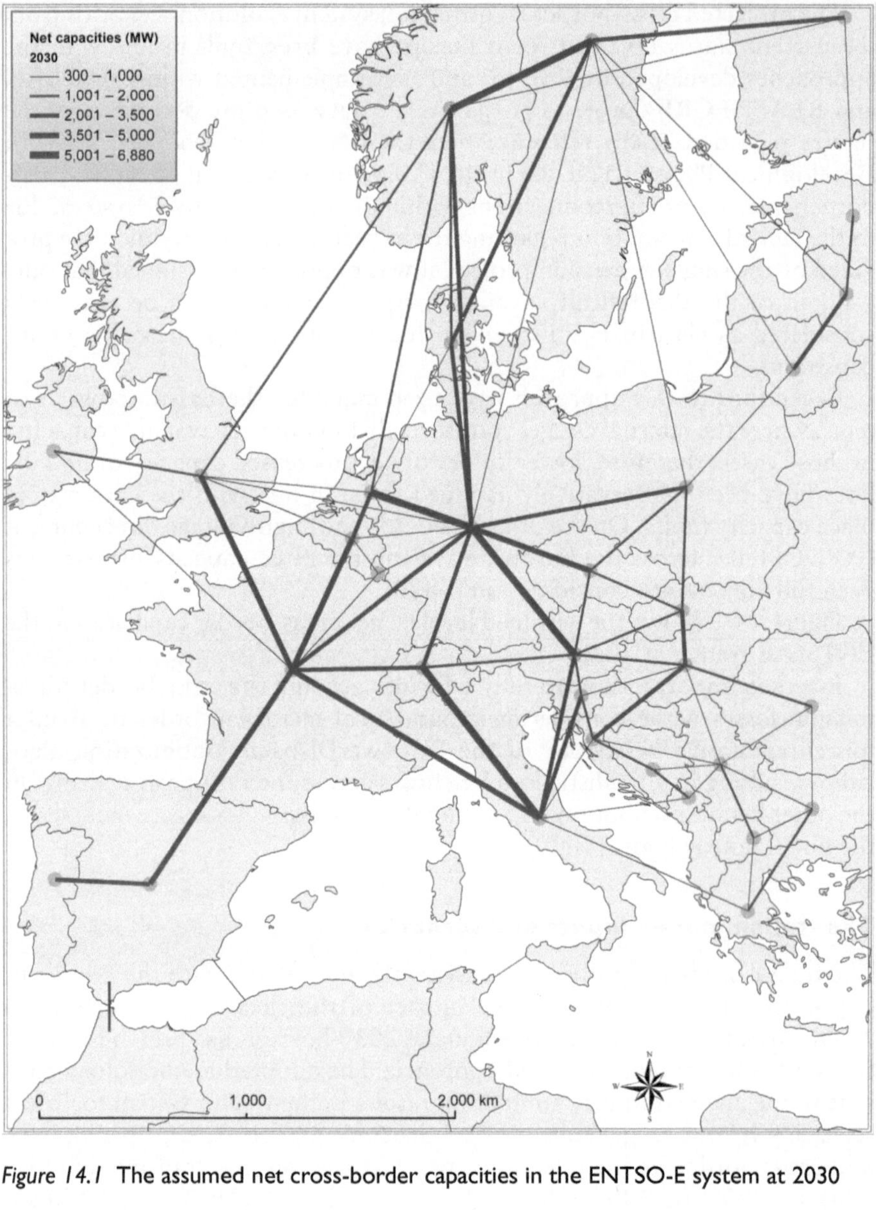

Figure 14.1 The assumed net cross-border capacities in the ENTSO-E system at 2030

The Italian study is carried out using the tool REMARK [7], developed by RSE. REMARK is able to conduct an OPF (Optimal Power Flow)-based analysis integrated by a non-sequential Monte Carlo methodology for static reliability investigation of complex liberalized electric systems divided in areas. This tool features full network representation by the simplified direct current model; power-flow controlling devices can be also incorporated in the model; variable wind generation is treated by the Monte Carlo methodology as well. REMARK is able to perform a quantitative assessment of the reliability and economic benefits, as well as other types of benefits, derived by transmission network expansion [7].

The Italian power system situation at 2010–2011 has been considered as starting point for building up the 2030 scenarios. The installed net generating capacity for Italy used for both 2030 European and Italian studies, as divided in the different electricity sources, is reported in Table 14.2[3] based on several sources of the data for the starting (2010–2011) situation as well as for the estimation of the future (2020 and 2030) evolution of the generation park in Italy [8][20]. In particular, the trend emerged during 2011 concerning the boosting penetration of PV has been fully considered towards the estimation of 2020 and 2030 potential generation. The high PV capacity assumed for 2020 and 2030 reflects this update. For the forecast of other RES (non-PV) penetration the 2020 targets set by the Italian National Renewable Energy Action Plan 2020 [15] have been taken into due account, whereas the estimation of future thermal generation capacity is based on the ENTSO-E forecast analyses [4][18], complemented by opportune assumptions.

In Italy, a continuously increasing generation quota, especially by RES (such as PV and biomass in addition to hydroelectric and geothermal sources), is installed on the downstream distribution grids. Because the

Table 14.2 Total installed net generating capacity in Italy (in MW)[4][15][18]

Electricity source	*2010*	*2011*	*2020*	*2030*
Hydroelectric	21,521	21,737	22,000	23,000
Geothermal	728	728	920	1,086
Solar (PV)	3,470	12,773	30,000	35,000
Solar (CSP)	0	0	600	1,000
Wind (onshore)	5,794	6,918	12,000	20,000
Wind (offshore)	0	0	680	1,000
Biomass	4,759	5,232	6,313	6,700
Nuclear	0	0	0	0
Thermal (coal)	6,600	6,600	7,000	7,000
Thermal (gas)	50,979	52,231	60,089	64,000
Thermal (oil)	8,179	8,179	4,235	4,000
Thermal (mixed)	6,811	6,870	5,063	7,113
Total capacity	**108,841**	**121,268**	**148,900**	**169,899**

focus of this study is on transmission and its grid-impacting generation, most downstream production units have been taken into account either directly or indirectly (via compensation balance of local loads) [8][20]. Concerning RES generation, it is modelled by REMARK according to specific hourly profiles (for each type of electricity sources) and is granted dispatch priority [7]. For the PV generation profile in Italy, the actual PV output has been taken into account by data available in [17]. On the other hand, the thermal (non-RES) generation is treated to simulate a merit order dispatch: it is then priced based on the variable generation costs adopted in the European study (see Table 14.1) and dispatched from the most convenient generation source up to the marginal one, given the system, zonal, interconnection and nodal constraints [7]. Further steps in the 2030 Italian generation scenario build-up have consisted first in the regional breakdown and then in the repartition among the connection voltage levels of the different generation capacities, where relevant. This evaluation has led to the choice of location of nodal injections by the various generation sources in the transmission grid [8][20].

The estimation of the national hourly demand for 2030 has been based starting from the 2010 data for the total energy consumption and the hourly load profiles specified for each region. To represent the scenario up to 2030, an average yearly growth rate has been applied to these regional profiles over the years, taking into account the available forecasts [8][20]. The total energy consumption in Italy at 2030 consistently corresponds to the total value assumed in the European analysis for Italy.

Concerning the transmission network evolution, the new interconnections foreseen and planned up to 2030 have been taken into account. These include the three submarine links of Italy with North Africa, namely with Algeria (via Sardinia) and with Tunisia and Libya (both via Sicily). Table 14.3[4] displays the list of existing and future cross-border transmission interconnections in Italy taken into consideration in the 2030 scenarios [8][12][20]. The corresponding net capacity values between Italy and bordering countries, as estimated at 2030 in the European study, are consistently used for the Italian specific case, too.

In order to evaluate the effect of cross-border flows (import and export) across Italy, a simplified, yet effective, modelling approach has been implemented by representing each cross-border interface towards an extended merit order dispatch. Within REMARK the bordering foreign buses have been modelled as connection points of both equivalent generating units and loads: the foreign bus generators are considered like thermal power plants participating in the merit order with the corresponding hourly marginal prices resulting from the European analysis. The exceptions to the approach of using equivalent generators concern the links with Malta and Corsica (France), where only a load is seen on the foreign side. Finally, opportune assumptions, also based on system evolutions in Albania, have been made to

Table 14.3 Interconnections at Italian borders considered in 2030 scenario (2012 update)

Node	*Country*	*Node*	*Country*	*Type*
TIVAT	Montenegro	VILLANOVA	Italy	HVDC (future)
ZUWARAH	Libya	CHIARAMONTE	Italy	HVDC (future)
EL HAOUARIA	Tunisia	PARTANNA	Italy	HVDC (future)
EL HADJAR	Algeria	RUMIANCA	Italy	HVDC (future)
ARACHTHOS	Greece	GALATINA	Italy	HVDC (existing)
CANDIA	Italy	KONJSKO	Croatia	HVDC (future)
DIVAČA	Slovenia	REDIPUGLIA	Italy	HVAC (existing)
OKROGLO	Slovenia	UDINE OVEST	Italy	HVAC (future)
DIVAČA	Slovenia	PADRICIANO	Italy	HVAC (existing)
VOLPAGO	Italy	LIENZ	Austria	HVAC (future)
CARDANO	Italy	THAUR	Austria	HVAC or HVDC (future)
GLORENZA	Italy	NAUDERS/WEST TIROL	Austria	HVAC (future)
BABICA	Albania	BRINDISI SUD ST.	Italy	HVDC (future)
ROBBIA	Switzerland	TIRANO/S. FIORANO	Italy	HVAC (existing)
CAGNO	Italy	MENDRISIO	Switzerland	HVAC (existing)
MUSIGNANO	Italy	LAVORGO	Switzerland	HVAC (existing)
MORBEGNO	Italy	LAVORGO	Switzerland	HVAC (future)
MESE	Italy	SOAZZA	Switzerland	HVAC (existing)
VERDERIO	Italy	SILS	Switzerland	HVDC (future)
MESE	Italy	GORDUNO	Switzerland	HVAC (existing)
AVISE	Italy	RIDDES	Switzerland	HVAC (existing)
VALPELLINE	Italy	RIDDES	Switzerland	HVAC (existing)
PONTE	Italy	AIROLO	Switzerland	HVAC (existing)
PALLANZENO	Italy	SERRA	Switzerland	HVAC (existing)
GRAND'ILE	France	PIOSSASCO	Italy	HVDC (future)
RONDISSONE	Italy	ALBERTVILLE	France	HVAC (existing)
VENAUS	Italy	VILLARODIN	France	HVAC (existing)
BROC CARROS	France	CAMPOROSSO	Italy	HVAC (existing)
MAGHTAB	Malta	RAGUSA	Italy	HVAC (future)
LUCCIANA	France (Corsica)	CODRONGIANUS	Italy	HVDC (existing)
LUCCIANA	France (Corsica)	SUVERETO	Italy	HVDC (existing)

include the corresponding Albanian generation and load in the 2030 model by an equivalent bus representation [8][20].

In sum, the Italian system model used in the study consists of 1,042 buses, 556 generating units, 856 lines and 479 transformers.

Case studies

For the interconnection expansion between North Africa and Europe, realistic estimations have been carried out [10][11], whereas the implementation of very ambitious plans (such as the ones in [9]) is considered to be feasible only after 2030. Different corridors are subject of analysis and assessment within this study, namely Spain–Morocco and Spain–Algeria as well as Italy–Tunisia, Italy–Algeria and Italy–Libya borders. In order to analyze the effects of power imports from Africa on the interconnected European electricity system and the Italian grid, three main scenarios, pessimistic (A), reference (B) and optimistic (C), are assumed in terms of the interconnectors between North Africa and Europe and their maximum transfer capacity (see Table 14.4 and Figure 14.2). In addition, a scenario with no electricity import from North Africa is applied to the model (baseline scenario D): this case sets in fact the base for a comparison of the three main scenarios.[5] Finally, a sensitivity analysis (based on the optimistic scenario for interconnection capacities) is introduced to assess a quite extreme situation (very optimistic scenario, C1) by lowering the North African electricity variable production cost (from €41.25/MWh to €10.52/MWh, see also Table 14.4) below the European electricity variable production cost for nuclear sources and therefore varying the merit order.

The European perspective

The European perspective on power imports from North Africa in 2030 is assessed in terms of the annual cross-border net exchanges and the hourly marginal energy source in each European country.

The main finding, independent of the three main 2030 scenarios, is that all the transmission links between North Africa and Europe result to be practically 100% fully loaded during all the hours of the year: this holds especially for the links with Italy, whereas in Spain for very few hours of the year the local cheaper energy sources (including hydro, wind, solar) and in

Table 14.4 Summary of the 2030 scenarios under study

Interconnector	*Pessimistic scenario (A)* *€41.25/MWh*	*Reference scenario (B)* *€41.25/MWh*	*Optimistic scenario (C)* *€41.25/MWh*	*Very optimistic scenario (C1)* *€10.52/MWh*	*Baseline scenario (D)*
Morocco–Spain	1,400 MW	2,000 MW	2,000 MW	2,000 MW	–
Tunisia–Italy	1,000 MW	1,000 MW	2,000 MW	2,000 MW	–
Algeria–Spain	1,000 MW	1,000 MW	2,000 MW	2,000 MW	–
Algeria–Italy	–	1,000 MW	2,000 MW	2,000 MW	–
Libya–Italy	–	1,000 MW	2,000 MW	2,000 MW	–

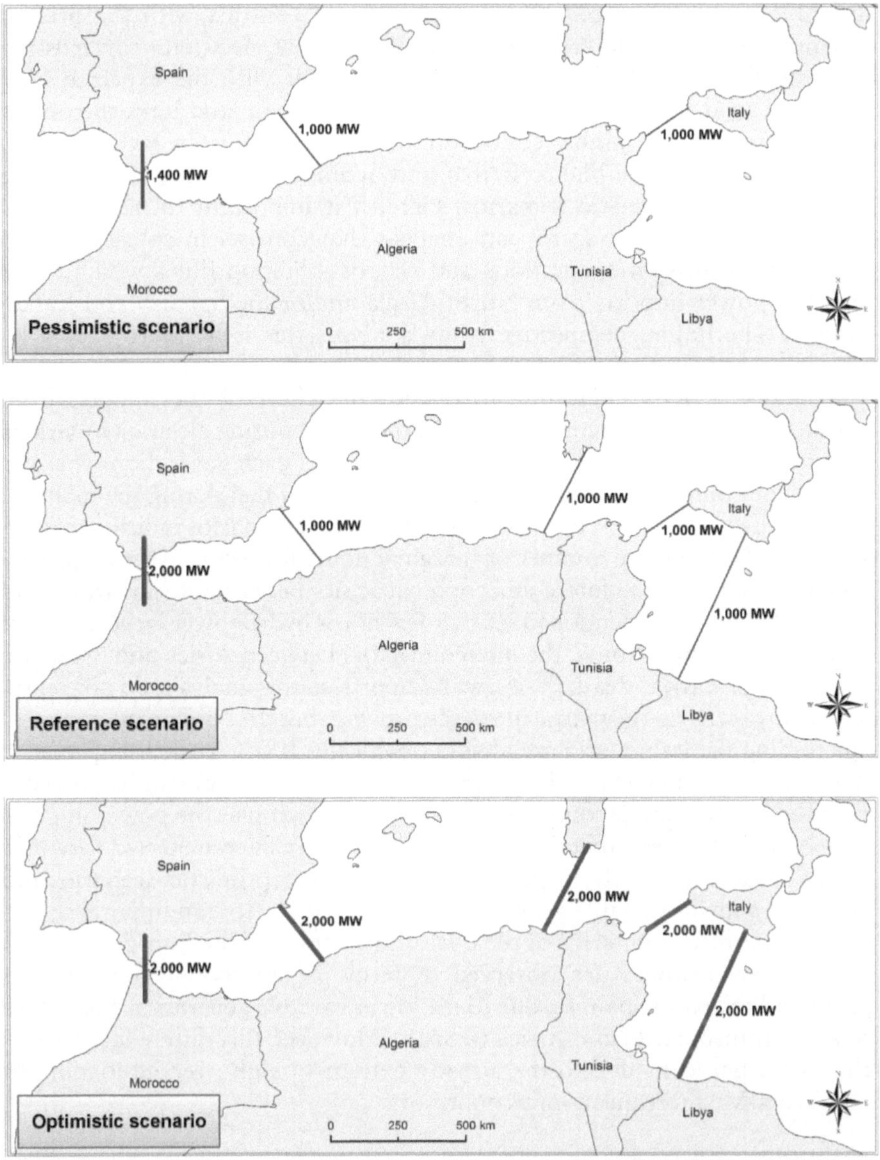

Figure 14.2 The three main 2030 scenarios for the interconnection between North Africa and Europe

some cases the convenient imports (from Portugal mainly) are sufficient to meet the national load and limit the power imports from Africa. Another important result is that the net power exchanges tend to follow the direction from south to north, that is from North Africa to Europe, for each of the three main 2030 scenarios. In addition, in contrast with the present situation, Spain and Italy become net exporters of electricity towards the other neighbouring European countries. Together with the expected high wind and solar installed generation capacities in Spain and Italy, the power imports from Africa highly contribute to this behaviour. In fact, the optimistic scenario shows higher Italian and Spanish power exports than the reference and pessimistic scenarios. Here, it is important to highlight the assumption behind the European analysis that consists in considering no internal transmission congestions and no possible loop flows which could limit the power imports from North Africa and/or the Spanish and Italian exports. The Italian perspective study analyzes this issue more deeply by focusing on the Italian grid detail.

In addition, it is interesting to discern the effect of the imports from North Africa on the Italian/Spanish and the neighbouring electricity markets reflected on the yearly hours of marginal prices per each generation type. For each of the countries considered, gas is the marginal fuel during the majority of the hours of the year. The results for the different scenarios indicate that for Spain and Portugal the number of marginal hours for North African generation increases with the interconnection capacities between North Africa and Europe. It should be mentioned that for few hours hydropower is the marginal source; during these hours, the interconnectors between Africa and Spain are not used. For Italy, instead, the African imports, in this analysis, do not represent in any scenario the marginal source proving that the African interconnectors feeding the Italian grid are always practically 100% loaded. Expectedly, changes in marginal generation sources throughout the year can be observed in all the countries interconnected to Italy and/or Spain as the power imports from North Africa are affecting the overall European interconnected network.

The sensitivity analysis performed using the optimistic scenario and assuming a lower variable generation cost for North African imports shows very similar results in terms of the overall behaviour of the European power system. Few changes are observed in terms of interconnection loading between Africa and Spain as, due to the lower variable generation cost, more power is transported from Africa to Spain. However, this difference remains almost insignificant due to the already extremely high average loading of such cross-Mediterranean interconnectors.

The Italian perspective

The same boundary conditions seen in the European study have been applied to the 2030 Italian case, subjected to the three main scenarios (A, B, C) together with the very optimistic scenario (C1) and the baseline scenario (D). The runs of REMARK tool have provided interesting results that are summarized in terms of exchange flows between market zones

Table 14.5 Exchange flows across zones

Exchange flows in GWh		*Scenario C*		*Scenario C1*	
From/To	From/To	→	←	→	←
NORTH	FRANCE	7,654	24,931	7,660	24,884
NORTH	SWITZERLAND	9,656	27,156	9,105	27,654
NORTH	AUSTRIA	9,578	10,481	9,553	10,291
NORTH	SLOVENIA	5,957	10,179	6,192	10,007
CENTER-NORTH	CROATIA	2,455	5,202	2,418	5,240
CENTER-SOUTH	MONTENEGRO	8,183	336	8,185	337
SOUTH	ALBANIA	2,462	1,817	2,528	1,748
SOUTH	GREECE	5,871	2,515	5,973	2,416
SARDINIA	ALGERIA	–	12,404	–	12,405
SICILY	TUNISIA	181	11,212	158	10,971
SICILY	LIBYA	104	13,788	110	14,071
SICILY	MALTA	2,186	–	2,186	–
NORTH	CENTER-NORTH	4,509	17,894	5,005	17,733
CENTER-NORTH	CORSICA	11	4,157	11	4,158
SARDINIA	CORSICA	5,023	2	5,024	2
CENTER-NORTH	CENTER-SOUTH	3,147	25,493	3,438	25,228
CENTER-SOUTH	SARDINIA	1	4,180	1	4,166
CENTER-SOUTH	SOUTH	–	49,419	–	49,137
SOUTH	SICILY	75	16,945	55	16,984

(Table 14.5 and Figure 14.3) and in terms of average marginal generation costs (Figure 14.4, color section, and Table 14.6) for the five cases. The Italian system is considered to be split in the following market zones: North, Center-North, Center-South, South, Sicily and Sardinia. In the run cases, a 2,000-MW net transfer capacity (NTC) has been considered on the South–Sicily interface; furthermore, between Sardinia and Center-North (through Corsica) the limit is given by the SA.CO.I. cable-link capacity, namely 600 MW, whereas between Sardinia and Center-South, the maximum capacity transfer amounts to 1,000 MW (due to SA.PE.I. link rating).

A first very important outcome of the analysis is that there is a clear, relevant flow of electricity from North Africa to Italy (Sicily and Sardinia). This import reaches its peak in scenarios C and C1 as expected, but, in relative terms, the scenario B sees the most favourable conditions. In the latter, in fact, during most of the yearly hours the interconnectors are fully loaded; however, it can be appreciated that for very few hours, Sicily is also able to export electricity to Tunisia and Libya. For the scenarios C and C1

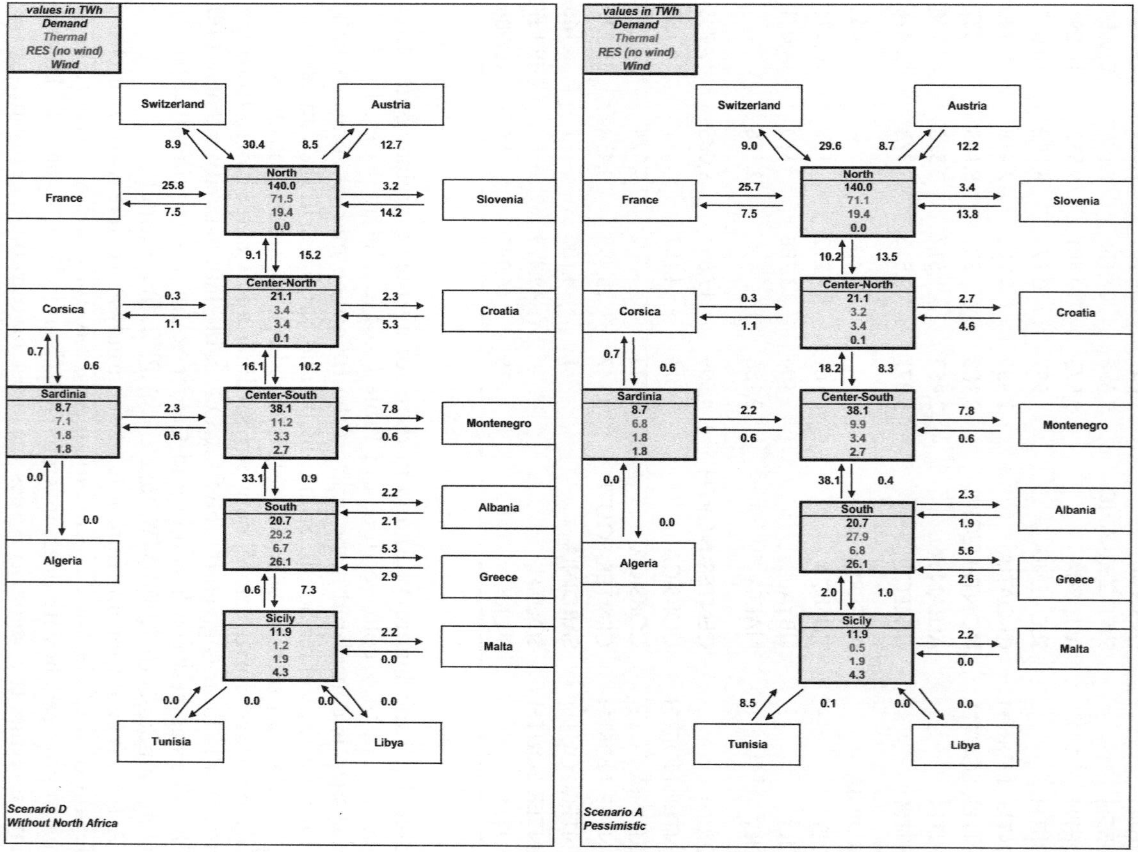

Figure 14.3a Flows and electrical power balances

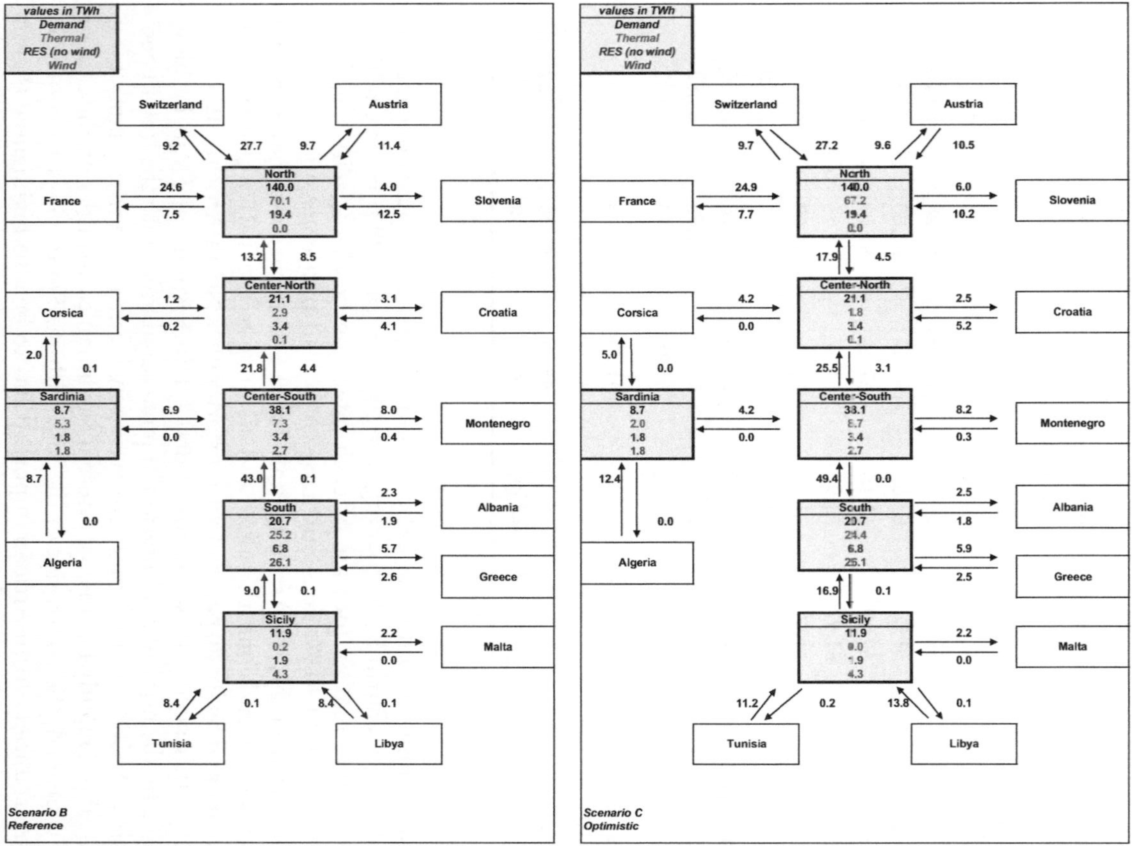

Figure 14.3b Flows and electrical power balances

Table 14.6 Locational marginal prices (average)

Locational marginal prices in €/MWh	*Scenario D*	*Scenario A*	*Scenario B*	*Scenario C*	*Scenario C1*
North	62.04	61.98	61.98	61.97	61.97
Center-North	61.59	61.88	61.87	61.86	61.86
Center-South	61.45	61.61	61.58	61.55	61.12
South	60.19	60.60	60.41	60.32	59.06
Sicily	60.21	60.65	60.44	41.94	12.23
Sardinia	61.87	61.87	61.22	52.03	37.59
France	62.28	62.25	62.28	62.28	62.26
Corsica	61.89	61.87	61.82	61.80	61.73
Switzerland	62.19	62.16	62.20	62.19	62.17
Austria	61.97	61.96	61.97	61.96	61.96
Slovenia	61.91	61.91	61.91	61.89	61.90
Croatia	61.89	61.89	61.90	61.88	61.88
Montenegro	62.32	62.42	62.35	62.06	62.06
Albania	61.86	61.88	61.87	61.85	61.83
Greece	61.08	61.42	61.28	61.14	60.43
Libya			41.25	41.25	10.52
Tunisia		41.11	41.25	41.25	10.52
Algeria			41.25	41.25	10.52
Malta	60.21	60.65	60.44	41.74	11.75

the maximum utilization of each corridor's capacity amounts to ca. 80%, whereas in scenario B, the loading capacity can reach 99% ca. Such difference implies that an increase of each interconnector's capacity (i.e. from 1,000 to 2,000 MW) does not necessarily lead to a consequent increase in import from North Africa to Italy for the facts that internal grid constraints exist and limit this flow and also due to local RES generation in Sardinia and Sicily dispatched with priority over North African electricity. In fact, this occurs in scenarios C and C1 in Sardinia and can be highlighted by the not full exploitation of inter-zonal links with the Italian peninsula (SA.CO.I. and SA.PE.I. are averagely loaded for a maximum of less than 50%), mainly because of internal Sardinian constraints. In the case of Sicily, instead, the constraint of the link with the South plays a major role in this reduced North African import. It has to be also highlighted that in scenarios C and C1, the largest African exporter is represented by Libya: its electricity also serves to supply the fixed 250 MW load in Malta.

Another important outcome of the study is that in all scenarios Italy is remaining a net electricity importer from the bordering countries of the

Alpine region (i.e. France, Switzerland, Austria, Slovenia). This difference with respect to the European study (with exception of the Italy-Slovenia interface) is due to the internal grid bottlenecks and cross-border constraints that limit the possibility that the RES generation, mostly concentrated in the southern and insular parts of the country, may flow to reach the northern borders to be exported. Concerning the flows on the Italian eastern borders with the Balkan region, the situation is partially reversed due to the large export to Montenegro that increases with the import from North Africa. Similar trends are observed with Albania and Greece, although the export magnitude is lower with respect to Montenegro. The same trend does not apply to Croatia, where the net flow has the sign of import in all cases, despite the fact that there are more hours in which Italy exports to Croatia, with the peak situation in scenario B.

Another important result is that, with the increased import from North Africa, the net flows of electricity are more and more directed from the islands and from the south of the country up to the north where large consumption is mainly concentrated. This trend can be seen from scenario B, where the net power flow between Center-North and North is reversed in direction with respect to the scenarios A and D: such trend confirms the impact of Italian RES generation as well as the African import influence.

Further elements of analysis can be derived from the results in Table 14.6 (whose details are plotted in Figure 14.4), in which the average yearly marginal prices are displayed for the different market zones as weighted with respect to loads. Considering a detailed comparison among the scenarios A, B, C and D within the Italian market zones, with the red colour associated to the minimum value of the scenario and the blue to the maximum, it is evident that the averagely weighted locational marginal prices in Table 14.6 are strongly dependent on the marginal generation source, which in most zones in all scenarios is represented by thermal (gas) production. The exceptions to this outcome are given by the Italian market zones of Sicily and Sardinia in which, especially in scenarios C and C1, the impact of import from North Africa is strong and makes the marginal prices much lower than in the rest of Italian zones in those scenarios. The same does not occur in the other Italian zones mainly due to internal grid bottlenecks. On the other hand, scenario A has led to a reduction of the zonal prices differences with respect to the scenario D: the price of Sardinia is unvaried whereas the zones of South and Sicily show a slight price increase. Furthermore, it can be additionally highlighted that also Malta's marginal price is strongly influenced by North African import, being even slightly lower than the average one in Sicily.

In the frame of this analysis, it is crucial to assess the possible weaknesses of the internal Italian network in terms of transmission grid behaviour in a way to identify relevant reinforcement options. In fact, several areas in the Italian system present a critical situation, especially in the scenarios B, C and

C1. These critical portions of the grid are mainly located in Sardinia, around the metropolitan areas of Rome, Naples, Milan, in Tuscany, in Center-South along the Tyrrhenian and Adriatic axes and at the interface between Center-South and South. Particularly the B scenario does stir up specific investigation with particular focus on the effect of the operation of phase-shifting transformers (PSTs) as a technical alternative to limit congestion. In total, five PSTs are considered within B scenario of which: two are located at Villanova in Center-South zone (to control the flows on both Adriatic axis lines), one is located at Foggia in South zone (to control the flow due to massive local RES generation) and two are located at Bisaccia in Center-South zone (to control the flow between the two Tyrrhenian and Adriatic axes). It can be observed by comparing the scenario B without PSTs and the reference scenario B as per Table 14.7 and Figure 14.5 (in the color section) that the impact of these PST devices in flow dispatch emerges, as they allow for example a larger export from Italy to Montenegro and also to Croatia

Table 14.7 Exchange flows in scenario B and its variants (in GWh)

		Scenario B without PST		*Scenario B*		*Scenario B with reinforcement Latina-Garigliano*	
From/To	From/To	→	←	→	←	→	←
North	France	7,573	24,541	7,526	24,597	7,544	25,007
North	Switzerland	10,434	23,711	9,230	27,688	9,048	27,641
North	Austria	9,571	12,155	9,662	11,358	9,242	11,710
North	Slovenia	3,887	12,497	4,039	12,501	4,739	11,305
Center-North	Croatia	1,886	5,069	3,054	4,078	2,987	4,309
Center-South	Montenegro	7,047	832	7,954	397	8,022	389
South	Albania	2,427	1,824	2,335	1,898	2,376	1,850
South	Greece	5,415	2,725	5,664	2,568	5,669	2,570
Sardinia	Algeria	–	8,672	–	8,672	–	8,671
Sicily	Tunisia	240	8,296	91	8,372	34	8,479
Sicily	Libya	269	8,285	55	8,391	58	8,390
Sicily	Malta	2,186	–	2,186	–	2,186	–
North	Center-North	7,041	15,394	8,513	13,200	8,478	13,268
Center-North	Corsica	165	1,595	158	1,210	149	1,450
Sardinia	Corsica	2,379	72	1,987	58	2,232	56
Center-North	Center-South	4,250	22,515	4,408	21,771	4,489	21,455
Center-South	Sardinia	12	6,452	1	6,899	1	6,524
Center-South	South	135	42,851	125	42,967	127	43,255
South	Sicily	424	9,329	77	9,036	33	9,146

and reduce the total dispatch costs of 2.1 million euros. On the other hand, doubling the transmission lines Latina-Garigliano located in Center-South between the areas of Rome and Naples can be considered as another reinforcement option to alleviate the congestions existing in this grid portion. The direct effect of this line's doubling in terms of total dispatch cost reduction would lead in a total saving of 4 million euros.

Considering the B scenario with its two variants (without PSTs and with reinforcement Latina-Garigliano) in terms of average marginal prices (see Figure 14.5, color section), it can be noticed how the former variant would lead to a bottleneck between the South and the Center-South, determining a net price difference; the latter variant has a higher price homogeneity.

Conclusion

The present chapter provides a techno-economic analysis based on the combination of two methodologies to preliminarily evaluate the North African electricity impact on the European system and the Italian grid in a 2030 horizon. A closer assessment of the Italian power system as a potential electricity hub in the Mediterranean region is carried out, paying particular attention to the behaviour of the Italian transmission grid. Three main scenarios of transmission capacity for the cross-Mediterranean interconnectors have been proposed and investigated. The combination of the European and Italian approaches has proved its feasibility to assess the impact of the North African electricity on the European system, and more specifically in the Italian case, providing a detailed picture of the inter-zonal flows and zonal prices. The major outcomes of this assessment consist in the confirmation of a sustained electricity flow from south to north with practically fully loaded cross-Mediterranean interconnections. On the other hand, the analysis proves that further investments on the Italian national transmission network may be necessary.

Notes

1. The views expressed in this chapter are the sole responsibility of the authors and do not necessarily reflect the views of the European Commission.
   This work has been partially financed by the Research Fund for the Italian Electrical System and Contract Agreement between RSE S.p.A and the Ministry of Economic Development.
2. This area is the most problematic one, both from the geopolitical and the development of grid infrastructures points of view.
3. Data related to biomass generation include also the ones for non-RES biomass; data related to gas-based generation include also the ones for derived gases, mixed gases and gas/oil mix.
4. HVAC stands for high voltage alternating current; HVDC stands for high voltage direct current. It has to be remarked that some links in Table 14.3 refer to existing interconnections to be expanded, like the one between Italy and Greece

(from 500 MW to 1,000 MW) or to be repowered, such as the one between Italy and France (Corsica) (from 300 MW to 600 MW). The Italy–Austria link to Thaur via Brenner tunnel may be based on HVAC via GIL (gas-insulated line) technology or on HVDC. The Italy–Malta link is based on 220 kV HVAC submarine cable. It has to be also said that on many cross-border links with the Alpine region (France, Switzerland, Austria, Slovenia) power flow controlling devices such as PSTs (phase-shifting transformers) exist or will be installed, to increase the cross-border capacity and overcome congestions.
5. It has to be highlighted that, notwithstanding the currently existing Morocco–Spain link, the rationale of baseline scenario D lies on the possibility to estimate the full effect of electricity flows from North Africa on the European system, by comparing the cases with and without the interconnectors bundle.

References

[1] European Commission, *Energy Infrastructure Priorities for 2020 and Beyond – A Blueprint for an Integrated European Energy Network*, COM(2010) 677 final, Nov. 2010.

[2] European Commission, *Energy Roadmap 2050*, COM(2011) 885, Dec. 2011.

[3] European Commission, Joint Research Centre, Smart Electricity Systems and Interoperability. http://ses.jrc.ec.europa.eu

[4] ENTSO-E, *Scenario Outlook & Adequacy Forecasts 2011–2025*, Feb. 2011. www.entsoe.eu

[5] Ricerca sul Sistema Energetico (RSE) S.p.A. www.rse-web.it

[6] G. van der Toorn, *Wind Power Capacity Data Collection*, Report D2.1 TradeWind project, Apr. 2007. www.trade-wind.eu/

[7] R. Calisti, M.V. Cazzol, *Tool for the Assessment of Benefits Given by the Expansion of Transmission Infrastructures*, Report D3.3.3 REALISEGRID project, June 2011. http://realisegrid.rse-web.it/

[8] G. Migliavacca, A. L'Abbate, R. Calisti, M. Soranno, C. Brancucci, C. Alecu, M. Vandenbergh, G. Fulli, *Solar Import from Africa: The Italian and the European Perspective*, 13th RSCAS-EUI Mediterranean Research Meeting 2012, Montecatini T., Italy, 21–24 Mar. 2012.

[9] DESERTEC Industrial Initiative (DII). http://dii-eumena.com/

[10] Medgrid Project. www.medgrid-psm.com/

[11] MED-EMIP, *MEDRING Update Study: Mediterranean Electricity Interconnections*, Apr. 2010. www.medemip.eu

[12] ENTSO-E, *Ten-Year Network Development Plan (TYNDP) 2010*, June 2010. www.entsoe.eu/major-projects/ten-year-network-development-plan/tyndp-2010/

[13] J. de Joode, O. Ozdemir, K. Veum, A. van der Welle, G. Migliavacca, A. Zani, A. L'Abbate, *Trans-National Infrastructure Developments on the Electricity and Gas Market*, Report D3.1 SUSPLAN project, Feb. 2011. www.susplan.eu/

[14] I. Losa, R. Calisti, A. L'Abbate, G. Migliavacca, C. Vergine, A. Sallati, *Application of the REALISEGRID Framework to Assess Technical-Economic and Strategic Benefits of Specific Transmission Projects*, Report D3.5.1 REALISEGRID project, July 2011. http://realisegrid.rse-web.it

[15] Italian Government, *National Renewable Energy Action Plan (NREAP) 2020*, 2010. http://ec.europa.eu/energy/renewables/transparency_platform/doc/dir_2009_0028_action_plan_italy.zip

[16] M. Korpås, L. Warland, J.O.G. Tande, K. Uhlen, K. Purchala, S. Wagemans, *Grid Modelling and Power System Data*, Report D3.2 TradeWind project, Dec. 2007. www.trade-wind.eu/
[17] M. Suri, T. Huld, E. Dunlop, H. Ossenbrink, *Potential of Solar Electricity Generation in the European Union Member States and Candidate Countries*, Solar Energy, 81, no. 10, pp. 1295–1305, 2007.
[18] ENTSO-E, *Scenario Outlook & Adequacy Forecasts 2012–2030*, July 2012. www.entsoe.eu/
[19] European Parliament and European Council, *Regulation (EU) No. 347/2013 of the European Parliament and of the Council of 17 April 2013 on guidelines for trans-European energy infrastructure and repealing Decision No. 1364/2006/EC and amending Regulations (EC) No. 713/2009, (EC) No. 714/2009 and (EC) No. 715/2009*, Official Journal of the European Union No. L 115, pp. 0039–0075, 25 Apr. 2013.
[20] C. Brancucci Martínez-Anido, A. L'Abbate, G. Migliavacca, R. Calisti, M. Soranno, G. Fulli, C. Alecu, L.J. de Vries, *Effects of North-African Electricity Import on the European and the Italian Power Systems: A Techno-Economic Analysis*, Electric Power Systems Research, 96, pp. 119–132, Mar. 2013.
[21] ENTSO-E, *Ten-Year Network Development Plan (TYNDP) 2012*, July 2012. www.entsoe.eu/major-projects/ten-year-network-development-plan/tyndp-2012/
[22] North Seas Countries' Offshore Grid Initiative, www.benelux.int/NSCOGI/

Chapter 15

Effects of the new 2020 strategy on regional energy initiatives and energy markets integration

Rosa María Fernández[1]

Last years have seen how the environment around energy issues has changed a lot through the European Union (EU). Even though it has been always claimed that there is a need for better integration, the measures taken did not seem to be leading Europe through the desired path. Goals such as sustainable development, emissions reduction or energy security of supply have been on the EU agenda for years, and the continuous modification of directives, for instance in the gas and electricity sectors, is a clear indicator of the unsuccessful outcome so far.

With the Kyoto Protocol, the move towards renewable energies seemed to have become the priority. In the current international economic crisis, assuring energy security of supply at affordable prices has come to the top of the list. Both items are related and are linked to the growing energy dependency affecting the EU, which was one of the reasons under the regional energy initiatives currently in place.

All in all, the *ad hoc* decisions over time show a lack of a real, common energy policy. But things seem to be changing, especially in light of the need for a new trend to take Europe finally out of the crisis, and after the entrance into force of the Lisbon Treaty (Official Journal of the European Communities [OJEC], 2007), which gave the commission for the first time a shared role to play with national governments in energy and environmental issues.

The energy sector is key for European economies and, at the same time, key for achieving a sustainable Europe and tackling climate change. Reducing energy consumption and using it in a more efficient way must be complemented and completed by the election of sustainable inputs (including sustainable energy sources) and sustainable production processes. These principles have been guiding EU decisions over time, but looking at some of the communications related to the energy sector in the last years, very critical with the slow progress achieved in all the areas, give us an idea of the difficulties to reach the desired results in a sector that not only affects the now 28 member states but also many other third countries, and which is also affected by many external factors.

Numerous measures affecting energy regulation have been approved since the 1990s, with a combination of goals: trying to provide security of supply, foster solidarity between countries and protect consumers (European Commission [EC], 2006). The final true goal would be having an integrated energy policy for Europe. But this is not an easy goal, given the disparity among situations in member states: Energy companies more or less open to competition, public or private ownership, availability of resources or institutional framework in favour (or not) of renewable energies, among others (Reiche & Bechberger, 2004).

The crisis has not helped change this situation. Investments in sustainable technologies and energy efficiency have been put aside because of the lack of resources and the limited access to finance. And regarding energy infrastructures, only projects considered of high priority will be carried out.

The EU must deal with a situation of lack of financial resources on the private sector, and a drastic decrease in public resources available, putting in risk the progress on research and development (R&D) activities on environment and sustainability. Additionally, the crisis has been affecting prices on the carbon markets, such as the European Trading Scheme (ETS). The transition period (2005–2007) of the ETS showed the need for stability in prices at a level high enough to guarantee the efficiency of the market and foster emissions reduction. We are right now in the opposite situation, worsened by a succession of negative events through 2011, such us the instability created in some Middle East countries or the nuclear disaster of Fukushima (Japan) in March 2011, with consequences still difficult to evaluate.

It is not surprising under these circumstances that the strategy outlined by the EU to get out of the crisis (EC, 2010a) gives especial relevance to the measures affecting the energy sector. One of Europe 2020 Strategy flagship initiatives is a "resource efficient Europe" (EC, 2011a), with the focus on decoupling economic growth from the use of resources, shifting towards a low-carbon economy, increasing the use of renewable energy sources, modernizing the transport sector and promoting energy efficiency.

The aim of this chapter is to analyze how the measures that will be implemented according to the Europe 2020 Strategy will affect energy markets integration and regional energy agreements, in light of the general goal of "smart, sustainable and inclusive growth".

We will not forget that before this strategy was put on the table, there were already some initiatives affecting third countries in the new directives coming from the Energy and Climate Change package approved in 2008. The new strategy reinforces the external dimension of the energy policy, and gives especial relevance to agreements with third countries, combining actions in both neighbourhood and development policies.

The chapter begins with an analysis of the external dimension of the EU energy policy and continues with a review of the bilateral and multilateral agreements in the area of energy cooperation that EU member states have signed. Bearing in mind all the data collected in those sections, the chapter concludes with an attempt to answer some questions about the possible future of the regional energy initiatives and their contribution to sustainable growth and energy markets integration.

The external dimension of the EU Energy Policy

The first thing we need to remark when talking about EU Energy Policy is that we hardly can say that such a policy exists. Despite the huge number of regulations and communications related to the energy sector coming from Europe, the fact is that *An Energy Policy for Europe* in 2007 (EC, 2007b) still highlights the need for that policy, which implies that there isn't any, and subsequent documents repeat the same discourse (EC, 2010a, 2010b). Europe needs an integrated energy policy, but it is an unfinished process with uneven progress.

The changing environment of the last years with unexpected events such as the crisis between Russia and Ukraine in the winter of 2008, leaving a considerable part of Europe without gas supply, made even more visible the high energy dependence (see Table 15.1) of EU countries and highlighted the need to search for alternative energy sources and alternative partnerships with third countries to guarantee security of supply.

In fact, security of supply has been present in all measures adopted within the context of the Energy and Climate Change Package. The Renewable Energies Directive (2009/28/EC: OJEC, 2009a) includes rules for energy imported from third countries, and the need to increase grid connections in and outside the EU, among other measures and targets. It is security of supply that guided the approval of a directive (2009/119/EC: OJEC, 2009b), imposing the obligation on member states to maintain minimum stocks of crude oil and/or petroleum products.

Energy dependence is not something to be solved in the short term, and it will take time to change patterns. The Europe 2020 Strategy claims for a shift towards renewable energy sources and forecasts a saving of €60 billion in oil and gas imports if goals are reached. But in the meantime, it is very important to improve interconnection with third countries to secure supply. It is one of the main reasons to promote initiatives such as Energy Community and Medreg. It should be nevertheless noticed that all the measures taken or planned, at least for the short and midterm, forecast an increase on energy consumption, and in fact, when reviewing individual member states' plans we still find that previsions on energy consumption or economic growth count on the continuation of the growing patterns of both global energy consumption and electricity demands (Department of Trade and Industry [DTI], 2006). This means that there is still a need to change behaviour towards reduction of energy consumption, and this is an area in which actions have been limited at least until the current international economic crisis appeared.

Table 15.1 EU energy dependence

	2000	*2001*	*2002*	*2003*	*2004*	*2005*	*2006*	*2007*	*2008*	*2009*	*2010*	*2011*
EU-27	46.7%	47.4%	47.6%	49%	50.2%	52.4%	53.7%	53%	54.6%	53.8%	52.6%	53.8%
Belgium	78.1%	80.6%	77.5%	79.6%	79.8%	80.1%	79.8%	77.1%	79.9%	74.3%	76.8%	72.9%
Italy	86.5%	83.3%	86%	83.9%	84.7%	84.4%	87%	85.1%	85.3%	82.8%	83.8%	81.3%
Denmark	–35.3%	–28.4%	–42%	–31.8%	–47.5%	–50.9%	–35.9%	–24.7%	–21.7%	–20.4%	–16.9%	–8.5%
Spain	76.7%	74.7%	78.5%	76.7%	77.6%	81.4%	81.2%	79.6%	81.3%	79.2%	76.8%	76.4%
France	51.6%	50.7%	51%	50.6%	50.8%	51.7%	51.5%	50.5%	50.9%	51%	49.1%	48.9%
UK	–17%	–9.4%	–12.5%	–6.5%	4.6%	13.4%	21.2%	20.5%	26.2%	26.2%	28.1%	36%

Source: Eurostat. Sustainable Development Indicators. Theme 6: Climate Change and Energy (last accessed 17.09.2013).

The Energy Green Paper of 2006 already remarked the need to reinforce the external dimension of an energy policy (EC, 2006) as the complement of the internal energy market. It advocated for measures started at communitarian level and then at national level, using a top-down approach. Some of the instruments mentioned to reach the desired goals include energy partnerships with producers, transit countries and other international actors.

Europe 2020 and the subsequent Energy 2020 (EC, 2010b) strategies, call for the integration of energy agreements not only as part of the Energy Policy but also of the Neighbourhood Policy, the Development Policy and the Common Foreign and Security Policy. Although initiatives such as Medreg are valued, it seems that there is a preference for the Energy Community approach, again a top-down approach in which partners must comply with the European energy acquis. It is also important to point out the ambitious target of creating not only an internal energy market but also a pan-European energy market.

There are many reasons to pursue this international cooperation on energy. The confirmation of the slow progress on targets such as energy efficiency for 2020, unlikely to be achieved (EC, 2011b), requires actions on as many fields as possible. Additionally, the measures already in place at the EU level have already started reducing greenhouse gas (GHG) emissions, and it will not be long before the EU will account for no more than 10% emissions worldwide. This makes it more necessary than ever for the EU to intensify efforts convincing other countries to get involved in the fight against climate change through resource efficiency policies, with energy-related ones the core of those policies.

The documents published during the last years by the EU regarding energy issues advocate for a medium- and a long-term approach when dealing with energy matters and for a united voice to represent the EU on decision making regarding projects that affect third countries. Bearing in mind the criticisms that from time to time appear at international level pointing out the lack of one unique voice defending EU ideas, consensus reached giving special importance to the external dimension of energy policy is of great relevance.

Nevertheless, the top-down approach widely used over time should be complemented by initiatives such as Medreg, whose flexibility allows quicker progress and seems to be more effective (Medreg, 2011) in terms of measures implementation and regulation harmonization.

Agreements with third countries will continue to be signed, given the evidence that energy issues are global issues and cannot be solved at national level and that cooperation is the best way to achieve positive results. Energy is an area that can help foster actions to reduce poverty in developing countries, usually rich in natural resources. These agreements are conceived as win–win situations. On one side, the EU guarantees security of supply, and on the other side, the third countries benefit from economic resources which contribute to increase their development levels.

There is still some controversy in the way that the external dimension of the energy policy should be considered, bearing in mind that the elements involved have multiple interrelations: market, governance and geopolitical elements seem to be given more or less importance depending on what is expected from the agreements in every moment (Herranz-Surrallés & Natorski, 2011). Some authors have criticized the lack of consideration of the geopolitical element in some cases (Youngs, 2007) or who is really benefitting from the agreements (Escribano-Frances & San Martin, 2011).

The failure in trying to get some countries adapted to the top-down approach, particularly Russia, can be a sign of the need for more flexible mechanisms, and initiatives such as Medreg can positively contribute to further achievements on this area. It is true that we talk about a voluntary (bottom-up) approach, but given that the regulators are the ones fostering the progress, the feasibility of a stable regulatory framework for energy appears more real, taking into account that actions include training and technological knowledge exchange (Medreg, 2011), which each country can apply locally with the necessary adaptations. We cannot forget either that although an independent initiative, Medreg is sponsored by the EU, so there will always be some type of surveillance coming from EU institutions to ensure that agreements do not contradict EU regulations.

In trying to answer questions such as why countries with no prospects of becoming EU member states still want to take part on these agreements, several reasons have been pointed out: Even if the chances of becoming a member state appear weak under current circumstances, potential partners may want to demonstrate their capability to be part of the club, so they try to identify themselves with the EU rules by adopting these rules as part of their agreements. Additionally, and particularly in the case of Eastern countries, they could also be searching for greater independence from Russian influence, so approaching the EU is a common-sense movement. Another good reason is the prospect of economic gains, as having common rules, norms and standards facilitates economic exchange with the EU (Prange, 2007). This complex set of reasons make us consider that even if there is a need for an integrated approach from the EU to agreements with third countries in light of a common energy policy, there may be also reasons for differentiated treatment towards each of the countries and the regional areas involved in negotiations.

Bilateral/multilateral energy agreements of EU member states

The decisions that one member state makes on its energy mix, energy suppliers and energy agreements have an effect on the rest of member states and on EU targets on energy efficiency or emissions reduction, as well as on its suppliers' economies and any other third country, because it has an influence over resources (oil/gas) reserves availability.

The EU advocates for the continuation of bilateral and multilateral agreements as a comprehensive approach to engage third countries on a broad range of aspects across sectors that touch on climate change issues (EC, 2011c), including energy.

To control the effects of these agreements and to have better coordination on the actions that EU member states undertake in trying to comply with their targets on GHG emissions reduction or their contribution to the internal energy market, the EC is now advocating for some short of ex-ante control of the agreements that each member state intends to sign with third countries (Official Journal of the European Communities [OJEC], 2012). The idea has the best of intentions, but it must be carefully analyzed.

On one side, exchange of information on energy agreements between member states and between member states and the EU Commission would have positive consequences if the EU could help the contracting member state avoid pressures coming from the third country (an energy producer) when signing clauses that could be considered abusive or against EU law. Additionally it would contribute to better coordination at EU level for implementing projects involving third countries and transit countries, especially those related to pipelines and grids connections.

On the other side, having to inform the commission before the agreement is signed and allowing modifications over a document that will have legal implications for the signatory parties could most likely cause delays of the whole process. The decision finally approving this matter tries to overcome this problem by establishing deadlines for the commission to answer, as well as not forcing member states to inform the commission before the agreement is signed but once it has been signed, unless there are doubts about compatibility with EU laws. Only under the premises of quickness, simplicity and flexibility, a legally binding mechanism would be preferred over voluntary measures, considered insufficient on this area: "*voluntary measures on their own have so far not proven to be sufficient to guarantee the type of exchange of information necessary to ensure that agreements concluded between Member States and third countries are lawful and do not threaten the proper operation and functioning of the Union internal energy market*" (EC, 2011f, emphasis mine).

Again, we find here another example of the top-down approach preferred by the EU for energy-related issues, and we should wonder at which point this new legal instrument will have an effect on voluntary initiatives in place, given that to some extent it seems they are being left out of scope.

Agreements between member states and third countries will continue existing and will continue being signed because they are considered very relevant for the new strategy of the EU of reinforcing the external dimension of the energy policy (EC, 2011e). If the approved mechanism works as planned, it will be a useful instrument to help guarantee security of supply, and it will prevent the need to start infringement procedures that would complicate the operation of the internal market.

In this case, it is important to remark that one of the purposes of the commission is to make member states have a European vision instead of a national vision when signing agreements. This is good in the sense that member states will have better information about the implications for other countries of their bilateral agreements, but it can find the reluctance of those who think that the EU is interfering with national issues, and we cannot forget that energy, particularly energy security of supply, has been largely considered a national issue (Andersen, 2000).

In order to consider the national vision versus European vision when signing energy agreements, we must take into account several factors. One of them is the particular situation of each member state with regard to the diversification of energy sources, both in terms of type of fuel and suppliers. Even if one country has a diversified mix of energy sources, it may be relying on a special partnership with a reduced number of suppliers, which is a risk equally important in order to guarantee security of supply. Over the last years, this has become an increasingly problematic factor due to the political instability of some Middle East countries, which is not forecasted to be solved in the short term.

Criticisms may also come from those who see this involvement at EU level as a way to influence the results in negotiations in favour of European companies, instead of thinking of the general good of the agreement as a whole, when for instance they try to gain market share in African countries against China (Youngs, 2007). But it is still soon to evaluate its results.

Another important factor is that when concluding agreements related to energy supply, it is normal that each country searches for partners geographically close, in order to reduce transport costs, especially if there is a need to build new grids. Geographical proximity can be also an indicator of a good neighbourhood relationship in other areas such as commerce, which could facilitate agreements in the area of energy. Historical and cultural links may also facilitate bilateral or regional agreements, and we can find that third countries may prefer having as counterpart one particular country instead of the EU as a whole. This is what happened, for instance, during climate change negotiations at the Copenhagen Summit in 2009 (Eppstein et al., 2010).

So if the EU wants to complete the internal energy market by 2014 (European Council, 2011) and wants to use that internal market as a model to be exported to third countries which can act as instrument to achieve its energy objectives, it first needs to make itself more attractive to potential partners, both politically and economically (Brunarska et al., 2011). Otherwise, instead of an integrated market, we could find several regional fragmented markets responding to national and regional interests.

The future of regional energy initiatives

Regional energy initiatives play a key role in the EU energy policy, giving importance to its external dimension every day. There are a number of initiatives in place and some others forthcoming, which ensure a future of more regional energy initiatives with wider scope.

The EU has signed energy cooperation agreements with Ukraine (2005) and with Azerbaijan and Kazakhstan (2006) and is continuously searching for new agreements: Algeria and Egypt, among others (Youngs, 2007). A good example could be the conversations with Algeria with regard to cooperation in the area of renewable energies (November 2011), which have given light to a memorandum of understanding (MOU) between two private consortiums, contributing to a more favourable regulatory framework between the EU and countries of the South Mediterranean area, with the goal to cover local electricity demand and 15% of the EU electricity demand by 2050 (EC, 2011j). A more comprehensive MOU for a strategic partnership in energy has been signed in July 2013 (EC, 2013b).

Other important initiatives include the Mediterranean Solar Plan (Resources and Logistics [RAL], 2010), with the general objective of creating 20 GW of new-generation capacity from solar and other renewable energy sources around the Mediterranean Sea by 2020. But the main issue for a successful outcome of this project keeps being the interconnection of grids, and the current international economic crisis adds a financial constraint to the natural barriers for the project development.

One of the best examples of the new approach that the EU wants to give to energy agreements is the plan to build a Trans Caspian Pipeline System to bring gas to Europe (EC, 2011h; 2013a). The negotiations between the EU, Azerbaijan and Turkmenistan have been pursued giving a mandate to the European Commission to act on behalf of the initially 27 member states, and for the first time, a treaty is proposed to support an infrastructure project. In the end, it will contribute to the completion of the Southern Corridor.

It would not make sense to try to list all the current agreements and initiatives between member states and third countries or the EU and third countries, but we can try to establish the different frameworks for the operability of these agreements. One big area would be the cooperation agreements between the EU and one individual third country which plays a relevant role at international level, as it is the case of Brazil, China, India, Iraq, Norway, South Africa, Ukraine or the United States (see the EU website for the external dimension of energy policy: http://ec.europa.eu/energy/international/bilateral_cooperation/bilateral_cooperation_en.htm). Russia–EU relations are also a special case, given the conflicting interests that have characterized negotiations in the last years and the problems that tension between Russia and transit countries of its past area of influence are causing for EU gas supply.

Bilateral agreements between a member state and a third country may have different reasons, and we can distinguish the case in which the third country is a supplier (energy producer) from the case in which the third country is a transit country or another energy importer. An example of the latter could be the collaborative projects between Italy and Turkey, such as the Blue Stream pipeline, the Samsun-Ceyhan pipeline or the ITGI (Interconnector Turkey–Greece–Italy) project (Ustun, 2011).

Regional initiatives have great relevance due to the big number of countries involved and the wide scope of the projects, not only in terms of long-run goals but in availability of financial resources, technical knowledge transfer, regulation harmonization, participation of private and public actors and so on. The Energy Community can be considered as the most important regional agreement on energy. Participants are the EU on one side and Albania, Bosnia and Herzegovina, Croatia (an EU member since July 2013), former Yugoslav republic of Macedonia, Moldova, Montenegro, Serbia, Ukraine and the United Nations Interim Administration Mission in Kosovo, on the other side, with Armenia, Georgia, Norway and Turkey as observers (Energy Community Secretariat, 2005). But this ambitious initiative has also its weakness. Knowing that its main purpose is to create a stable regulatory and market framework, through the implementation of key elements of EU legislation, the continuous changes on that legislation make it difficult to happen. The summary of its first five functioning years gives an idea of the progress achieved so far and the pending issues, particularly the implementation of the formally adopted regulation and the boost of the necessary investments for infrastructures development (EC, 2011i), with the commission calling for the enforcement of the measures implementation and the conditionality of the financial resources to the effective implementation of the EU acquis (EC, 2011g).

On the positive side, Energy Community has facilitated collaboration between some of the countries that are contracting parties. Proof of this is shown in the projects approved in Macedonia for improving power networks with Albania, Serbia and Kosovo and in some initiatives from Bulgaria for creating an "electrical energy bourse" in which Greece, Serbia and Macedonia would participate (Taleski, 2011). It is not surprising then that the EU tries to apply the same approach to other geographical areas, particularly the Mediterranean, with the creation of a Southern-Energy Community (EC, 2011d).

But projects such as the Mediterranean Solar Plan, for instance, raise doubts about how it may influence on Morocco development, because it is required that some conditions are fulfilled, such as the participation of local labour force in plant construction and not only in maintenance activities or the establishment of feed-in-tariffs which do not undermine Moroccan companies in favour of European ones (Escribano-Frances & San Martín, 2011). A similar approach could be used in the case of Turkey, where there is a limited influence of the EU in the reform of the energy sector but it is not so clear at which point it can influence on institutional reforms (Carafa, 2011).

To some extent there is an overlap of actions and initiatives affecting some geographical areas. The Mediterranean Solar Plan and Medreg, for instance, combine regulatory and voluntary approaches in the Mediterranean area, which gives a proof of the importance that certain partners have for the EU. In this case, Medreg presents a complementary framework to gain progress on energy markets integration, with action plans for the short term (Medreg, 2011) that allows measuring the effectiveness of implementation and quicker adaptation to changing environments.

In light of all the initiatives mentioned, which are established with a long-term vision, it is clear that they will continue in the coming years and we will see new agreements and initiatives involving more countries and probably extending the geographical and political areas of influence approved. It is still to be seen which approach these future initiatives may take, but given the preference for regulations, mandatory and binding measures, the top-down approach seems to be the path to be followed by future actions.

Conclusion

The international economic and financial crisis, the difficulties on reaching a binding climate change agreement to reduce GHG emissions, the geopolitical situation in the Middle East countries and the events of the last years have contributed to convert energy issues in a global political issue. In such a complex situation, the EU is trying to reinforce its presence in international markets and convince potential partners of adopting its regulatory framework, with the aim of guaranteeing energy security of supply in the context of a sustainable development to become a low-carbon economy in the long term.

The promotion of renewable energies and the targets approved for every member state regarding the participation of renewable energies in total energy consumption for 2020, reinforced by directives and regulations which remark the importance of the external dimension of the EU Energy Policy, create the perfect framework to boost the implementation of regional energy initiatives with third countries. Given the fact that energy producers are usually developing countries, justification of the need for these agreements in those countries comes from the potential to foster economic growth and poverty reduction.

But we have seen that the contribution of these agreements to those aims is conditioned by the particular implementation practices in each case, and unless carefully planned, they may only satisfy the EU needs of security of supply.

It has also been established that there are different approaches to reach energy agreements, but the EU prefers the top-down one, which involves the purpose of attracting third countries interested in adopting EU regulations and frameworks, with the aim of increasing security and the rule of law on agreements signature. There is a clear intent to complete the internal energy market and the ambition to reach a pan-European market, but we should wonder if this is too ambitious a purpose, bearing in mind that particularly in the case of bilateral agreements the national interests usually prevail over the general EU interest. The decision from the EU establishing an information exchange mechanism for energy agreements tries to correct this issue.

Some of the problems presented by this approach have been also pointed out, together with the possibility to combine this top-down approach with voluntary bottom-up initiatives which, although considered insufficient by EU authorities, can perfectly complement and complete the positive outcome of regional energy initiatives.

As a concluding remark and given all the facts presented through the chapter, it is the opinion of this author that regional energy initiatives will have a prominent role in the coming years for the EU energy policy, and we are already seeing the external dimension playing a major role each day, with the area of renewable energies acting as one of the prominent links. But there is no need to take the EU acquis as the most accurate way to approach energy agreements. Only if the main goal is to satisfy EU needs we can see the regulatory approach as the preferred one, and even if it is the most used nowadays, we need to wait and see the results that this approach will give in the medium and long term. There is plenty of room for voluntary initiatives that can produce positive outcomes in terms of markets integration, through their flexibility and adaptation to third countries' environment and a process of mutual learning from third countries could attract those countries to the EU vision. In the end, all this means that both top-down and bottom-up approaches are needed and complementary.

Note

1. The findings, interpretations and conclusions expressed in this chapter are those of the author and do not reflect the positions of the institution.

References

Andersen, S.S. (2000). *EU Energy Policy: Interest Interaction and Supranational Authority*. ARENA Working Papers WP 00/5, University of Oslo. ARENA Centre for European Studies. Oslo.

Brunarska, Z., Jarosiewicz, A., Loskot-Strachota, A. & Wisniewska, I. (2011). "Between Energy Security and Energy Market Integration. Guidelines for the Future Development of the EU's External Energy Policy in Europe's Neighbourhood". Center for Eastern Studies, Warsaw, June.

Carafa, L. (2011). "How Far Does the EU Influence Energy Sector Reform in Turkey?" In *European Energy Policy: The Environmental Dimension*, F. Morata & I. Solorio (Eds.). Barcelona. Universidad Autónoma de Barcelona, pp. 231–250.

Department of Trade and Industry (DTI) (2006). *The Energy Challenge. Energy Review Report 2006*. London: HM Government.

Energy Community Secretariat (2005). *Treaty establishing the Energy Community*. Extracted from the Energy Community website: www.energy-community.org/portal/page/portal/ENC_HOME/ENERGY_COMMUNITY/Legal/Treaty

Eppstein, G., Gerlach, S. & Huser, M. (2010). "The EU's Impact in International Climate Change Negotiations. The Case of Copenhagen". The Diplomatic System of the European Union, Maastricht University, Maastricht. Retrieved 28.12.2011 from http://dseu.lboro.ac.uk/members/Maastricht/DSEU%20Epstein%20Gerlach%20Huser%20paper.pdf.

Escribano-Frances, G. & San Martin, E. (2011). "Morocco and the Mediterranean Solar Plan: A Driver for the Development of Whom?" In *European Energy Policy: The Environmental Dimension*, F. Morata & I. Solorio (Eds.). Barcelona. Universidad Autónoma de Barcelona, pp. 209–230.

European Commission (2006). *Green Paper. A European Strategy for Sustainable, Competitive and Secure Energy*. Brussels 8.3.2006. COM (2006) 105 final.

European Commission (2007a). *Communication from the Commission to the Council and the European Parliament. Prospects for the internal gas and electricity market*. Brussels. 10 January 2007. COM (2006) 841 final.

European Commission (2007b). *Communication from the Commission to the European Council and the European Parliament. An Energy Policy for Europe*. Brussels. 10 January 2007. COM (2007) 1 final.

European Commission (2010a). *Communication from the Commission. Europe 2020. A Strategy for Smart, Sustainable and Inclusive Growth*. Brussels. 3 March 2010. COM (2010) 2020.

European Commission (2010b). *Communication from the Commission to the European Parliament, the Council, the European Economic and Social Committee and the Committee of the Regions. Energy 2020. A Strategy for Competitive, Sustainable and Secure Energy*. Brussels. 10 November 2010. COM (2010) 639 final.

European Commission (2011a). *Communication from the Commission to the European Parliament, the Council, the European Economic and Social Committee and the Committee of the Regions. A Resource-Efficient Europe – Flagship Initiative under the Europe 2020 Strategy*. Brussels. 26 January 2011. COM (2011) 21.

European Commission (2011b). *Communication from the Commission to the European Parliament, the Council, the European Economic and Social Committee and the Committee of the Regions. Energy Efficiency Plan 2011*. Brussels. 8.3.2011. COM (2011) 109 final.

European Commission (2011c). *Communication from the Commission to the European Parliament, the Council, the European Economic and Social Committee and the Committee of the Regions. A Roadmap for Moving to a Competitive Low Carbon Economy in 2050*. Brussels. 8 March 2011. COM (2011) 112 final.

European Commission (2011d). *Joint Communication to the European Parliament, the Council, the European Economic and Social Committee and the Committee of the Regions. A Partnership for Democracy and Shared Prosperity with the Southern Mediterranean*. Brussels. 8 March 2011. COM (2011) 200 final.

European Commission (2011e). *Communication from the Commission to the European Parliament, the Council, the European Economic and Social Committee and the Committee of the Regions. On Security of Energy Supply and International Cooperation – "The EU Energy Policy: Engaging with Partners beyond Our Borders"*. Brussels. 7 September 2011. COM (2011) 539 final.

European Commission (2011f). *Proposal for a Decision of the European Parliament and the Council Setting Up an Information Exchange Mechanism with regard to*

Intergovernmental Agreements between Member States and Third Countries in the Field of Energy. Brussels. 7 September 2011. COM (2011) 540 final.

European Commission (2011g). Press Summary. Energy Community – Commission calls for reforms to boost investments. 10 March 2011.

European Commission (2011h). Press Release. EU starts negotiations on Caspian pipeline to bring gas to Europe. Brussels, 12 September 2011. IP/11/1023.

European Commission (2011i). Press Release. Energy Community – Five years of regional cooperation. Brussels, 21 October 2011. IP/11/1223.

European Commission (2011j). Press Release. Commission welcomes Desertec and Medgrid cooperation on solar energy in North Africa and the Middle East. Brussels, 24 November 2011. IP/11/1448.

European Commission (2013a). Press Release. EU Commission welcomes decision on gas pipeline: Door opener for direct link to Caspian Sea. Brussels, 28 June 2013. IP/13/623.

European Commission (2013b). Memorandum of Understanding on the establishment of a strategic partnership in the energy sector between the European Union and the People's Democratic Republic of Algeria. 07.07.2013 (in French).

European Council (2011). Conclusions European Council. 4 February 2011.

Eurostat. *Sustainable Development Indicators. Theme 6: Climate Change and Energy.* http://epp.eurostat.ec.europa.eu/portal/page/portal/sdi/indicators/theme6 (last accessed 17.09.2013).

Helm, D. (2003). *Energy, the State, and the Market. British Energy Policy since 1979.* Oxford. Oxford University Press.

Herranz-Surrallés, A. & Natorski, M. (2011). "Energy Security in the EU's Relations with Eastern Neighbours: Driver or Brake for a Comprehensive External Energy Policy?" In *European Energy Policy: The Environmental Dimension*, F. Morata & I. Solorio (Eds.). Barcelona. Universidad Autónoma de Barcelona, pp. 157–188.

Medreg (2011). *MEDREG Response to EC Public Consultation on the External Dimension of the EU Energy Policy.* 4 March 2011.

Official Journal of the European Communities (2007). *Treaty of Lisbon amending the Treaty on European Union and the Treaty establishing the European Community, Signed at Lisbon, 13 December 2007.* C306 of 17 December 2007.

Official Journal of the European Communities (2009a). *Directive 2009/28/EC of the European Parliament and of the Council of 23 April 2009 on the Promotion of the Use of Energy from Renewable Sources and Amending and subsequently Repealing Directives 2001/77/EC and 2003/30/EC.* 5 June 2009.

Official Journal of the European Communities (2009b). *Council Directive 2009/119/EC of 14 September 2009 imposing an Obligation on Member States to Maintain Minimum Stocks of Crude Oil and/or Petroleum Products.* 9 October 2009.

Official Journal of the European Communities (2012). *Decision No 994/2012/EU of the European Parliament and the Council, of 25 October 2012, Establishing an Information Exchange Mechanism with regard to Intergovernmental Agreements between Member States and Third Countries in the Field of Energy.* 27 October 2012.

Prange, H. (2007). "A Ring of Friends . . . ? EU External Energy Policy and Regulatory Boundary Shifting towards its Eastern Neighbourhood". Paper prepared for the 6th CEEISA Convention "Global and Regional Governance – European Perspectives and Beyond", Panel "Energy Security and EU's Eastern Neighborhood", 24–26 May 2007, Wroclaw, Poland. Retrieved on 14 February 2012 from www.ceeisaconf.uni.wroc.pl/wordy/papers%201%20session/Prange.doc.

Reiche, D. & Bechberger, M. (2004). "Policy Differences in the Promotion of Renewable Energies in the EU Member States". *Energy Policy* 32, pp. 843–849.

Resources and Logistics (RAL) (2010). "Identification Mission for the Mediterranean Solar Plan. Final Report". FWC Beneficiaries Lot 4 –N° 2008/168828. Project funded by the European Union, ENPI – Neighbourhood – Mediterranean & Eastern Europe.

Taleski, M. (2011). "Macedonia Strives to Improve Regional Energy Networking". *Southeast European Times*. 4 November. Retrieved on 14 February 2012 from www.setimes.com/cocoon/setimes/xhtml/en_GB/features/setimes/features/2011/11/04/feature-04

Ustun, C. (2011). "Energy Cooperation between Import Dependent Countries: Cases of Italy and Turkey". *Perceptions* XVI, no 1, pp 71–89.

Youngs, R. (2007). "Europe's External Energy Policy: Between Geopolitics and the Market". CEPS Working Document No. 278/November 2007. Centre for European Policy Studies, Brussels.

Chapter 16

The diffusion of renewable energy technologies in the Mediterranean basin

State of the art and perspectives

*Alessandro Sapio**

Having lost its prominence in trade flows in the aftermath of Columbus's great discovery, the Mediterranean is enjoying a revived centrality thanks to the burgeoning demand for green energy. European policymakers and businesspeople are increasingly alert to the investment prospects offered by the huge endowments of solar and wind resources that characterize the countries in the Mediterranean basin.

Although advanced countries possess the financial and technological means to abate greenhouse gas emissions, the natural resources of emerging countries are powerful weapons in the fight against global warming. The ensuing mutual dependence of advanced and emerging countries vis-à-vis the perils of climate change is exemplified by the geo-economic map of the Mediterranean, with its north–south patterns of economic development and resource endowments. The immense, virtual reservoir of solar power called the Sahara as well as the rich wind resources in the North African coastlines are largely untapped.

The European Union (EU), however, is keen to find solutions to the insufficient development of renewables in the Middle East and North Africa (MENA) area, as it depends on external gas and oil supplies for more than half of its energy requirements. As testified by the events of January 2009, when the Russian dispute with Ukraine led to major cut-offs of gas supplies, as well as by the turmoil caused by the Arab Spring revolutions, the energy security and independence of Europe are threatened by diplomatic clashes. A direct approach to the security of supply problem has been attempted by the European Commission through the Southern Gas Corridor, in hope to curtail the Russian bargaining power by increasing the number of gas suppliers.[1] However, the recent setback of the Nabucco-West Pipeline and the limited expected impact of the alternative Trans Adriatic Pipeline further motivate the quest for alternative energy sources. The high human and economic losses due to nuclear disasters (Three Mile Island, Chernobyl, Fukushima) and fears of nuclear proliferation are pushing European governments to de-commission their national endowments of nuclear power plants, ruling out a potential solution. Together with continued public support to European renewables, the large-scale development of solar and wind power in MENA countries is thus seen as a most welcome way out of the European energy security problems – despite persisting niches of skepticism concerning the prospective economic efficiency of renewable energy technologies (RETs) and the burden of renewable energy (RE) subsidies on public budgets.

This chapter outlines the main existing empirical findings on the diffusion of renewables and their effects on the Mediterranean electricity systems. It also offers an overview of the interdependent roles that regulatory harmonization and physical integration can play in driving the development of renewables in the area.

Renewable energy sources in the Mediterranean: The status quo

The Mediterranean offers a fascinating blend of resource-rich and resource-needing countries. The spatial distribution of natural, technological and financial resources creates mutual dependence patterns and promising trade opportunities: the affluent north could help the deployment of the plentiful endowments of solar and wind power available in the south.

In the southern Mediterranean, the outlook for green power generation is highly promising. With an average solar irradiance of 2,400 kWh/m^2/year, solar power is an obvious candidate for the development of the RE sector in the MENA area. No European country comes close to such a high solar power potential. The Sahara Desert spans an area of about 3.9 million square kilometres; part of it could be used for concentrating solar power (CSP) installations. The Negev and Arava Deserts in southern Israel offer further land exploitable for CSP, for about 3% of their 15,600 square kilometres (Yasner 2012). Thanks to the intense solar irradiance, CSP facilities could produce at relatively low average costs (Trieb and Müller-Steinhagen 2007). Photovoltaic (PV) installations on the rooftops of densely populated cities could add to the existing hybrid solar power facilities, which in the last decades have helped countries such as Morocco achieve a nearly complete rural electrification (Tsikalakis et al. 2011).

Besides the solar power potential, average wind speeds exceeding 10 m/s in some locations on the Egyptian Red Sea coast and on the Atlantic coast of Morocco, higher than in most southern European locations, testify to the large potential for wind power generation in MENA countries. Significantly high wind power potentials are also found on the Algerian, Tunisian, Libyan and Turkish coasts.[2] Geothermal energy supplies from promising locations on the Golan Heights and in Turkey may compete with those from the Italian site of Larderello. Significant biomass resources are found in Israel and Turkey. The latter also boasts an economically feasible hydro potential of 42,000 MW (Boluk 2013).

But currently, the main sources of wealth for MENA countries are represented by fossil fuels. The southern Mediterranean hosts major producers and exporters of fossil fuels, such as Algeria and Libya. About 57% of the world oil reserves and 41% of the world natural gas resources can be found in that area (Nicolini and Porcheri 2012). Along with shale gas, renewables could help these countries cope with the nearing peak in hydrocarbon production.[3]

On the opposite side of the Mediterranean, one finds high-income countries that, although depending on imports of fossil fuels, are important sources of financial flows to the MENA area, through either development aid, public–private partnerships, or foreign direct investments (FDIs), due to their superior financial development.[4] Furthermore, Lanjouw and Mody (1996) and Dechezleprêtre et al. (2012) among others have shown that advanced countries are also the main sources of green innovations.[5]

Guided by the preceding picture, one could argue that financial flows and technology transfers from high-income countries, combined with the rich natural resource endowments of the MENA area, would yield plenty of renewable energy outputs. The diffusion process of RETs in the southern Mediterranean, however, has not taken off.[6]

In Table 16.1, for each country in the Mediterranean basin, two measures of RET diffusion are reported for the year 2011. One is the RET penetration rate, namely the share of renewable energy capacity in total electricity capacity.[7] In the table, this metric is disaggregated by technology. The other proxy for diffusion is the per capita non-hydro RE capacity.[8]

The overall penetration of renewable sources was highest in South-East European countries, such as Albania (90.3%), Bosnia and Herzegovina (49.1%) and Croatia (47.6%). However, this is the effect of heavy reliance on hydropower. Once hydro is factored out, the diffusion of non-hydro renewables was highest in Spain, where it accounted for slightly more than one-fourth of total generation capacity, as well as in Italy and Greece, with penetration rates between 10% and 20%. In North Africa, the largest presence of non-hydro RE capacity was in Morocco (4.5%) and in Egypt (2%); the performance of Turkey was better (5.4%), but overall, RE penetration rates are much lower than in Europe. Looking at the individual technologies, Spain stands out as the leader in wind power (21.3%) but it is outpaced by Italy in solar power (10.8%). Greece and France fare better in wind (9.9% and 5.2% respectively), but much less so in solar. The wind penetration rate in Italy was 5.8%. The data on the MENA area reveal that Morocco and Egypt produced almost all of their non-hydro RE via wind power. Biomass production is characterized by generally smaller penetration rates, with Italy (2.2%) as the leader. Geothermal penetration rates are negligible. The data on per capita non-hydro RE capacity tell a very similar story: EU countries hold the lion's share, with Spain, Italy, France and Greece as the top four countries. Notice that the non-hydro RE capacity in Israel is relatively large

Table 16.1 Diffusion of renewable energy technologies in the Mediterranean: renewable penetration rates and non-hydro renewable per-capita capacity, for 2011

Countries	*RE penetration rates (%)*						*Per capita non-hydro RE capacity (kW per capita)*
	RE	*non-hydro RE*	*Solar, tide & wave*	*Wind*	*Geothermal*	*Biomass & waste*	
Albania	90.3	–	–	–	–	–	–
Algeria	2.2	–	–	–	–	–	–
Bosnia and Herzegovina	49.2	–	–	–	–	–	–
Croatia	47.6	3.3	–	3.1	0.1	0.1	0.031
Cyprus	13.1	13.1	0.9	12.1	–	0.1	0.081
Egypt	12.1	2.0	–	2.0	–	–	0.007
France	22.9	8.8	2.4	5.2	0.1	1.1	0.176
Greece	29.5	14.2	3.8	9.9	–	0.5	0.208
Israel	1.6	1.5	1.4	0.03	–	0.07	0.028
Italy	31.6	19.5	10.8	5.8	0.7	2.2	0.383
Lebanon	11.9	–	–	–	–	–	–
Libya	–	–	–	–	–	–	–
Malta	1.2	1.2	1.2	–	–	–	0.017
Morocco	24.9	4.5	–	4.5	–	–	0.009
Slovenia (2010)	35.5	1.9	0.4	-	–	1.5	0.029
Spain	39.6	26.5	4.2	21.2	0.1	1	0.586
Syria	10.8	–	–	–	–	–	0.00005
Tunisia	2.9	1.3	–	1.3	–	–	0.005
Turkey	37.2	5.4	–	3.4	0.2	1.8	0.040

Source: Author's calculations on US Energy Information Administration and World Bank data.

with respect to its population, if compared with countries characterized by higher penetration rates, such as Egypt, Morocco and Turkey.

It is worth remarking that, currently, larger RE supplies from MENA would not represent credible sources for the energy needs of the northern Mediterranean anyway, because there is only one interconnection link between MENA and EU countries, namely the 400 kV line between Morocco and Spain, with a capacity of 1,400 MW. A 400-kV undersea transmission line between Tunisia and Sicily, the Elmed cable with a 1,000-MW capacity, has been planned. Other projects under consideration include links between Algeria and Sardinia, Algeria and Spain and Libya and Sicily (Nicolini and Porcheri 2012).[9]

Behind the Mediterranean patterns of renewable energy diffusion

The evidence of asymmetric RET diffusion patterns along the north–south Mediterranean axis is, apparently, a puzzle: the countries that could produce more renewable energy are those that most strikingly lag behind. Reviewing the policy-oriented econometric literature on the determinants of RET diffusion rates is the natural starting point on the way to solving the puzzle. The weakness of regulatory settings in MENA countries seems to be a major barrier to the development of their green energy industries. However, this section goes on to argue that the policy-diffusion line of causation in emerging countries (e.g. MENA) is less straightforward than in advanced economies (such as the EU), because it is mediated by the influence of external policy principles, as well as by the solution to local and external shortages in complementary transmission infrastructures.

Reviewing the evidence

To anticipate somewhat the findings reviewed below, the emerging picture is one in which promotion policies stand out as the neatest determinants of RET diffusion, even though not all policies have been scrutinized with the same accuracy, not all tools are equally effective and some tools only stimulate the diffusion of certain technologies. The impact of cross-country income and financial development is not robust, whereas technological progress in the green energy sector seems to be a driving force in the diffusion of RETs.[10] Let us go through the evidence in some more detail.

A number of works highlight the primary role of support policies. The cross-sectional econometric study by Menz and Vachon (2006) focuses on the policies adopted in 37 US states between 1998 and 2003 to encourage the development of wind power, and finds that both the sheer adoption and the "policy experience" (i.e. the number of years elapsed since policy adoption) in renewable portfolio standards and mandatory green power options have favorably affected the installed wind capacity, the absolute growth in wind capacity and the number of large wind power installations, whereas public benefit funds have yielded no significant effects. These results are confirmed by Yin and Powers (2010) who more appropriately account, in a panel data model, for dynamic relationships and feedbacks and include socio-economic variables, which, however, lack statistical significance. Jenner et al. (2012) separately analyze the growth in solar photovoltaics and wind capacity in 26 EU countries between 1992 and 2008, to find that the growth of both RE sources is positively associated with the share of gas in electricity production, whereas feed-in tariffs have been beneficial only in the wind power sector. The effects of socio-economic variables (per capita gross domestic product [GDP], per capita energy consumption) are not statistically significant. In del Rio and Tarancon (2012), support policies per se are not as effective as the removal of administrative barriers and the stability of the regulatory setting in a panel of EU countries. Pfeiffer and Mulder (2013) confirm this result on a panel of 108 developing countries from 1980 and 2010, along with a positive effect of income and schooling, while they cast doubts on the effectiveness of the Kyoto Protocol and observe no effect from financial development. These latter results run against the findings by Brunnschweiler (2010) of significant Kyoto and financial development effects on RET diffusion in 119 non–Organisation for Economic Co-operation and Development (OECD) countries.

Indirect effects of support policy measures through technical change are signaled in Popp et al. (2011), who find that knowledge stocks concerning green energy, measured as patent counts, positively affect the per capita RE capacity in 26 OECD countries from 1991 to 2004. A stronger effect is found with patent indicators that account for the quality of the innovation, for example patents awarded in multiple countries. However, these effects are smaller than those associated with the adoption of the Kyoto Protocol, underlining the direct role of support policies.

Finally, Narbel (2013) analyzed a sample of 107 middle- and high-income countries, with data for the 2007–2009 period, showing that higher dependence on coal fosters the adoption of RETs both for climate change and for energy security reasons, whereas with respect to gas dependence, the two policy objectives tend to counterbalance each other.[11]

Diverging energy policy principles

As the evidence suggests, RE promotion policies (or lack thereof) are major candidates to explain the distance between EU and MENA countries in RET diffusion rates.[12] Indeed, policy principles and tools in RE support widely differ across the Mediterranean.

In the north, the energy policies of the countries in the EU bloc have been harmonized upon the approval of the Third Energy Package, according to which EU members adopt National Renewable Energy Action Plans to reach legally binding targets coherent with the EU-wide goals of 20% penetration rate in gross final energy consumption, 20% energy saving, and 20% greenhouse gas emissions reduction by 2020. The EU energy policy can be described through four pillars (Cambini and Franzi 2013a): (1) independent regulatory agencies (IRAs), (2) incentive-based tariffs, (3) third-party-access regime and (4) network unbundling. Kitzing et al. (2012) have identified three trends in the EU policy towards renewables: the increasing use of price-control instruments (such as feed-in tariffs and feed-in premia) instead of quantitative limits; the differentiation of policy types by installation size, with a preference for feed-in tariffs designed to encourage small installations; the "stacking" of multiple instruments. Although cross-country diversity in practices is still persistent, the process of convergence in the design of the policy tools across the EU is slowly progressing, as suggested in the historical outline of RE promotion offered by Haas et al. (2011).

National support to green energy in the southern shore departs from the sheer efficiency logic that permeates the EU framework, in order to pursue strategic and socio-economic goals. Indeed, the challenges faced by the energy industries in MENA countries are unusual for Europe, including the use of energy for desalination, the constraints given by water scarcity, the requirements of a rapidly growing population, the goals of rural electrification, the socialization of fossil fuel export revenues and the paradoxes of plenty that entrap resource-rich countries.

Seen as a political instrument, tariff setting in MENA is the responsibility of the executive power, only seldom entrusted to independent, technocratic bodies, as with IRAs established in Egypt, Israel, Jordan and Turkey. Feed-in tariffs have been introduced in Algeria, Israel and Turkey, but their levels are seen as insufficient to stimulate the diffusion of renewables (see Ata [2013] for the case of Turkey). The low level of end-user and feed-in tariffs, along with widespread fuel consumption subsidies, diminish the competitiveness of renewable energy with respect to conventional sources (Trieb et al. 2011; Nicolini and Porcheri 2012).

Another key feature of the MENA energy industries is represented by the dominant position of state-owned public utilities. Political resistance to change is, however, understandable, because vertical unbundling would open up the door to foreign private investors that, by virtue of their superior competence, may threaten the national interests of MENA countries. Egypt (Suding 2011) and Morocco (Fritzsche et al. 2011) are reforming their energy industries in the direction of greater openness to competition, whereas the Turkish electricity market was liberalized back in 1982 (Baris and Kucukali 2012). Morocco is developing an advanced status with the EU for energy cooperation (Escribano and Solorio 2012). Some measures favorable to public–private partnerships have been adopted in Tunisia, as well as law provisions in Algeria allowing the involvement of private actors in CSP projects. Yet, all this must be weighed against the fact that MENA countries aim to help the creation of local knowledge bases in RETs and to develop their own national solar industries, as testified by the Algerian, Egyptian and Moroccan renewable energy plans (Kilpeläinen 2013).

In interpreting differences in the principles and tools for RE support as the main drivers of the observed gaps in RET diffusion, some fundamental asymmetries between advanced and emerging countries need to be accounted for.

In high-income countries, policy measures are mainly aimed at stimulating domestic investments, allowed by relatively efficient credit markets and high saving rates. Instead, the diffusion of RETs in emerging countries depends on foreign investments. In designing green energy promotion policies, thus, emerging countries may need – or be required – to create interfaces with the regulatory settings of high-income countries, at least to an extent that guarantees acceptable risk–return profiles to foreign investors.

Further, economic agents exposed to RE promotion policies in high-income economies can deploy infrastructures that are already in operation or that only require relatively modest reinforcements. Emerging countries, instead, need to use policy levers to provide citizens with basic services, as well as to fill the gaps in transport infrastructures and in developing a sufficient absorptive capacity, that is the ability to identify, assimilate and commercially exploit new knowledge (Cohen and Levinthal 1990). Hence, the sets of policy measures that are empirically effective in MENA countries and in the EU are bound to be different, and the impact of MENA promotion policies is, by necessity, less timely and more likely to show up with significant delay. The external dimension of energy policies and the issue of infrastructural gaps is dealt with in the upcoming subsections.

The limping Europeanization of green energy policy

The existing diversity in RE support approaches and intensities around the Mediterranean is viewed by the European Commission (EC) as detrimental for the achievement of the EU energy policy targets. The so-called external energy policy of the EU, as promoted in the EU Energy 2020 Strategy (COM(2010) 639) and reinforced in the subsequent EC Partnership for Democracy and Shared Prosperity with the Southern Mediterranean (COM(2011) 200), pursues the integration of the national energy markets around the Mediterranean under the aegis of the *acquis communautaire*. Europeanization may be seen by MENA countries as an opportunity to shape support policies that are attractive for foreign investors.

As theorized by Cambini and Franzi (2013a), the EC can promote the alignment of the energy policies of neighbor countries to the EU framework via direct pressure and network-based influence. Direct pressure in some cases is enacted through more or less explicit conditionality clauses, such as those imposed on candidate member countries (Croatia until recently and Turkey). Indirect pressure is conveyed through institutions where EU sits together with non-EU countries, such as MedReg and the Union for the Mediterranean (UfM). In particular, capacity-building and knowledge-transfer projects financed by the EC can play a prominent role in shaping new understandings on RE promotion in MENA countries.[13]

The existing heterogeneity in policy types and designs around the Mediterranean suggests that the Europeanization process is still far from achieving its goals. A number of difficulties appear along the path to regulatory convergence.

For one, rules harmonization is not a sufficient condition to encourage investments. Recent evidence suggests that reforms in MENA countries often do not go beyond mere "cosmetic" changes, lacking substantial handovers of regulatory powers from the executive. Through a survey, Cambini and Franzi (2013b) evaluate the independence of the IRAs established by MENA and other Mediterranean countries in terms of decision-making autonomy, organizational autonomy and agencies' accountability. The survey respondents include energy regulators of Croatia, Egypt, Israel, Jordan and Turkey and the state-owned utilities of Morocco and Tunisia. The regulators are found to possess autonomy on some relatively minor issues, such as consumer protection, dispute settlements and licenses but notably not on tariff setting, a function retained by the executive.

Another key barrier to the Europeanization of RE support policies lies in the divergence of goals and interests across the Mediterranean, which are bound to foster regulatory convergence within more limited geographical areas (Kausch and Youngs 2009). Faced with the prospects of rising domestic demand for energy, MENA countries are increasingly aware of RETs as a solution to their own energy dependence problems (Mason and Kumetat 2011), breaking the complementarity between EU and MENA countries that triggered the Mediterranean Solar Plan. In light of this, regulatory integration among (subsets of) MENA countries is not an unlikely scenario, and following Escribano (2010) and Rubino (2011), it may even be desirable in view of the commonality of interests that coagulate around energy corridors. Not by chance, direct pressure from the EU is found to be less effective in countries such as Turkey and Algeria (Cambini and Franzi 2013a), reflecting the key positions of those countries in the geopolitical chessboard. MENA countries are starting to cooperate on the design of RE and energy efficiency policy tools through institutions such as the Middle East and North Africa Renewable Energy Conference (MENAREC); discussions concerning regional integration plans are also progressing, as witnessed by the focus of the Arab Maghreb Union on the national electricity markets in the Maghreb, as well as by the Eight Countries Interconnection project involving Egypt, Iraq, Jordan, Lebanon, Libya, Palestine, Syria and Turkey.

Leaping the infrastructural hurdle

Filling the gaps in transmission infrastructures is most definitely a key step towards fostering the development of RETs in the Mediterranean basin. Energy policymaking in MENA countries needs to prioritize the design of institutional arrangements in order to cope with the causal ambiguity between RE and transmission investments known as the "chicken and egg" dilemma (Vajjhala et al. 2008).[14]

The appeal and the very economic feasibility of HVDC cables to connect the grids of EU and MENA countries would be very limited without alleviating the between- and within-country bottlenecks that plague both sides of the Mediterranean. The transmission shortages and lack of synchronization between the Maghreb and Libya do not allow the benefits from the Spain–Morocco interconnection to spread across the whole MENA area. Israel can be still considered as an "electric island" due to lack of interconnection with bordering countries (Yasner 2012). The lack of transport infrastructures to bring wind power produced on the Red Sea coast to the large urban centers along the Nile and the Mediterranean coast of Egypt is yet another barrier to the diffusion of RETs in the MENA area. The reach of northbound energy flows across the Mediterranean is limited also by the heavily congested corridors that are found in Europe, for example in the Pyrenees and at the Greek borders, let alone the bottlenecks within Italy (see L'Abbate et al. 2014).

The bottom line is that, while still relying on the high-income world for financing RE investments within their own borders, MENA countries also depend on the physical investments in transport capacity made within foreign power networks. The grid reinforcements required in the northern Mediterranean would nevertheless be primarily financed by public actors, in light of the grid externalities that cause the private returns to reinforcements to fall below the social returns. Hence, diplomatic relationships may be far more critical than market forces for jumping the infrastructural hurdle.

Effects of renewables on Mediterranean electricity systems

While the increasing adoption of clean technologies is bound to trigger positive environmental externalities, it also yields effects that are more readily quantifiable, for example on electricity price levels and volatilities, as well as on the costs for operating and updating the transmission system. Detecting and measuring these effects is useful in order to get timely feedbacks on the performance of the support measures.

Price reduction effects

Because of negligible marginal costs and priority dispatch rules, greater RE supply causes the displacement of high marginal cost, fossil-fueled power plants, thereby yielding price reduction effects at least in the short run. Empirical evidence on price reduction effects due to wind power is provided, among others, by Moesgaard and Morthorst (2007) on western Denmark between 2004 and 2007, by Traber and Kemfert (2011) on Germany in 2007 and 2008 and by Woo et al. (2011) on the Texas electricity market between 2007 and 2010. Closer to a Mediterranean focus, Saenz de Miera et al. (2008) compare Spanish electricity prices in three consecutive days with similar levels of electricity demand and simulate the market solution in the absence of wind generation using data for 2005, 2006 and 2007. The price reduction effect in the Spanish market is found to be greater than the increase in the costs for the consumers arising from feed-in tariffs. The simulations in L'Abbate et al. (2014) also show that, with enough transmission capacity between the Italian islands and North Africa (five lines, 2,000 MW each) and with the current internal bottlenecks, average electricity prices in Sardinia and Sicily would stay significantly below those in the other Italian market zones.

Theoretical works have demonstrated that price reduction effects of wind power penetration are exacerbated if thermal power plants are run by oligopolistic companies. Indeed, when wind is low, generating companies that possess market power face a higher residual demand; therefore, they find it optimal to withhold capacity, pushing up the electricity price. Similarly, the exercise of market power may keep electricity prices well above the marginal costs of the market-clearing unit even after substantial increases in wind capacity. This is an outcome of the auction model by Banal-Estañol and Rupérez Micola (2011), in which electricity demand is served by two symmetric high-cost plants and a low-cost, wind power plant. The intermittency of wind power creates uncertainty on the market equilibrium solution, forcing power generating companies to hedge that uncertainty by inducing positive price-cost margins.[15]

An alternative reason for why electricity prices can remain above marginal costs is provided by Guerci and Sapio (2012). Looking at the patterns of network congestion generated by a simulation model of the Italian electricity market, the authors find that high wind penetration comes at the cost of more frequent separation between the southern regions, rich in wind, and the "windless" northern regions. The ensuing market power opportunities partly offset the price reduction effects of high wind penetration. These results can be relevant for Egypt and Morocco, because the most promising sites for wind power production in those countries are placed quite remotely from the most densely populated areas.[16]

It is worth noting that, in the long run, electricity prices may even increase, as power generating companies have the opportunity to adjust their portfolios by investing in peaking plants with higher marginal costs (Nicolosi and Fürsch 2009; Steggals et al. 2011). However, upward price pressures may manifest themselves even in the short run. Billette de Villemeur and Pineau (2012) have theoretically shown that the integration between a competitive market (in the importing country) and a regulated monopoly (in the exporting country) may cause the electricity prices to increase on the exporting side of the transmission line. Indeed, the regulated monopolist has incentives to export and thus restrains the energy supply at home.

The latter theoretical result is particularly appealing for the analysis of renewable energy trade patterns across the Mediterranean. If the direction of RE flows is from MENA countries to the EU, as predicted by the Mediterranean Solar Plan and confirmed by L'Abbate et al. (2014) in this book, the implication is that electricity prices may grow in the southern Mediterranean shore, where public utilities still monopolize the industry. Clearly, this would make a strong case for liberalization of the electricity industry in the MENA countries.

However, there is still considerable uncertainty regarding which scenario will prevail. Brand and Zingerle's (2011) simulations over the 2025 horizon show that, even after the integration into the European grid, Maghreb countries will remain net importers if electricity from Spain remains cheaper. The stronger demographic growth in the south may add momentum to this trend (Trieb and Müller-Steinhagen 2007). Hence, research efforts are needed to assess how the Mediterranean configuration of electricity prices may be affected by renewable energy trade flows when the monopolist is hosted by the importing country.

Volatility and risk management capabilities

Downward pressure on wholesale prices, much desired by electricity users, comes at the cost of higher electricity price volatility, at least in the short run. This is due to the fluctuations of RE outputs that can be extremely wide in the case of wind. In the oligopoly model proposed by Green and Vasilakos (2010), power generating companies view wind sales as a shift in their residual demand functions; consequently, increased wind penetration adds to the uncertainty in the residual demand they face, yielding volatile electricity prices.

One key implication of magnified volatility is that the demand for risk management tools in MENA countries is bound to increase as the RET diffusion trajectories unfold (Woo et al. 2011). In particular, higher volatility may trigger the set up of market segments catering to risk traders. Risk management, however, involves high human specificities that may discourage the participation of small power-generating companies in critical transactions, such as power balancing, creating an effective lower bound to market concentration.[17] In light of the limited financial development of MENA countries, the increased electricity price volatility would aggravate their dependence on foreign financial intermediaries, by adding the demand for risk management services to the current dependence on FDIs and international aid.

Energy storage and investments in transmission capacity provide alternative, non-financial solutions to risk management. Improved transmission links act as buffers with respect to RE fluctuations by smoothing them on a wider geographical scale. However, decisions on transmission expansions and the (uncertain) outcomes of research and development (R&D) on energy storage are beyond the reach of individual generating companies, let alone those dominating the MENA countries, that often lack both the incentives and the capabilities to affect the technological frontier in energy infrastructures.

The cost of complementary infrastructures

Increasing RE penetration may cause losses in the average technical efficiency of a country's power generation portfolio. Due to the displacement effect, thermal units tend to operate for fewer hours, to the detriment of their efficiency rates. Adding to this, the fluctuations of wind power call for the availability of flexible conventional plants, such as combined-cycle gas turbines (CCGT), to ensure the adequacy and security of supply. Volatility, in this respect, may even be good, as it would provide the financial incentives to delay the economic obsolescence of the gas-fired plants used for reserve services (Albadi and El-Saadany 2010; Steggals et al. 2011). The sign of the net effect on the demand for flexible thermal units is still an empirical question. A positive net effect would testify to the complementarity of natural gas and renewables, as already suggested by some empirical evidence on the diffusion of RETs (see Jenner et al. 2012). This would place countries such as Algeria and Egypt at the forefront of RE development in the Mediterranean, in light of their plentiful natural gas reserves. The simulation results in Traber and Kemfert (2011) are rather pessimistic in this respect, as their findings suggest that the load factors of flexible gas-fired plants would be disproportionately reduced.

Further complementary investments may be required in the perspective of higher RE penetration rates. The simulations in Guerci and Sapio (2011) show that, as wind power approaches its potential, congestion frequencies may considerably increase across the southern Apennines and between Calabria and Sicily. The inadequacy of the inherited transmission grids vis-à-vis the challenges of growing RE penetration rates finds a quantitative counterpart in the frequency of forced outages of RE plants, ordered by transmission system operators in the face of threats for reliability of supply. Curtailments of wind turbines in the southern Apennines between Campania and Apulia have caused the loss of about 10.7% of wind energy in 2009, 5.3% in 2010, 2.3% in 2011 and 1.1% in 2012 (GSE Statistical Reports, 2009, 2010, 2011, 2012; Lo Schiavo et al. 2013). The simulation study by Caralis et al. (2011), calibrated on the Greek power system, has explored the pros and cons of the parallel integration of distributed PV generation and wind power, a highly relevant issue for MENA countries too, thanks to their potential for diversified green energy portfolios. PV facilities rely on sunlight, which is more predictable than wind and, in comparison with gas-fired back-ups, its efficient scale is smaller and is less burdensome on the power grid. In line with these insights, simulation results show that at moderate penetration rates, distributed PV generation facilitates the integration of wind power by minimizing the wind turbine curtailments and the associated efficiency losses.

Conclusion

In the Mediterranean basin, welfare improvements may come from adequate flows of renewable energy, finance and technology among mutually dependent countries. Despite the wealth of solar and wind resources, the RE sector has not blossomed yet in the MENA area. The intensity and direction of RE support measures in the southern Mediterranean do not seem to be conducive to high-RET diffusion rates. Having illustrated the status of RE support measures and the main barriers to RET diffusion, some challenges and questions for future research can be outlined.

One area for future theoretical and empirical work concerns the causal relationship between the level of a country's economic development and the type and intensity of its efforts at greening the economy. In econometric models of RET diffusion, economic development indicators (e.g. GDP per capita) and policy measures enter as independent determinants. In other words, estimates can be used to compute how RETs' diffusion would change if emerging countries adopted the same policy tools and designs as in the high-income world. This reflects the belief that institutions can be costlessly replicated without a careful assessment of the subtle interdependencies that characterize economic systems, such as the institutional complementarities (Aoki 2001) among the legal, financial and educational systems, and between these and the technological base of a country.

Goals and priorities that drive the design of energy policies in MENA countries are shaped by a distribution of economic and political power that is vastly different from what can be observed in EU countries, as dramatically testified by the revolutionary movements and civil wars that have been upsetting the region in recent years. Basic issues that have already been solved in Europe, such as the widespread access to energy, are still far from finding satisfaction in most MENA countries; others (e.g. desalination) are unknown in the northern shore. These issues are even more crucial in countries, such as Algeria, Libya and Egypt, richly endowed with hydrocarbon reserves, because Dutch disease phenomena are always behind the corner.

Solving the Mediterranean RET diffusion puzzle may also benefit from understanding spatial causation processes in policymaking. As clear from financial and technological dependence patterns, leaping the infrastructural hurdle requires cooperation across borders. Investments in RE facilities and in transport capacity are complementary; the value of transmission investment projects in a country depends on the undertaking of complementary transmission investments in other countries. Hence, beyond the effects of explicit cooperation agreements, there may be cross-border spill-overs from infrastructural investments. The effectiveness of support policies and the very choice of policy types and designs may not be neutral with respect to such external influence.

Finally, the existing literature on the effects of increasing renewables penetration raises a number of issues. Supposing new interconnectors are built across the Mediterranean, the spatial distribution of electricity prices after a significant increase in renewables penetration will depend on the prevailing direction of the energy flows and on the persistence of state-owned monopolies in MENA countries, but we still lack a theoretical understanding of these relationships. Greater reliance on intermittent power sources, moreover, means higher electricity price volatility, which may imply an increasing demand for risk management institutions in the MENA electricity sectors, aggravating their dependence on external flows of capital and competences. A final issue is whether the cheap availability of natural gas in some MENA countries will push local governments to offset the negative impact of renewable penetration on the efficient use of gas-fired units.

Notes

* The findings, interpretations and conclusions expressed in this chapter are those of the author and do not reflect the positions of the institution.

1. See the European Commission's Communication "Second Strategic Energy Review. Communication from the Commission to the European Parliament, the Council, the European Economic and Social Committee and the Committee of the Regions", COM(2008) 781 final, http://eur-lex.europa.eu/LexUriServ/LexUriServ.do?uri=COM:2008:0781:FIN:EN:DOC
2. Baris and Kucukali (2012) report an estimate of a 20,000 MW potential for wind power generation in Turkey.
3. It is worth noting that there are no nuclear facilities for civilian use in North Africa, although Egypt, the holder of significant uranium deposits, has commissioned a 1,200 MW nuclear power plant. Turkey is planning to invest in nuclear power as well, whereas nuclear investments by other North African countries would come at the price of increased fuel dependence (Supersberger and Führer 2011).
4. The data presented in Čihák et al. (2012, Appendix II) allows us to grasp the financial development gap between EU and MENA countries. Mukasa et al. (2013) offer a survey on financing wind power investments in Africa.
5. Considering only Mediterranean countries, 92.8% of fractional patents have been awarded to inventors from southern EU countries by the European Patent Office between 1999 and 2008, with France and Italy acting as the leaders, whereas outside the EU, only Israel (10 patents) and, more recently, Turkey (4 patents between 2006 and 2008) have shown some dynamism. Source: author's elaborations on OECD patent statistics data.
6. See Stoneman (1991) and Geroski (2000) for comprehensive works on the economics of technology diffusion, whereas Popp et al. (2009) illustrate the evidence on the diffusion of environmentally friendly technologies in developed countries, and Neuhoff (2005) focuses on the deployment of RETs.
7. The penetration rate can be alternatively measured as the share of renewable energy capacity in total electricity demand. The two definitions may not be equivalent due to electricity imports and exports. This discrepancy can

be ignored here, as the electricity trade with and among MENA countries is severely limited by lack of interconnection capacity (see the section "Leaping the infrastructural hurdle").

8. The source of data on power generation capacities is the US Energy Information Administration. Per capita non-hydro RE capacities are computed by using World Bank population data.
9. Based on extensive information on solar radiation levels, land cover and use, land forms, population density, location of water sources and protected areas, availability of roads and transmission corridors, Ummel and Wheeler (2008) have identified five sites for solar fields (in Morocco, Libya, Egypt, Jordan and Saudi Arabia), as well as two undersea lines (Morocco to Spain through the Strait of Gibraltar; Tunisia to mainland Italy through Sardinia and Corsica), plus an overland link from Jordan to Turkey through Syria. May (2005) has illustrated three examples of high-voltage direct current (HVDC) lines connecting promising sites for CSP generation in MENA (in Algeria, Libya and Egypt) with three major European demand centers (Aachen, Milan and Wien).
10. Although the evidence chiefly concerns the economically developed world, it may nevertheless provide hints as to the direction that research on RET diffusion in the Mediterranean may take. Performing econometric analyses on the diffusion of green energy technologies in MENA countries is nevertheless difficult, due to paucity of data and the low variance in RET diffusion rates across the area.
11. In a panel of 19 EU countries covering the 1998–2009 period, Marques et al. (2010) show that the cumulated number of RE policies is associated with larger wind overcapacity, that is, with lower capacity factors. While not strictly on the sheer diffusion of RETs, these results are relevant in defining what lies behind the diffusion of the most technically efficient RETs.
12. Technical change is itself guided or at least heavily influenced by public policies, as witnessed by the dramatic increase in green technology patenting after the adoption of the Kyoto Protocol. See Popp (2012) and Popp et al. (2009).
13. Bottom-up forces also contribute to the Europeanization of energy policy around the Mediterranean, for example lobbying to favor a market-based approach to the development of the energy sector and the inflow of FDIs from the EU.
14. Interconnectors open up wider markets for the energy produced by means of RETs, thereby boosting the expected returns from investing in RE facilities. At the same time, the returns from transmission investments are discouraged by uncertainty over the future localization of renewable energy plants and the associated expected impact on congestion patterns.
15. The authors also explore the effects of nuclear power, finding that the implied price-cost margin would be lower than that implied by the integration of wind power, due to nuclear power being dispatchable.
16. Examples include Zafarana and Hurghada in Egypt, as well as Tarfaya and El-Aaiún in Morocco.
17. See also Sioshansi and Oren (2007) for empirically inspired conjectures on the effects of transaction costs on electricity bidding strategies.

References

Albadi, M. H., and E. F. El-Saadany (2010). Overview of wind power intermittency impacts on power systems. *Electric Power Systems Research* 80.6: 627–632.

Aoki, Masahiko (2001). *Toward a comparative institutional analysis*. MIT Press.

Ata, Raşit (2013). The current situation of wind energy in Turkey. *Journal of Energy*, Article ID 794095. doi:10.1155/2013/794095

Banal-Estañol, Albert, and Augusto Rupérez Micola (2011). Behavioural simulations in spot electricity markets. *European Journal of Operational Research* 214.1: 147–159.

Baris, Kemal, and Serhat Kucukali (2012). Availability of renewable energy sources in Turkey: Current situation, potential, government policies and the EU perspective. *Energy Policy* 42: 377–391.

Billette de Villemeur, Etienne, and Pierre-Olivier Pineau (2012). Regulation and electricity market integration: When trade introduces inefficiencies. *Energy Economics* 34.2: 529–535.

Boluk, Gulden (2013). Renewable energy: Policy issues and economic implications in Turkey. *International Journal of Energy Economics and Policy* 3.2: 153–167.

Brand, Bernhard, and Jonas Zingerle (2011). The renewable energy targets of the Maghreb countries: Impact on electricity supply and conventional power markets. *Energy Policy* 39.8: 4411–4419.

Brunnschweiler, Christa N. (2010). Finance for renewable energy: An empirical analysis of developing and transition economies. *Environment and Development Economics* 15.3: 241–274.

Cambini, Carlo, and Donata Franzi (2013a). Assessing the EU pressure for rules change: Perception by southern Mediterranean energy regulators. *Mediterranean Politics* 19.1: 59–81.

Cambini, Carlo, and Donata Franzi (2013b). Independent regulatory agencies and rules harmonization for the electricity sector and renewables in the Mediterranean region. *Energy Policy* 60: 179–191.

Caralis, George, S. Delikaraoglu, and Arthouros Zervos (2011). Towards the optimum mix between wind and PV capacity in the Greek power system. In *European Wind Energy Conference and Exhibition – Scientific Proceedings*, Brussels, pp. 75–79.

Čihák, Martin, Asli Demirgüç-Kunt, Erik Feyen, and Ross Levine (2012). *Benchmarking financial systems around the world.* Policy Research Working Paper 6175, World Bank.

Cohen, Wesley M., and Daniel A. Levinthal (1990). Absorptive capacity: A new perspective on learning and innovation. *Administrative Science Quarterly* 35.1: 128–152.

Dechezleprêtre, Antoine, Matthieu Glachant, Ivan Hascic, Nick Johnstone, and Yann Meniere (2012). Invention and transfer of climate change mitigation technologies: A global analysis. *Review of Environmental Economics and Policy* 5.1: 109–130.

del Rio, Pablo, and Miguel-Angel Tarancon (2012). Analysing the determinants of on-shore wind capacity additions in the EU: An econometric study. *Applied Energy* 95: 12–21.

Escribano, Gonzalo (2010). Convergence towards differentiation: The case of Mediterranean energy corridors. *Mediterranean Politics* 15.2: 211–229.

Escribano, Gonzalo, and Israel Solorio (2012). *Morocco and the European Union: Bridging opportunities through the Mediterranean solar plan.* An overlook of Morocco's perspectives. Paper presented at XXIV Convegno SISP, Venezia.

Fritzsche, Kerstin, Driss Zejli, and Dennis Tanzler (2011). The relevance of global energy governance for Arab countries: The case of Morocco. *Energy Policy* 39.8: 4497–4506.

Geroski, Paul A. (2000). Models of technology diffusion. *Research Policy* 29.4: 603–625.

Green, Richard, and Nicholas Vasilakos (2010). Market behaviour with large amounts of intermittent generation. *Energy Policy* 38.7: 3211–3220.

GSE (2009). *Rapporto Statistico 2009 – Impianti a fonti rinnovabili - Settore Elettrico*. Roma: Gestore Servizi Energetici.

GSE (2010). *Rapporto Statistico 2010 – Impianti a fonti rinnovabili - Settore Elettrico*. Roma: Gestore Servizi Energetici.

GSE (2011). *Rapporto Statistico 2011 – Impianti a fonti rinnovabili - Settore Elettrico*. Roma: Gestore Servizi Energetici.

GSE (2012). *Rapporto Statistico 2012 – Impianti a fonti rinnovabili - Settore Elettrico*. Roma: Gestore Servizi Energetici.

Guerci, Eric, and Alessandro Sapio (2012). High wind penetration in an agent-based model of the electricity market. *Revue de l'OFCE* 5: 415–447.

Haas, Reinhard, Christian Panzer, Gustav Resch, Mario Ragwitz, Gemma Reece, and Anne Held (2011). A historical review of promotion strategies for electricity from renewable energy sources in EU countries. *Renewable and Sustainable Energy Reviews* 15.2: 1003–1034.

Jenner, Steffen, Felix Groba, and Joe Indvik (2012). Assessing the strength and effectiveness of renewable electricity feed-in tariffs in European Union countries. *Energy Policy* 52: 385–401.

Kausch, Kristina, and Richard Youngs (2009). The end of the Euro-Mediterranean vision. *International Affairs* 85.5: 963–975.

Kilpeläinen, Sarah (2013). Energy relations between the European Union and North Africa. *Journal of Contemporary European Research* 9.2: 346–359.

Kitzing, Lena, Catherine Mitchell, and Poul Erik Morthorst (2012). Renewable energy policies in Europe: Converging or diverging? *Energy Policy* 51: 192–201.

L'Abbate, A., Migliavacca, G., Calisti, R., Brancucci, C., Chaouachi, A., and G. Fulli (2014). Electricity exchanges with North Africa at 2030: The European and Italian approaches, in Carlo Cambini and Alessandro Rubino, eds. *Regional Energy Initiatives in the Mediterranean Basin* (New York: Routledge).

Lanjouw, Jean Olson, and Ashoka Mody (1996). Innovation and the inter-national diffusion of environmentally responsive technology. *Research Policy* 25.4: 549–571.

Lo Schiavo, Luca, Maurizio Delfanti, Elena Fumagalli, and Valeria Olivieri (2013). Changing the regulation for regulating the change: Innovation-driven regulatory developments for smart grids, smart metering and e-mobility in Italy. *Energy Policy* 57: 506–517.

Marques, Antonio C., Jose A. Fuinhas, and J. R. Pires Manso (2010). Motivations driving renewable energy in European countries: A panel data approach. *Energy Policy* 38.11: 6877–6885.

Mason, Michael, and Dennis Kumetat (2011). At the crossroads: Energy futures for North Africa. *Energy Policy* 39.8: 4407–4410.

May, Nadine (2005). Eco-balance of a solar electricity transmission from North Africa to Europe. Diploma Thesis, Technical University of Braunschweig.

Menz, Fredric C., and Stephan Vachon (2006). The effectiveness of different policy regimes for promoting wind power: Experiences from the states. *Energy Policy* 34.14: 1786–1796.

Moesgaard, Rune, and Poul Erik Morthorst (2007). *The effect of wind power on spot market prices*. Fredriksberg, Danish Wind Industry Association-Risoe/DTU EWEC.

Mukasa, A.D., Mutambatsere, E., Arvanitis, Y., Triki, T. (2013). *Development of wind energy in Africa*. Abidjan, African Development Bank.

Narbel, Patrick A. (2013). What is really behind the adoption of new renewable electricity generating technologies? *Energy for Sustainable Development* 17.4: 386–390.

Neuhoff, Karsten (2005). Large-scale deployment of renewables for electricity generation. *Oxford Review of Economic Policy* 21.1: 88–110.

Nicolini, Marcella, and Simona Porcheri (2012). *The energy sector in Mediterranean and MENA countries*. FEEM Working Paper no. 73/2012, Fondazione Eni Enrico Mattei, Milano.

Nicolosi, Marco, and Michaela Fürsch (2009). The impact of an increasing share of RES-E on the conventional power market. The example of Germany. *Zeitschrift für Energiewirtschaft* 33.3: 246–254.

Pfeiffer, Birte, and Peter Mulder (2013). Explaining the diffusion of renewable energy technology in developing countries. *Energy Economics* 40: 285–296.

Popp, David (2012). *The role of technological change in green growth*. National Bureau of Economic Research No. w18506, Washington, DC.

Popp, David, Ivan Hascic, and Neelakshi Medhi (2011). Technology and the diffusion of renewable energy. *Energy Economics* 33.4: 648–662.

Popp, David, Richard G. Newell, and Adam B. Jaffe (2009). *Energy, the environment, and technological change*. National Bureau of Economic Research No. w14832, Washington, DC.

Rubino, Alessandro (2011). *A Mediterranean electricity co-operation strategy*. Vision and rationale, Competition and Regulation in Network Industry (CRNI), Brussels.

Saenz de Miera, Gonzalo, Pablo del Rio Gonzalez, and Ignacio Vizcaino (2008). Analysing the impact of renewable electricity support schemes on power prices: The case of wind electricity in Spain. *Energy Policy* 36.9: 3345–3359.

Sioshansi, Ramteen, and Shmuel Oren (2007). How good are supply function equilibrium models: An empirical analysis of the ERCOT balancing market. *Journal of Regulatory Economics* 31.1: 1–35.

Steggals, Will, Robert Gross, and Philip Heptonstall (2011). Winds of change: How high wind penetrations will affect investment incentives in the GB electricity sector. *Energy Policy* 39.3: 1389–1396.

Stoneman, P. (1991). Technological diffusion: The viewpoint of economic theory, in P. Mathias and J. Davis, eds., *Innovation and technology in Europe: From the eighteenth century to the present day*, The Nature of Industrialization Series (Oxford and Cambridge: Blackwell), pp. 162–184.

Suding, Paul H. (2011). Struggling between resources-based and sustainable development schemes. An analysis of Egypt's recent energy policy. *Energy Policy* 39.8: 4431–4444.

Supersberger, Nikolaus, and Laura Führer. (2011). Integration of renewable energies and nuclear power into North African energy systems: An analysis of energy import and export effects. *Energy Policy* 39.8: 4458–4465.

Traber, Thure, and Claudia Kemfert. (2011). Gone with the wind? Electricity market prices and incentives to invest in thermal power plants under increasing wind energy supply. *Energy Economics* 33.2: 249–256.

Trieb, Franz, and Hans Müller-Steinhagen (2007). Europe-Middle East North Africa cooperation for sustainable electricity and water. *Sustainability Science* 2.2: 205–219.

Trieb, Franz, Hans Müller-Steinhagen, and Jürgen Kern (2011). Financing concentrating solar power in the Middle East and North Africa. Subsidy or investment? *Energy Policy* 39.1: 307–317.

Tsikalakis, Antonis, T. Tomtsi, N. D. Hatziargyriou, A. Poullikkas, C. Malamatenios, E. Giakoumelos, O. Cherkaoui Jaouad et al. (2011). Review of best practices of solar electricity resources applications in selected Middle East and North Africa (MENA) countries. *Renewable and Sustainable Energy Reviews* 15.6: 2838–2849.

Ummel, Kevin, and David Wheeler (2008). *Desert power: The economics of solar thermal electricity for Europe, North Africa, and the Middle East.* Center for Global Development Working Paper 156, Washington, DC.

Vajjhala, Shalini, Anthony Paul, Richard Sweeney, and Karen Palmer (2008). *Green corridors: Linking interregional transmission expansion and renewable energy policies.* RFF Discussion Paper no. 08–06, Resources for the Future, Washington, DC.

Woo, Chi-Keung, I. Horowitz, J. Moore, and A. Pacheco (2011). The impact of wind generation on the electricity spot-market price level and variance: The Texas experience. *Energy Policy* 39.7: 3939–3944.

Yasner, Rebecca A. (2012). Maximizing renewable electricity in Israel: Energy security, environmental impact, and economic development. Senior Honors Thesis, Carnegie Mellon University, Pittsburgh, PA.

Yin, Haitao, and Nicholas Powers (2010). Do state renewable portfolio standards promote in-state renewable generation? *Energy Policy* 38.2: 1140–1149.

Contributors

Editors

Carlo Cambini is Associate Professor of Industrial Organization at the Politecnico di Torino, research fellow at the IEFE – Bocconi, Milan, and economic member of the Scientific Committee of the Florence School of Regulation, Robert Schuman Centre for Advanced Studies – European University Institute (EUI). He received the Young Economist Essay Award at the 1999 EARIE conference in Lausanne, Switzerland. His research focuses on industrial economics, regulatory economics and competition policy with a strong background in telecommunications, transportation and energy markets. He has being published in leading scientific international journals such as the *RAND Journal of Economics*, the *Journal of Industrial Economics*, the *International Journal of Industrial Organization*, the *Journal of Economics & Management Strategy* and the *Journal of Regulatory Economics*. He acted as consultant for national agencies, such as the Italian Regulatory Authority for Communications (AGCOM), the Authority of Energy (AEEG), the Italian Treasury and local governments. He was also invited as a keynote speaker in conferences organized by the Australian Regulatory and Competition Authority (ACCC) and by the Commerce Commission in New Zealand.

Alessandro Rubino is a Lecturer in Economics at Bari University. His research interests are in the fields of regulation, industrial organization, energy market integration and demand side participation. Alessandro obtained his PhD in Economics from Siena University. Before completing his PhD, he worked at Ofgem as Regulatory Economist. His main tasks were related to the analysis of the impact of the Large Combustion Plants Directive (LCPD, 2001/80/EC) on the generation portfolio in UK. In addition Alessandro has worked on the design and implementation of the incentive scheme for the national transmission system operator for gas and electricity. From June 2009 to July 2012 Alessandro has worked as Research Assistant at the Florence School of Regulation (European

University Institute) within the Robert Schuman Centre, a research unit specialized in applied economics. He has been junior expert in the EC founded project "Paving the Way for the Mediterranean Solar Plan" and coordinator of the International Energy Regulation Network (IERN). From July 2012 to January 2014 Alessandro was head of capacity building and knowledge dissemination at the Enel Foundation.

Authors

Miriam Bardolet is a qualified lawyer in Spain and France, specializing in energy law and holds an LLM in International Environmental Law and Policy from University College London. Since 2010, she is the legal counsel of the team Regulation and Markets at Dii GmbH. In her current position, Miriam has been involved in the regulatory analysis and design for Dii's long-term strategy "Desert Power 2050: Getting Started". Miriam is an expert for power sector regulation in MENA (including, Morocco, Tunisia, Algeria, Jordan, Egypt and Saudi Arabia) as well as for the design of cooperation mechanisms for renewables projects of EU member states with third countries (Article 9, RES-E Directive). Prior to joining Dii, Miriam was an associated lawyer with Cuatrecasas Gonçalves Pereira and Freshfields Bruckhaus Deringer, where she worked on mergers and acquisitions transactions as well as acquisition, sale, development, financing and construction of energy projects.

Ernesto Bonafé works at the Secretariat of the Energy Charter Treaty in Brussels. His work focuses on legal and policy issues of energy markets at national, regional and international level. Previously he was legal expert in the EU project "Paving the Way for the Mediterranean Solar Plan" and coordinated the energy training at the Florence School of Regulation, where he teaches regulatory authorities at the annual training. He holds a PhD in Energy Policy and EU Law from the European University Institute of Florence. He worked as legal counsel in various law firms within Europe. He got his law degree at the University of Valencia (Spain) and holds an LLM in European Law at the Université Libre de Bruxelles.

Ana-Maria Boromisa is Head of Department for International Economic and Political Relations and President of the Scientific Council at the Institute for Development and International Relations (IRMO) in Zagreb. She teaches the economics of European integration at the University of Zagreb, and occasionally lectures at various training courses (e.g. regional trainings organized by the World Trade Organization). Her research interests include energy and environment in an international context. Ana-Maria Boromisa holds a PhD in Economics (University of Zagreb), a Masters degree in European Studies (College of Europe) and her first degree is in Electrical Engineering (University of Zagreb).

Carlo Brancucci Martínez-Anido was employed for three years as a scientific researcher at the European Commission – Joint Research Centre – Institute for Energy and Transport, Petten, the Netherlands. His research focused on energy system modelling and techno-economic assessments of the European power system. Carlo's main achievement has been the development of EUPowerDispatch, a minimum-cost dispatch model of the European electricity system. In September 2013 he defended his PhD thesis at Delft University of Technology, the Netherlands. He holds a Masters degree in Aeronautical Engineering from the University of Bristol, UK (2009).

Dirk Buschle has been Deputy Director of the Energy Community Secretariat since 2011 and its Legal Counsel since 2007. Prior to his current position, he was Head of Cabinet of the President of the Court of Justice of the European Free Trade Association (EFTA) in Luxembourg. Dirk Buschle graduated from Constance University, Germany, and earned his PhD at St. Gallen University in Switzerland, where he is Deputy Director of the Institute for European and International Business Law. He has widely published in different areas of European law, speaks regularly at conferences and has lectured at the Universities of Reykjavik, Constance and St. Gallen as a visiting professor. He co-chairs the annual St. Gallen International Energy Forum as well as the Vienna Energy Law Forum.

Roberto Calisti received his electrical engineering diploma in 1983 and programmer qualification in 1987. He joined the Electrical Research Centre of ENEL, Italy, in 1985 where he has been active in the Test and Instruments section. Since 1996 he has been working on power system reliability modelling. In 2000 he joined CESI, Milan, Italy and then he has been since 2006 with CESI RICERCA (now: RSE – Ricerca sul Sistema Energetico), Milan, Italy, where he is engaged in Power Systems Development department. He has actively contributed to the development of different grid planning and reliability tools and models. He is involved in several Italian and European projects.

Bagdagul Kaya Caner is an Energy Expert in the Energy Market Regulatory Authority of Turkey since 2004. She has worked on regulatory/legal issues on both domestic and international levels concerning energy, especially natural gas markets. She participated in preparation of secondary legislation that will be applied in Turkish natural gas sector. She scrutinized contracts relating to sales and transport of natural gas. She is a lawyer by training and was admitted to Ankara Bar Association in 2003. She has participated in the working groups of the Energy Community, the Energy Regulators Regional Association, the Association of Mediterranean Regulators for Electricity and Gas and the European Regulators' Group for Electricity and Gas, as a Turkish representative. She gave lectures on Turkish natural gas law, EU energy law in Ankara University

Banking and Commercial Law Institute and the Ankara Bar Association. She holds a Master of Laws degree from the Queen Mary, University of London; her Masters thesis on "Renewable Energy and International Investment Protection Regime" was awarded with distinction. She has post-graduate certificates from the Florence School of Regulation–European University Institute and the Institute of Public Utilities–Michigan State University. She is also a PhD candidate in the Ankara University Law School, researching investment arbitration in the energy sector. She has published several articles on issues such as investment arbitration, competition in energy markets, energy regulation and carbon trading.

Aymen Chaouachi holds BS and MS degrees in Electrical Engineering from the École Nationale Supérieure d'Ingénieurs, Tunis, Tunisia, received in 2004 and 2006, respectively, and a PhD degree in Electrical Engineering from the Tokyo University of Agriculture and Technology, Japan, received in 2011. He worked as a lecturer in the Tunisian Ministry of Education, and he has been involved in several photovoltaic (PV) research projects with the Tunisian Research and Technology Center of Energy. Before joining the European Commission – Joint Research Centre – Institute for Energy and Transport, Petten, The Netherlands, he was with Hitachi Ltd. Power System Company, Japan. His research interests include renewable energy, power system control and stability, smart grids, microgrids and electrical vehicles.

Rosa María Fernández currently teaches Environmental Economics and Sustainable Development at the University of Birmingham. Rosa holds a degree in economics from the University of Oviedo (Spain) and joined in 2006 UNED (Spain National Distance Education University) as Lecturer. She defended her PhD in 2010, on "European Union and Climate Change. Main Economic Effects". She has been a visiting researcher at the University of Exeter (UK) and York University (Toronto). Her publications focus on climate change, energy policy and corporate social responsibility. Rosa has been an associate tutor at the University of Bradford.

Donata Franzi is a consultant on EU funds for research at the University of Modena and Reggio Emilia. After receiving his Bachelor degree in International Relations and his Masters degree in International Development, Donata started consulting on EU projects drafting and management, gaining a wide knowledge of EU programs in the field of energy and environment, transport, structural funds. She got a PhD in Political Systems and Institutional Change in 2012 at the IMT Institute for Advanced Studies in Lucca (Italy). Her main research interests are EU governance, rules harmonization processes and the utilities sector.

Gianluca Fulli leads the "Smart Electricity Systems and Interoperability" group at the European Commission – Joint Research Centre – Institute for Energy and Transport, Petten, The Netherlands, conducting research and providing scientific support to EU policies on power systems and smart grids. He is an electrical engineer (1998, Università La Sapienza, Rome, Italy) with combined industrial and research expertise on smart grids techno-economic assessment and power systems planning, operations and technologies. He previously worked for the Italian transmission system operator (TSO) where he was responsible for cross-border interconnectors planning and new generation connection, with NGT UK, and with IBM, Italy. He is involved in several research projects and EU initiatives on smart/power grids.

Natasha Hroneska is a former student of the College of Europe in Bruges, Department of EU International Relations and Diplomacy Studies (Albert Einstein Promotion). Currently she works at the Ministry of Foreign Affairs of the Republic of Macedonia. Before joining the Ministry, she worked as an energy policy analyst in Analytica, a think-tank in Skopje, and for different international organizations. She is a graduate in Political Sciences from the University Ss. Cyril and Methodius, and also holds a Masters degree in Southeast European Studies from the National University of Athens.

Josefine Kuhlmann, LL.M., works as a legal expert at the Austrian national regulatory authority for electricity and gas. She wrote her dissertation about the Energy Community and its implications for external relations of the EU during her time as a university assistant at the Institute for European and International Law at the Vienna University of Economics and Business. She gained valuable insights during an internship as an assistant to the legal counsel of the Energy Community. She graduated in Law and International Business Relations following studies in Vienna, Graz, Eisenstadt, and Krakow.

Angelo L'Abbate graduated in Electrical Engineering at the Politecnico di Bari, Italy, in 1999. In 2003–2004 he received his PhD in Electrical Energy Systems at the Politecnico di Bari, Italy, in partnership with the University of Dortmund, Germany. In 2004–2005 he was an active researcher at the Mediterranean Agency for Remote Sensing (MARS), Benevento, Italy, and at the University of Ljubljana, Slovenia, as CNR-NATO Fellow. From 2005 through 2008, he was at the European Commission – Joint Research Centre – Institute for Energy, Petten, the Netherlands. Since 2009 he has been working at CESI RICERCA (now RSE – Ricerca sul Sistema Energetico), Milan, Italy. His fields of interest include modeling and planning of power T&D systems, RES and DG integration, FACTS and HVDC. He is involved in several Italian and European projects.

Gianluigi Migliavacca graduated in Electronic Engineering at the Polytechnic University of Milan in 1991 and joined the Automation Research

Center of ENEL, Italy, in 1994, working on mathematical modeling and dynamic simulation of thermal power plants. In 2000 he joined CESI, Milan, Italy and since 2006, he has been with CESI Ricerca (now RSE – Ricerca sul Sistema Energetico), Milan, Italy, where he is now Head of the Transmission Network Planning research group. He carries out research on mathematical modelling and regulatory issues concerning deregulated electricity markets, cross-border trade issues and transmission planning. During 2005 he was a consultant at the Italian regulator on the development of a common energy market in South-East Europe. He is involved in several Italian and European projects.

Simone Mori is Executive Vice President in Enel Holding in charge of Regulation, Environment and Innovation. After receiving a degree in physics and an MBA, Simone Mori held several positions at the Enel Group. He is also Head of Enel Carbon Strategy Division. He is a member of the board of directors of Enel Trade, Enel Foundation and Cesi. He is also Vice President of Assoelettrica (Italian Association of Electricity Companies) and Senior Transatlantic Fellow at the German Marshall Fund of the United States. Simone Mori teaches Economics and Management of Energy Sector at "La Sapienza" and Luiss Universities in Rome.

Alessandro Ortis is Honorary President of MEDREG and EBRD Senior Adviser. From December 2003 through February 2011, Alessandro Ortis was concurrently the Chairman of the Italian Regulatory Authority for Electricity and Gas, the Vice President of the Council of European Energy Regulators (CEER), the President of the Association of the Mediterranean Regulators for Electricity and Gas (MEDREG) as well as Chairman of the International Energy Regulation Network (IERN). He also served as Director General for Energy and Mineral Resources at the Italian Ministry for Productive Activities, President of the Cassa Conguaglio (Equalisation Fund) for the Electricity Sector, a member of the Governing Board of the International Energy Agency, a member of Italy's Higher Council for Public Works and a member of the Technical-Scientific Commission at the Italian Department for Civil Protection. Alessandro Ortis also was Vice President of ENEL (the Italian Electric Group; 1987–1992), President of EURELECTRIC (the European Electric Industry Association), Chairman of the International Energy Agency's Group of Experts for the electricity sector and Lecturer in industrial production management (Trieste University). After attending the Nunziatella Military College, he took a degree in nuclear engineering at Politecnico di Milano University; he is also a graduate of Bocconi University's Business School. He was awarded the Honours of Knight of the Grand Cross of the Italian Republic and Chevalier de la Légion d'Honneur de la République Française.

Roman Petrov holds an LL.M in EU Law (Durham University, UK, 1998), a PhD in Law (National Academy of Science of Ukraine, 2000) and a PhD in Law (Queen Mary, University of London, UK, 2005). He conducted post-doctoral research as Max Weber Fellow at the European University Institute (Italy, 2006–2008) and had visiting research fellowships at the University of Heidelberg (Germany), the University of Oxford (UK) and Ghent University (Belgium). Dr. Petrov is Founder and the first elected president of the Ukrainian European Studies Association. Currently Dr. Petrov is Jean Monnet Chair in EU Law and Head of the Jean Monnet Centre of Excellence at the National University's Kyiv-Mohyla Academy in Ukraine. Areas of Dr. Petrov's research and teaching include EU law, EU external relations law, approximation and harmonization of legislation in the EU, rights of third-country nationals in the EU, legal aspects of regional integration in the post-Soviet area.

Sergio Piazzi is a Senior Executive Officer of the United Nations. In April 2007, while retaining his UN rank and status, he was appointed as the first Secretary General of the Parliamentary Assembly of the Mediterranean. As a UN officer, in the past, he served in Romania, the Persian Gulf, Rwanda, Poland, Turkey and Indonesia, among others. He developed the UN/NATO policy framework for cooperation in civil emergency management in the context of the Partnership for Peace (PfP), and in 2000, he organized the Fribourg Forum, the UN Ministerial Conference on European Cooperation and Coordination in Crisis Management, and contributed to the launching of the Africa 2000 programme. In 2004, he negotiated the UN/EU treaty on cooperation in crisis operations.

Peter Polajnar works in the Directorate-General for Enlargement of the European Commission. He is responsible for economic reporting on the Western Balkans and Turkey. He is engaged in accession negotiations, for which he drafts EU positions on industrial and commercial policy. In this context, he observes the functioning of the Central European Free Trade Agreement and other similar economic initiatives in the Western Balkans. He studied economics at the University of Ljubljana and received a Masters degree in European Studies from College of Europe Bruges/Warsaw. His theses dealt with innovation systems. He has continued publishing on innovation, competitiveness and appropriate governance since then.

Alessandro Sapio is Assistant Professor of Economics at the Parthenope University of Naples since 2007. He obtained a PhD in Economics and Management from Sant'Anna School of Advanced Studies, Pisa in 2005, and has been Visiting Research Fellow at the Energy Markets Group of the London Business School. His research interests include the empirical and theoretical analysis of electricity markets and of the capital structure of small and medium enterprises. Dr. Sapio has published his research in international journals such as *Energy Policy*, *Industrial and Corporate Change*, *Small Business Economics* and the *Journal of Evolutionary Economics*.

Milica Uvalic is Full Professor of Economics, University of Perugia (Italy), formerly a member of the UN Committee for Development Policy, Public Policy Scholar at the Woodrow Wilson Center (Washington D.C.) and Assistant Minister in the first post-Milosevic federal government of Yugoslavia. He has published extensively on transition economies (particularly the Balkans), on macroeconomic policies, privatization, regional cooperation and EU enlargement. Milica's recent books include *Serbia's Transition; Towards a Better Future* (Palgrave-Macmillan, 2010) and *Western Balkans' Accession to the European Union; Political and Economic Challenges* (Beograd, 2010) (co-edited with Bozidar Cerovic).

Francesca Pia Vantaggiato is Research Associate and a PhD candidate in Politics at the Centre for Competition Policy (University of East Anglia, Norwich, UK). She holds an MA in Economics obtained in July 2013 from the University of Parma. In 2011, she obtained an MA in European Studies from the same university, following the achievement of a postgraduate high-level diploma in EU Studies from the European College of Parma Foundation. Her research interests concern institutional analysis, regulatory authorities and regulation, energy policy and EU politics and integration. She has been working with energy regulators and carrying out research on energy regulation for the past three years while working as Research Assistant at the Robert Schuman Centre (Florence School of Regulation) of the European University Institute (EUI, Florence). Moreover, in 2009 she graduated in Conference Interpreting at the School of Modern Languages for Interpreters and Translators of the University of Bologna, after obtaining her BA degree in Interpreting and Translation at the same institution. She speaks English, Russian, French and Spanish.

Index